Environmental Modeling

Ekkehard Holzbecher

Environmental Modeling

Using MATLAB®

Springer

Priv.-Doz. Dr.-Ing. Dipl.-Math. Ekkehard Holzbecher
Georg-August Universität Göttingen
Goldschmidtstr. 3
37077 Göttingen
eholzbe@gwdg.de

Additional material to this book can be downloaded from http://extra.springer.com.

ISBN 978-3-662-50053-8 ISBN 978-3-642-22042-5 (eBook)
DOI 10.1007/978-3-642-22042-5
Springer Heidelberg Dordrecht London New York

Softcover reprint of the hardcover 2nd edition 2012

Printed on acid-free paper

Springer is part of Springer Science+Business Media (www.springer.com)

Dedicated to
my children Gesa and Gero

Foreword

The book has two aims:

(a) To introduce basic concepts of environmental modeling and
(b) To exercise the application of current mathematical software packages

To the target group belong all natural scientists who are dealing with the environment: engineers from process and chemical engineering, physicists, chemists, biologists, biochemists, hydrogeologists, geochemists, ecologists . . .!

As the book is concerned with modeling, it inevitably demands some mathematical insight. The book is designed to

1. Be a door opener to the field for novices without any background knowledge of environmental modeling and of MATLAB®, and
2. To surprise those, who have some expertise, with advanced methods which they have not been aware of

For this book MATLAB® was chosen as the computer tool for modeling, because

1. It is powerful, and
2. It is available at most academic institutions, at all universities and at the research departments of companies

Other mathematical products could have been selected from the market, which would perform similarly well for most application problems presented in the various chapters. But MATLAB® is rather unique in it's strong capabilities in numerical linear algebra.

There are 20 chapters in the book. The first chapters are concerned with environmental processes and their simulation: (1) transport, consisting of advection, diffusion and dispersion, (2) sorption, (3) decay or degradation, (4) reaction, either kinetic or thermodynamic. Following aim (b) there are sub-chapters inserted for the introduction of MATLAB® modeling techniques. The first part of the book ends with chapters on ordinary differential equations and parameter estimation (inverse modeling).

The second part of the book starts with chapters on flow modeling. Flow, if present, is an important, but mostly also complex part within an environmental compartment. Core MATLAB® allows simple flow set-ups only. Therefore the focus is on potential flow, which has applications in hydro (water) and aero (air)-dynamics as well as in porous media (seepage and groundwater). Concepts of MATLAB® are deepened within these chapters. At the very end special topics appear: image processing and geo-referencing, graphs, linear systems, the phase space and graphical user interfaces.

Berlin

Foreword to the Second Edition

The consistently commendatory and positive reactions that I obtained unanimously for the first edition of 'Environmental Modeling' encouraged me to work on a second improved and extended version of the book and the accompanying software, which is available herewith. The reader may allow me to cite some surely exaggerating voices from the Internet:

> Excellent work. It will be more helpful for the younger researchers also for the sr. scientists for understanding basics and applications of MATLAB in environmental engg. It is THE BEST book.
> I love this book, because you wrote it in a programming manner and I love programming, so I learnt advection and diffusion excellent. Because after I read the equation I modeled it in Matlab and saw the results. And it remains in my mind. I want to thank you because of writing this book. It helps a lot to the students and researchers to learn environmental modeling deeply.

Special thanks to my students at the Georg-August Universität, Göttingen, and at Freie Universität, Berlin, who gave me clues on how the mathematical viewpoint, taken in this book, is conceived by an audience that is usually not especially trained in topics as mathematical physics. Some of the improvements directly result from the work with the students.

A new chapter was added, in which an introduction into numerical methods is given – an important topic that was missing in the first edition, as I was told by some readers.

Special focus has been laid to extend the capability to use 'Environmental Modeling' as a reference book. The list of keywords in the MATLAB® command index, although not covering the complete list of commands, has been extended significantly. Personally I think that this is the major improvement in relation to the first edition. I hope that in that way the book will help readers and modelers to understand the commands quickly and to apply them correctly.

Final thanks to all people at the mentioned universities, the publishers at Springer Verlag, Heidelberg, and the people of the bookprogram of MathWorks.

Göttingen

Preface

"Environmental Modeling using MATLAB®" by Ekkehard Holzbecher is an excellent publication and a novel approach covering the intersection of two important, growing worlds – the world of environmental modeling and of mathematical software.

Environmental modeling is a science that uses mathematics and computers to simulate physical and chemical phenomena in the environment (e.g., environmental pollution). This science was initially based on pen-and-paper calculations using simple equations. In the last 50 years, with the development of digital computers, environmental models have become more and more complex, requiring often numerical solutions for systems of partial differential equations.

Mathematical software, such as MATLAB®, has been developed in the last two decades. These packages have been particularly successful for users of personal computers. Mathematical software provides a set of tools for solving equations both analytically and numerically. This is a major improvement in comparison to the programming tools (e.g., FORTRAN) previously used by scientists. Mathematical software offers extremely valuable and cost-effective tools that improve the productivity of the programmer by at least an order of magnitude. The use of these tools also minimizes the risk of programming errors. In addition, mathematical software offers unique visualization tools that allow the user to immediately visualize and often animate simulation results.

Scientists who become familiar with a tool like MATLAB® will never go back to previous ways of computer programming.

The book "Environmental Modeling using MATLAB®" provides a clear, comprehensive, and very instructive introduction to the science of environmental modeling, and more importantly, includes the MATLAB® codes for the actual solutions to the environmental equations. MATLAB® codes are listed in the book and also included as more complete versions in an attached CD[1].

---

[1] The first edition of the book included a CD. Readers of the second edition obtain the accompanying software via Internet.

I highly recommend this book to both beginners and expert environmental professionals. The book will be particularly useful to those scientists who have postponed learning and using mathematical software. This book will open a new world to them!

Paolo Zannetti
President, The EnviroComp Institute
Editor of Book Series on Environmental Modeling

Acknowledgements

A book on such a wide topic as this one includes experiences and knowledge shared with lots of other people. I ask for understanding that I mention no names here. To be fair, giving one name would make it necessary to add numerous others. As one and only exception I want to mention my wife Susanne for reading the first edition of the book, correcting stylistic errors, providing improved formulations and for mental support.

Thanks to Humboldt University Berlin (HUB), the Leibniz Institute of Freshwater Ecology and Inland Fisheries (IGB), to the Weierstrass Institute for Applied Analysis and Stochastics (WIAS), to the Geosciences of Freie Universität Berlin (FUB) and of Georg-August Universität Göttingen. Special thanks to the students who attended the modeling courses in Berlin and Göttingen, for their questions and for their ideas. Without them I would not have become so familiar with the software. Parts of the book are based on seminar scripts.

Especially I want to thank The MathWorks, Inc (http://www.mathworks.com/) for providing the most recent versions of the MATLAB® software as well as other support by including the book in the MathWorks Book Program.

Contents

Part I
Primer to Modeling with MATLAB®

Chapter 1
Introduction

1.1 Environmental Modeling Using MATLAB®

There are various types of models in the environmental sciences, and surely there is no unique opinion about the essence of an environmental model. Differences may mainly concern the scope of the models and the modeling methods. Concerning the scope, this book is relatively open; i.e. examples from different branches of environmental science and technology are included, mainly from the hydrosphere and the geosphere, and also from the biosphere and the atmosphere. However, the examples are selected for demonstration purposes and can in no way represent the vast variety of phenomena and approaches, which can be met in publications and studies of all types of environmental systems.

Concerning the methods, the book does not represent the entire field either. In this book modeling is *process-oriented* and *deterministic*. These two terms characterize almost all presented methods, which, according to many opinions, represent the most important approach to understand environmental systems. There are environmental problems, for which other approaches not tackled here work more successfully. Statistical or stochastic methods are not mentioned, for example. Data processing, either graphical or numerical, as for example in Geo-Information Systems (GIS), appears rudimentary in this book.

Processes are in the focus of the presented approach. In the modeling concept of this book processes can be of physical, chemical or biological nature. The reproduction of biological species is a process, death is another; degradation of biochemical species, or decay of radioactive species are other examples. Some relevant processes are explained in detail: diffusion, dispersion, advection, sorption, reactions, kinetic and/or thermodynamic and others.

A view into journal or book publications shows that models of the treated kind, process-oriented and deterministic, are applied to different environmental compartments, to different phases and to different scales, as well as to multi-phase and multi-scale problems. There are models of the entire globe, of earth atmosphere and oceans, of the global atmosphere, of the sea, of rivers, lakes and glaciers, of

E. Holzbecher, *Environmental Modeling*,
DOI 10.1007/978-3-642-22042-5_1, © Springer-Verlag Berlin Heidelberg 2012

watersheds, of the soil, of terrestrial or aquatic sediments, of aquifers, and of parts of streams, and so forth. There are models of technical devices for environmental purposes, in addition. Experimental set-ups in laboratories are simulated in order to understand relevant processes.

The methods presented in this book are deterministic, throughout. A search for any statistics would be in vain. The description of processes is translated into mathematical terms. Often the approach leads to differential equations, which are conditions concerning the change of a variable, like concentration or population density, in space and time. Nowadays the solution of such equations is not as tedious as in former times. Using core MATLAB®, problems in 0 and 1 space dimensions can be solved comfortably. Core MATLAB® is also convenient for solving 2 and 3-dimensional problems with analytical solutions. For more complex modeling in more than one dimension, toolboxes, especially the MATLAB® partial differential toolbox, can be recommended.

The aim of the book is to introduce basic concepts of environmental modeling. Starting from basic concepts the problems are transformed into mathematical formulations. Strategies for the solution of the mathematical problems on the computer are outlined. The main aim of the book is to communicate the entire path of such a modeling approach. At some points algorithmic details will be omitted for the general aim. Who is interested strictly in computer algorithms, will be better served with a book on numerics, applied mathematics or computational methods. It is important that the modeler has a basic understanding of the underlying numerics. There is no need, however, to dive so deep into the algorithms that one would be able to program them oneself. In fact, it is an advantage of the chosen software that modeling tasks, which could be handled only by people with profound programming knowledge and skills, become now available to a wider audience.

Who is addressed? In a broader sense everyone is addressed, who is dealing with or is interested in the simulation of environmental systems on a computer. In a considerable part of the book concepts of environmental modeling are introduced, starting from basic principles, tackling differential equations and numerical solutions. In another similarly big part of the book special implementations are introduced and described. If someone is very familiar with another mathematical software, the book may be of help too, as most of the described models can also be realized using other maths computer programs.

There are several good and excellent books on environmental modeling and on MATLAB®. Richter (1985) deals with ecological systems and with time dependencies (but no space dependencies), as well as Deaton and Winebrake (1999) using STELLA®.[1] Shampine et al. (2003) also present MATLAB® modeling of ordinary differential equations; concerning applications they do not address environmental modeling particularly; concerning methods, they do not address partial differential equations. Gander and Hrebicek (1997) offer little to

[1] See: http://www.iseesystems.com/.

the specialized environmental modeler, although some of the presented mathematical tools could be applied to environmental problems. Christakos et al. (2002) focus on the connection of time dependent simulation and GIS using MATLAB®. McCuen (2002) treats statistical methods (which do not appear here) for modeling hydrologic change. Cantrell and Cosner (2003) examine spatial ecology via reaction-diffusion models, without reference to any specific software package. Lynch (2005) addresses scientists and engineers in his general introduction to numerical methods without preference for any specific software and with few references to applied environmental modeling. In his introduction to MATLAB® Kiusalaas (2005) addresses engineers in general. The topic of Zimmerman (2004) is chemical process simulation using FEMLAB[2] code. Hornberger and Wiberg (2005) have the hydrologist's perspective on numerical methods. Trauth (2010) focuses on image- and data-processing, as well as statistical methods for geoscientists. Finlayson (2006) deals with the chemical aspects and gives an introduction to MATLAB® as one of several modeling tools. All these books[3] differ concerning scope and methods; and none of them has the same constellation of scope and methods, as it is presented in this book.

The book is divided into 20 chapters which differ concerning scope and complexity. The first ten chapters form a primer on fundamental concepts and basic environmental modeling. All of the model examples presented are 0- or 1-dimensional. In the further ten chapters more complex models, as for example spatial 2D, are outlined with an explanation of the underlying methods. Concepts of flow modeling are introduced.

In this book the focus on basic '*core*' MATLAB®[4] is intended. There is the hope to address a wider audience, as not all readers may have access to the complete palette of MATLAB® toolboxes. On the other hand, there are lots of powerful commands in core MATLAB® and novice users might be confused being confronted with more specialized tools. It turns out that this is not a severe restriction, as most basic tasks, which are of interest to the environmental modeler, can be performed using core MATLAB®. For advanced higher dimensional and coupled problems the MATLAB® partial differential equation toolbox has to be used, or COMSOL Multiphysics alternatively. COMSOL has developed a multi-physics software environment, which can be applied with MATLAB® in the background.

Although other mathematical codes have developed a similar extension from a special purpose module to a toolbox for mathematical calculations in general, matrix manipulation is the backbone and stronghold of the MATLAB® package

---

[2] Now COMSOL; see: http://www.comsol.com/.

[3] There are numerous other books on MATLAB®, which could not all be checked by the author. The reader can get a list on the MathWorks Website http://www.mathworks.com/support/books.

[4] For this book we mainly used the most up-to-date version of MATLAB®. The latest version was release R2011b. However, there are some references back to version 7 that was used in the first edition of the book. Most of the commands described in the book should work equally independent of the MATLAB® version.

and explains its strong competitiveness. Therefore Sect. 1.2 gives a brief reminder of basic matrix operations.

The book is accompanied by software containing advanced and final versions of the program files described in the text. The Mathworks logo

appears where MATLAB® files of the accompanying software are referenced. The terms 'modeling' and 'simulation' are synonymously in the concerned scientific and technical literature. However, the term 'model' appears to be more general, encompassing all types of attempts to capture one or more aspects of a real system, and is therefore preferred in this book. The term 'simulation' also fits to the presented approach, as it suggests that processes which are relevant for the behavior of a system are included in the computer simulation. In the sequel the term is used for time-dependent dynamics.

The book contains relatively simple models throughout. It is not the case that complex models constructed by MATLAB® don't exist, but they are not appropriate for an introduction into modeling techniques. For such an aim models should be as simple as possible, even more, when novice modelers are addressed.

Usually the extensive work with a model leads to renewed extensions, which turn simple models into complex ones almost as a rule. Not all models are improved by doing this. Jørgensen (1994) envisages the connection between model complexity and knowledge, gained by the model, as shown in Fig. 1.1. Simple models can be improved by extensions, but there is a certain peak position after which further extensions do not add to the knowledge – rather quite the contrary. An improved model design increases the quality of the model (lets take gained knowledge as a quality measure), but further extensions of the improved model may finally lead to a situation in which the increase of model complexity is counter-productive.

The model evaluation study of Constanza and Sklar (1985) provides a plot similar to Fig. 1.1, but with 'articulation' on the x-axis and 'effectiveness' on the

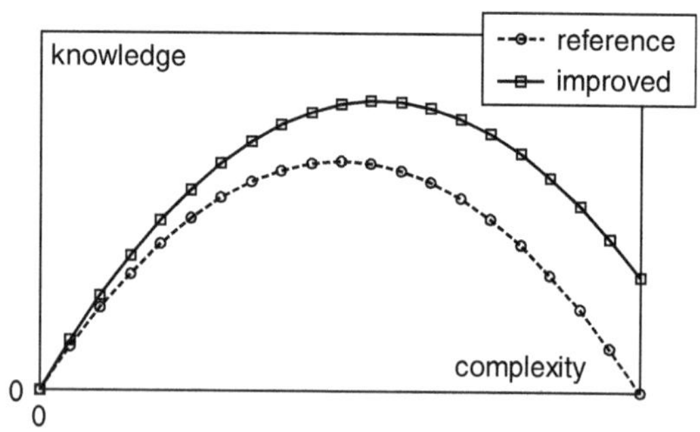

Fig. 1.1 Model evaluation: knowledge gained vs. complexity

y-axis. Chwif and Barretto (2000) envisage a similar picture, putting 'level of detail' on the x-axis and 'model confidence' on the y-axis. All these terms can be taken as different terms for the complexity of a model on one side and its performance on the other side.

The method, how to construct complex models, is another topic which is left out in the book. The major drawback of complex models is the increased number of parameters, sometimes to a drastical extend. The situation may be worsened by the fact that many new parameters are usually difficult to obtain or have to be determined by parameter estimation runs with the model. Another drawback may appear, if the model becomes very sensitive to one or more parameters, i.e. that relatively small changes of a parameter induce a tremendous effect on the output results. A complex model which depends sensitively on numerous unknown parameters can surely not be used as a predictive tool.

However, complex models have their justification. Whether they can be successful also depends on the architecture, design and construction itself, especially on the analytical and/or numerical techniques.

A complex model concerning sediment phosphorus and nitrogen processes is presented by Harper (2000): the SNAPP model is constructed in MATLAB® and contains even a graphical user interface. As another example Luff et al. (2001) present a MATLAB library to calculate pH distributions in marine systems. Kumblad et al. (2003) construct an ecosystem model of the environmental transport and fate of carbon-14 in a bay of the Baltic Sea, just to give another example. A complex MATLAB® surface fluid flow model for rivers, streams and estuaries is presented by Martin and Gorelick (2005).

It is not the aim of modeling to set up complex models. The opposite of that statement is a more suitable goal: the aim of modeling is to find simple models that explain some aspects of a real system. Unfortunately that aim turns out to be a tricky one, because every real system appears to be complex, as long as there is ample knowledge about the driving mechanisms. Moreover, if a system is complex, a simple model can explain a few aspects at the most and that may not be enough to solve a real problem.

1.2 Introduction to MATLAB®

MATLAB® is a mathematical software, originated and mainly developed by mathematicians (Moler 2004). The name envisages a laboratory for matrix calculations, where the mathematical term of a matrix refers to an array of numbers such as

$$\mathbf{A} = \begin{pmatrix} 1 & 2 \\ 3 & 4 \end{pmatrix} \tag{1.1}$$

Linear algebra is the name of the mathematical field in which calculations with matrices are treated. Some basic terms are listed in the appendix of this chapter.

While MATLAB® was designed for numerical linear algebra in the beginning, it has become a tool for all types of mathematical calculations in the meantime. Nowadays, MATLAB® has been applied in nearly every field of scientific or technical calculations. In the academic branch there is almost no university where MATLAB® is not available.

With MATLAB® innumerable types of mathematical operations can be performed. Of course, numerous linear algebra calculations are available, such as inversion of matrices, eigenvalue and eigenvector determination, which can be applied to perform various tasks, for example, the solution of systems of linear equations. One may perform basic statistics, numerical differentiation and integration, evaluate all types of functions, solve dynamical systems and partial differential equations, estimate parameters and so forth. All this is part of core MATLAB®, a collection of basic mathematical tools.[5]

Before some details of linear algebra are examined, an introduction into the work with MATLAB® is necessary. This should be read by novices, but can be skipped by those who have already worked with the program.

1.2.1 Getting Started with MATLAB®

When MATLAB® is opened, the user obtains a graphical user interface on the display, as it is shown in Fig. 1.2, containing several windows. The main window, to start with, is the 'Command Window', where commands are given and answered. In the command window the MATLAB® prompt '>>' stands at the position where the user command is shown on the display, during and after entering.

In order to start type the command:

```
a = 2
```

Press the return button and the program gives an answer, here with the information that a variable `a` was created in the machine containing the value `2`:

```
a =
     2
```

A new prompt appears after the answer of the system, in order to enable the user to give the next command. Note that only the line after the last prompt in the command window can be used for a new command. The former lines remain in the command

[5] Core MATLAB® can be extended by numerous toolboxes for special purposes, for details see: http://www.mathworks.com/products/. Most interesting for environmental modelling, as it is treated here, are the optimization toolbox and the partial differential equations toolbox.

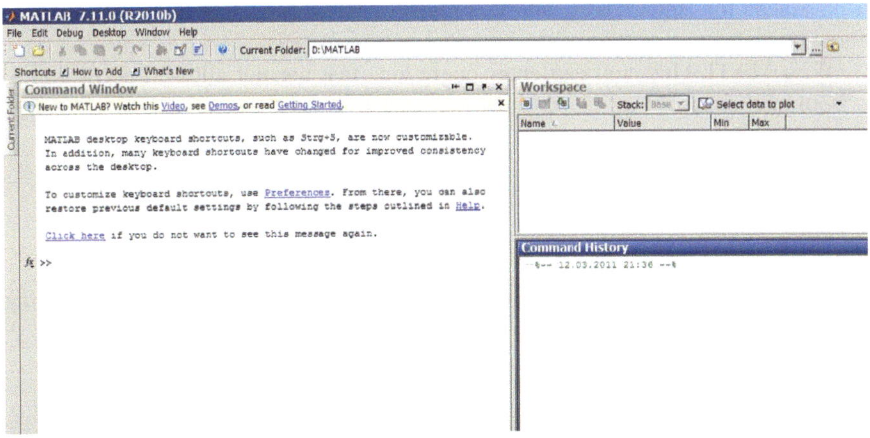

Fig. 1.2 Appearance of MATLAB® graphical user interface

window to allow the user to have an overview on the previous work and the produced answers. Confirm that the `;` as closing character of the command, for example

```
a = 2;
```

prevents that the answer is shown in the command window. Variables and their values are stored during a session, if you do not chose to delete them. As we can work with these variables as we want, we can imagine easily that the mathematical software offers a lot more than a simple calculator.

Variable names may be quite long. Type

```
namelengthmax
```

(and enter) to see what is the limit on your installation. Also note that MATLAB® is case-sensitive: in general `C` is not the same as `c`. As you see, several variables are predefined and may better not be used. A very prominent example is:

```
pi
```

Try also:

```
i
ans =
        0 + 1.0000i
```

to learn that MATLAB® works with imaginary numbers, with the imaginary unit `i` as $\sqrt{-1}$. In order to check this, try:

```
sqrt(-1)
sqrt
```

`sqrt` stands for 'square root' and is a reserved name for a function – more about functions later.

With MATLAB® you can enter operations that are not allowed according to school mathematics:

```
5/0
Warning: Divide by zero.
ans =
   Inf
```

If there is not a warning, as shown, on your computer, switch on the warning option, by:

```
warning on
```

Note that `Inf` is the value of the 'ans' variable. Infinity is thus a value and be used as such:

```
5/Inf
ans =
     0
```

or:

```
 5*Inf
ans =
  -Inf
```

but:

```
Inf/Inf
ans =
   NaN
```

`NaN` stands for 'Not a Number'; remember that ∞/∞ has no unique value. Try also the following expressions:

```
0/0
(+Inf)+(-Inf)
0*Inf
0^0
```

Standard output has few significant digits. With

```
format long
```

one can switch to a representation with more digits. Try it with `pi` now! Note that this concerns only how values are displayed on the screen. Internally the default data type is 'double.' It corresponds to a 64 bit[6] representation on most computers, in contrast to 'single' or 'float,' which corresponds to a 32 bit representation only.

The data type of a variable can be asked for by the 'class' command. Example: what is the data type of `pi`?

```
class (pi)
  ans =
double
```

You convert from default double to float using the 'single' command:

```
single(pi)
ans =
   3.1415927
```

Note that there are eight significant digits. That holds for all singles, as will become clear soon. Example:

```
single(sqrt(10000000))
ans =
  3.1623e+003
```

From the MATLAB® response observe that the number consists of two parts: the significant or mantissa (in front of the 'e') and the exponent (behind the 'e'). The 'e' obviously indicates the start of the exponent. All doubles or singles are represented this way. Thus it is clear that

```
10e5
ans =
      1000000
```

or:

```
10e14
ans =
  1.0000e+015
```

Internally both the mantissa and the exponent are stored as binary numbers, i.e. represented not to the basis 10, but to the basis 2. This is nearby, as the binary system knows only 0 and 1 as digits, which corresponds to the two states of a bit. Real numbers are represented on the computer internally by floating point numbers of different type; most important are the just mentioned *double* and *single*.

MATLAB® constructs the single and double data type according to IEEE[7] Standard 754. A single is stored in 32 bits. Any value stored as a double requires 64 bits, formatted as shown in the Table 1.1:

[6] A binary digit on a computer, which can take only two states: 0 or 1.

[7] Institute of Electrical and Electronics Engineers.

Table 1.1 Bit usage for double

Bits	Usage
63	Sign (0 = positive, 1 = negative)
62 to 52	Exponent, biased by 1,023
51 to 0	Fraction f of the number 1.f

Table 1.2 Standard operator symbols

Symbol	Operation
+	Addition
−	Subtraction
*	Multiplication
/	Division
^	Power
( )	Specify evaluation order

Every number that has more than 53 binary digits, which corresponds to 17 decimal digits,[8] can not be represented exactly on a computer, as a double. Note that 53 binary digits are considered although the significant requires 52 bits only. This is due to the convention that every number is converted to a form which has a single non-zero digit in front of the dot. Example: 0.01 is represented as $1.0 \cdot 10^{-2}$ in the decimal system, but as $1.1001100110011001101{\cdot}2^{-4}$ in the binary system on the computer.

As there is only one non-zero number in the binary system (1), the first significant digit is always the same and does not have to be considered in the number representation. The idea goes back the very early days of computers, when it was already utilized by Zuse.[9]

We have to realize that despite of its enormous computation capacity numbers can only be represented with a certain accuracy. Only certain numbers are represented exactly. An accuracy measure is given by the distance from 1.0 to the next largest double-precision number, that is 2^{-52} and is called within MATLAB® as

```
eps
ans =
  2.2204e-016
```

Due to rounding the error of a result after mathematical operations may be significantly below the 17-digit limit for doubles, resp. the 8-digit limit for singles. We will demonstrate this in the following exercise. Numbers represented exactly in the decimal system, may be represented with an error in the binary system (Table 1.2).

[8] For the conversion recall that $2^{10} = 1024 \approx 10^3$.

[9] Konrad Zuse (1910–1995), German computer pioneer.

Table 1.3 Number representation in MATLAB®

Characteristics	Double	Single
Storage [bits]	64	32
Storage [byte]	8	4
Mantissa storage incl. sign [bits]	53	24
Exponent storage incl. sign [bits]	11	8
Max. significant decimals, mantissa	16	8
Max. significant decimals, exponent	3	2
Max. exponent of binaries	1,023	127
Min. exponent of binaries	−1,022	−126
Maximum float (absolute value)	$1.7977{\cdot}10^{308}$	$3.4028{\cdot}10^{38}$
Minimum float (absolute value)	$2.2251{\cdot}10^{-308}$	$1.1755{\cdot}10^{-38}$
Machine accuracy (of 1)	$2.2204{\cdot}10^{-16}$	$1.1921{\cdot}10^{-7}$

Due to the limited number of digits in the exponent there are minimum and maximum positive and negative numbers, that are not $\pm\infty$ or ± 0. From the 11 bits for the exponent one is needed for the sign, leaving the range to represent $2^{10} \approx 1{,}024$ numbers, i.e. from 0 to 1,023. The MATLAB® functions `realmax` and `realmin` return the maximum and minimum values that you can represent with the double data type. The range for double is −1.79769e+308 to −2.22507e−308 and 2.22507e−308 to 1.79769e+308. For further characteristics of floating point numbers see the following Table 1.3.

The command window is good for an introduction into MATLAB®. Finally, the work with M-files replaces extensive operating in the command window (see Chapter 2.5). Nevertheless, for certain tasks, the command window will remain the most direct and simple way to compute with MATLAB®.

Aside from the command window, the user may select numerous other views of the desktop. The different options are depicted in Fig. 1.3. Very important is the workspace view, where all variables of the current session are visible and directly available. The workspace of the just started session, shown in Fig. 1.2, is depicted on the left side of the figure. The workspace appears only if the view is selected in the 'Desktop' submenu, as shown in Fig. 1.3. Using `who` or `whos` in the command window is an alternative way to access the workspace (and its contents).

Here, `a` is the only variable in the workspace which is of 'double' type and of 1×1 size (a single variable and not a 'real' matrix). A double-click on the block-panel symbol, left of the variable name in the workspace, delivers an array editor, in which the contents of variables can be viewed directly. In the simple example case the result is given in Fig. 1.4. With the array editor it is not only possible to view variables, but also to change them. The user can easily explore the use of the editor on her/his own.

To mention is the 'command history' view, in which all commands are listed. An example with one command only is depicted in Fig. 1.5. The user can initiate the repeated command, mostly with some workspace variables changed, by double-click in the command history window. This is a shortcut to the alternative method to

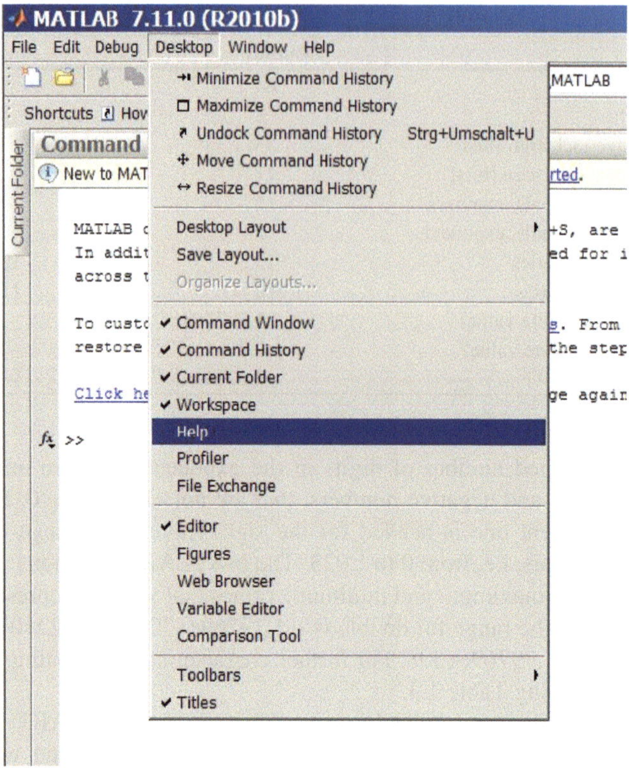

Fig. 1.3 Submenu-entries of desktop main entry, listing all possible views of the desktop

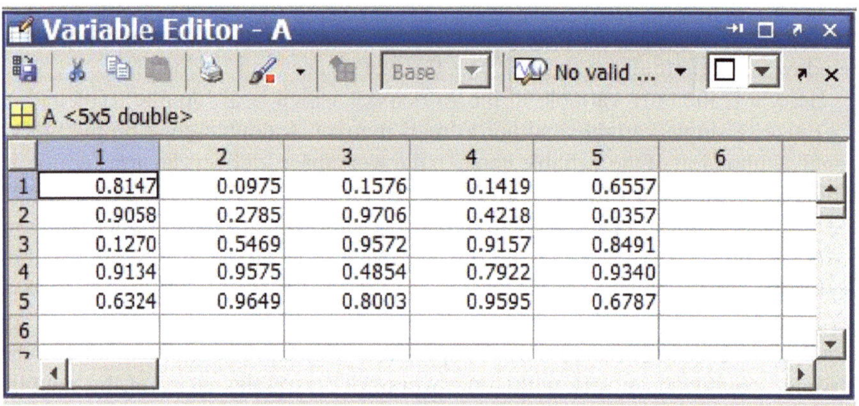

Fig. 1.4 MATLAB® array editor

Fig. 1.5 MATLAB® command history view

copy a former command in the history view and to paste it in the command window. The user may wish to perform some alterations in the command and can do that easily, before the command is executed after pressing the return button.

In the command window the up-arrow and down-arrow keys of the keyboards offer an alternative, allowing a sequential loop through former commands.

1.2.2 Matrices in MATLAB®

The name 'MATLAB' is a combination of 'matrix' and 'laboratory.' With respect to the suite of various mathematical tools, which are made available by recent versions of the software, one might think the origin of the MATLAB® software is numerical linear algebra. A comprehensive treatment of matrix algebra is given by Robbin (1995).

A *matrix* is a 2-dimensional array of numbers, for which examples are given right below. Matrices can be specified directly by the user. Entries in lines are separated by blanks; lines are separated by ';'.

```
A = [1 2 3; 4 5 6]
```

```
A =
     1     2     3
     4     5     6
```

The example matrix has two rows and three columns. Matrix dimensions are 2 and 3. `A` is a 2 $\times$ 3 matrix. It is thus non-square, as a *square matrix* has the same number of rows and columns. Once a matrix is constructed, its elements can be called by using usual round brackets, which is exemplified by:

```
A(2,1)
```

```
ans =
     4
```

The element in the second row and first column of **A** is 4. The first counting index is 1, in contrast to conventions in other programming languages.

As no variable is used in the command, MATLAB® uses the notation `ans =` in order to indicate an answer to the given command. Sub-matrices of a matrix can be called by using ':', as the following example illustrates:

```
A(2,2:3)

ans =
     5     6
```

Elements in the second line, second and third column are given in the answer. The ':' without any numbers is used to indicate the entire range. In the example, the entire first column of **A** is given

```
A(:,1)

ans =
     1
     4
```

There are several special commands to input special types of matrices. *Vectors* are multi-element matrices, for which either the number of rows or the number of columns is 1. Row vectors with constant increment can be specified as follows:

```
v = [2:0.5:5]

v =
  Columns 1 through 3
    2.0000    2.5000    3.0000
  Columns 4 through 6
    3.5000    4.0000    4.5000
  Column 7
    5.0000
```

`v` is a row vector, containing all values between 2 and 5 with increment 0.5.[10] A column can easily be obtained by using the *transponation* operation, which in MATLAB® is performed by the `'`:

```
  v'

ans =
    2.0000
    2.5000
    3.0000
    3.5000
    4.0000
    4.5000
      5.0000
```

[10]The *comma* in common numbers is a *dot* in all mathematics software products, thus also in MATLAB®.

The last element of a vector can also be reached by using the `end` keyword in the following way:

```
v(end)
ans =
     5
```

or

```
w = v'; w(end)
ans =
     5
```

The size of a matrix, i.e. the numbers of columns and rows, is given by the `size` command:

```
size(v)
ans =
     1    7
```

`length` returns the bigger of both:

```
length(v)
ans =
     7
```

The empty array is specified by:

```
empty = []
empty =
     []
```

Matrices containing 1s are given by:

```
B = ones(2,3)
B =
     1     1     1
     1     1     1
```

Matrices containing zeros are produced analogously:

```
B = zeros(3,1)
B =
     0
     0
     0
```

How matrices containing a constant, different from 0 and 1, can be obtained easily, is demonstrated by the following command:

```
C = 4.5*ones(3,5)

C =
    4.5000    4.5000    4.5000    4.5000    4.5000
    4.5000    4.5000    4.5000    4.5000    4.5000
    4.5000    4.5000    4.5000    4.5000    4.5000
```

The `ones`–matrix is multiplied by a single value, a so called *scalar*, here 4.5. The `*` stands for multiplication. As will be explained in more details in the next Section, there are several multiplication operations in linear algebra and in MATLAB®. In the previous command line the `*` stands for *scalar multiplication*, where all elements of the matrix are multiplied by the same scalar value.

The command for random matrices is

```
C = rand(2,5)

C =
    0.9501    0.6068    0.8913    0.4565    0.8214
    0.2311    0.4860    0.7621    0.0185    0.4447
```

Random values between 0 and 1 are entries of the matrix. If there is only one integer argument in the preceding matrix types, a square matrix results:

```
D = rand(2)

D =
    0.9501    0.6068
    0.2311    0.4860
```

As mentioned above matrices are 2-dimensional arrays. Single numbers can be regarded as 1-dimensional arrays. MATLAB® can, of course, handle arrays of higher dimensions. We demonstrate this by introducing the `randn` command:

```
E = randn(2,4,2)

E(:,:,1) =
    0.0000    1.0950    0.4282    0.7310
   -0.3179   -1.8740    0.8956    0.5779
E(:,:,2) =
    0.0403    0.5689   -0.3775   -1.4751
    0.6771   -0.2556   -0.2959   -0.2340
```

which is a 3-dimensional array of random numbers with mean value $\mu = 0$ and standard deviation $\sigma = 1$. In the same manner, all previous matrix generating commands can be applied to obtain higher dimensional arrays if the number of arguments in the call exceeds 2. Multi-dimensional arrays can be viewed using the array editor, but they cannot be edited within the editor. In order to do this, address single elements from the command window, or specify 2-dimensional sub-arrays:

```
E1 = E(:,:,2)
```

and edit those.

1.2.3 Basic Matrix Operations

It is expected that readers are already familiar with matrix operations and basics of linear algebra. The purpose of the following is (1) to be a reminder for those, to whom matrices are not (yet) part of daily practice and (2) to introduce the notation used in the following chapters of this book.

A matrix is a 2-dimensional array of numbers. A matrix has a certain number of lines and columns, and the single numbers in the matrix are called elements (sometimes the term 'entries' is used here as alternative). The matrix in (1.1) has two lines and two columns, and the element in the second line and first column is 3. A single number can be conceived as special case of a matrix with one line and one column. Thus matrix algebra is a generalization of the usual calculations with single numbers. However, in order to distinguish 'real' arrays from single numbers, bold letters are used for matrices and vectors.[11]

Basic operations as known from single numbers can be generalized for matrices. Matrices can be added. The sum of the matrices **A** and **B**

$$\mathbf{A} = \begin{pmatrix} a_{11} & a_{12} & \dots & a_{1m} \\ a_{21} & a_{22} & \dots & a_{2m} \\ \dots & \dots & \dots & \dots \\ a_{n1} & a_{n2} & \dots & a_{nm} \end{pmatrix} \text{ and } \mathbf{B} = \begin{pmatrix} b_{11} & b_{12} & \dots & b_{1m} \\ b_{21} & b_{22} & \dots & b_{2m} \\ \dots & \dots & \dots & \dots \\ b_{n1} & b_{n2} & \dots & b_{nm} \end{pmatrix} \tag{1.2}$$

is given by:

$$\mathbf{A} + \mathbf{B} = \begin{pmatrix} a_{11} + b_{11} & a_{12} + b_{12} & \dots & a_{1m} + b_{1m} \\ a_{21} + b_{21} & a_{22} + b_{22} & \dots & a_{2m} + b_{2m} \\ \dots & \dots & \dots & \dots \\ a_{n1} + b_{n1} & a_{n2} + b_{n2} & \dots & a_{nm} + b_{nm} \end{pmatrix} \tag{1.3}$$

In order to add two matrices, both need to have the same number of lines and columns. In each element of the matrix **A** + **B**, the sum of the corresponding elements of **A** and **B** appears. One may also say that in order to obtain the element

[11] A vector is a matrix consisting of one line or one column only. Terms as line-vectors or column-vectors are used, too.

in the i-th row and j-th column of **A** + **B**, the elements in the i-th row and j-th column of **A** and **B** have to be added:

$$(\mathbf{A}+\mathbf{B})_{ij} = a_{ij} + b_{ij} \tag{1.4}$$

Example in MATLAB®:

```
A = [1 2; 3 4]; B = [-1 0; 1 2];
  C = A+B

  C =
         0      2
         4      6
```

When the number of columns or the number of lines do not coincide, MATLAB® produces an error:

```
D = [5 6 7 8];
  A+D

  ??? Error using ==> +
  Matrix dimensions must agree.
```

Clearly the subtraction of matrices is defined analogously. One may also formally introduce subtraction by the definition that subtraction of **B** is the addition of **–B**. As one may expect **–B** contains the negative of the elements of **B** and is the inverse of **B** with respect to the addition operation. The generalizations of matrix multiplication and division are slightly more complex.

It was already mentioned that there are several multiplication operations. Correspondingly there are several division operations. Aside from scalar multiplication, there are several matrix multiplications. The standard matrix multiplication for the two matrices **A** and **B**, given by

$$\mathbf{A} = \begin{pmatrix} a_{11} & a_{12} & \dots & a_{1k} \\ a_{21} & a_{22} & \dots & a_{2k} \\ \dots & \dots & \dots & \dots \\ a_{n1} & a_{n2} & \dots & a_{nk} \end{pmatrix} \text{ and } \mathbf{B} = \begin{pmatrix} b_{11} & b_{12} & \dots & b_{1m} \\ b_{21} & b_{22} & \dots & b_{2m} \\ \dots & \dots & \dots & \dots \\ b_{k1} & b_{k2} & \dots & b_{km} \end{pmatrix} \tag{1.5}$$

in order to obtain a new matrix **A** · **B**, is defined by the following formula:

$$(\mathbf{A}\cdot\mathbf{B})_{ij} = \sum_{l=1}^{k} a_{ik} b_{kj} \tag{1.6}$$

This is a formula for the element in the i-th row and j-th column of the matrix **AB**. Matrices can be multiplied if the first matrix has the same number of columns as the second matrix has columns (inner dimension). In formula (1.6) that number is k. Elements in lines of the first matrix are multiplied with columns of the second matrix, and the products are summed in order to obtain an entry in the result matrix **A** · **B**.

Example in MATLAB:

```
C = A*B
```

```
C =
     1     4
     1     8
```

If the inner dimensions of the matrices do not agree an error message results. Matrix multiplication is a generalization of the multiplication of single numbers. Clearly, if the product $\mathbf{A} \cdot \mathbf{B}$ is possible, the product $\mathbf{B} \cdot \mathbf{A}$ is only possible if **A** and **B** are square matrices. Even then the identity $\mathbf{A} \cdot \mathbf{B} = \mathbf{B} \cdot \mathbf{A}$is not valid generally (see exercises below).

The multiplication, described by formula (1.6), is the standard multiplication of matrices, denoted by a '·'-dot in the formulae and by a `*` in MATLAB® commands. Analogously to the definition of addition, given in (1.4), there exists also an element-wise multiplication:

$$(\mathbf{A} \cdot \mathbf{B})_{ij} = a_{ij} b_{ij} \tag{1.7}$$

In order to perform this multiplication, matrices **A** and **B** need to have the same number of rows and columns. In formulae element-wise multiplication is denoted by `.*` in MATLAB® commands, distinguishing element-wise operation from the standard matrix multiplication. In formulae we use the '·'-dot or omit the operator symbol entirely. There are scalar multiplication and vector product as further operations which are explained below.

Division of matrices can be defined for both multiplications. To start with the simple case: element-wise division is performed with element values. In MATLAB® element-wise division is denoted by `./`. Element-wise division with the same matrix delivers a matrix containing 1 in each entry, which is the unit matrix with respect to element-wise multiplication.

Example in MATLAB®:

```
C = A./B
```

```
Warning: Divide by zero.
(Type "warning off MATLAB:divideByZero" to suppress this warning.)
C =
    -1   Inf
     3     2
```

Obviously, in three entries the element-wise division is performed. In the second entry of the first row `Inf` stands for *infinity*, which is the result of a division by zero.[12]

[12] In contrast to school knowledge, division by zero is allowed in MATLAB®. The result is infinity. MATLAB® shows a warning (but no error) in order to remind the user that such an operation may result in some errors in further operations.

Example in MATLAB®:

```
C = A./A

C =
     1     1
     1     1
```

The matrix with entries 1 everywhere is the unit matrix of pointwise matrix multiplication.

The unit matrix with respect to matrix multiplication is given by:

$$\mathbf{I} = \begin{pmatrix} 1 & 0 & \dots & 0 \\ 0 & 1 & \dots & 0 \\ \dots & \dots & \dots & \dots \\ 0 & 0 & \dots & 1 \end{pmatrix} \tag{1.8}$$

This matrix is a diagonal matrix, as there are non-zero elements only in the main diagonal from the top left to the bottom right. The unit matrix within the matrix algebra corresponds to the 1 in usual multiplications using single numbers. In MATLAB® it is delivered by the `eye`-command. `eye(n)` produces the unit matrix with `n` lines and `n` columns.

```
eye(3)

ans =
     1     0     0
     0     1     0
     0     0     1
```

Formally, the division can be introduced similarly to the definition of subtraction given above: division by **B** is the multiplication with the inverse of **B**, denoted as $\mathbf{B}^{-1}$ in mathematical notation. In MATLAB®, the inverse of a matrix is denoted as `inv(B)` and is defined by the formula:

$$\mathbf{B} \cdot \mathbf{B}^{-1} = \mathbf{I} \tag{1.9}$$

It can be checked easily that the matrix multiplication in (1.9) can only be performed for square matrices **B**. However, this is not the only requirement; the matrix needs to be *regular* in order to be *invertible*. The regularity of matrices is a standard topic in textbooks on linear algebra and will thus not be deepened here. Matrices for which no inverse exists are also denoted as *singular*.

The inverse of a matrix is unique, i.e. there is only one matrix with the property (1.9). If the matrix **B** is regular, the division is defined by the expression $\mathbf{A} \cdot \mathbf{B}^{-1}$.

Example in MATLAB®:

```
X = B*inv(A)

X =
    2.0000   -1.0000
    1.0000         0
```

X is the solution of the linear system $\mathbf{XA} = \mathbf{B}$. The MATLAB® user will usually use the `/`-operator for the same expression: `B/A`. However, the division procedure is not unique, because matrix multiplication does not have the commutation property to which we are used to from calculating with single numbers (see exercise below). There are two types of divisions depending on the order in which the product is performed. The `/` operator has to be used if $\mathbf{A}^{-1}$ is the second operator; if $\mathbf{A}^{-1}$ is the first operator, the `\` (backslash) is necessary.

Example in MATLAB: instead of the expression `inv(A)*B` one may use:

```
Y = A\B

Y =
    3.0000    2.0000
   -2.0000   -1.0000
```

Y is the solution of the linear system $\mathbf{A} \cdot \mathbf{X} = \mathbf{B}$. One speaks of *left-division* for the latter case and of *right-division*' in the former case. Both division operators can also be used for non-square matrices (see Sidebar 1.1).

The power operation is defined for matrices as well. There is a power operator for matrix multiplication and another for pointwise operation. In MATLAB® these are written by `^` for matrix multiplication and `.^` for pointwise multiplication. In formulae the power notation is used for matrix multiplication.

Example in MATLAB®:

```
A^2

ans =
     7    10
    15    22

A.^2

ans =
     1     4
     9    16
```

It was already mentioned that matrices can be multiplied by a single number, by a so-called scalar. The *scalar multiplication* is demonstrated by the example:

```
c = 2;
c*A

ans =
     2     4
     6     8
```

All elements of the matrix are multiplied by the scalar. Note that in MATLAB® the `*` is also used for scalar multiplication. MATLAB® distinguishes between scalar multiplication and standard matrix multiplication automatically (both are written by the same `*` operator). The type of the operands gives the unique clue which of both operations is meant. From the dimension of the operators the program finds out, whether matrix or scalar multiplication is meant. In formulae we will usually use no symbol to indicate scalar multiplication. Sometimes we use the '·' in formulae in order to separate the factors.

Sidebar 1.1: Over- and Underdetermined Systems
Both divisions, / and \ , are also possible for non-square matrices. In the latter case the linear system $\mathbf{A} \cdot \mathbf{X} = \mathbf{B}$ is over- or underdetermined. In the case of an overdetermined system the equations usually can not be fulfilled exactly. Then the solution $\mathbf{X}$ is computed in the sense of least square minimization, i.e. $\mathbf{X}$ minimises the expression $\mathbf{A} \cdot \mathbf{X} - \mathbf{B}$. In the following example the right side of the system is a column vector $\mathbf{b}$: The solution $\mathbf{x}$ is also a column vector.

```
x1 = 1; y1 = 1;
x2 = 2; y2 = 1.5;
x3 = 3; y3 = 2.2;

A = [x1 x2 x3; 1 1 1 ]';
b = [y1 y2 y3]';
x = A\b
A*x-b

x =
    0.6000
    0.3667
ans =
   -0.0333
    0.0667
   -0.0333
```

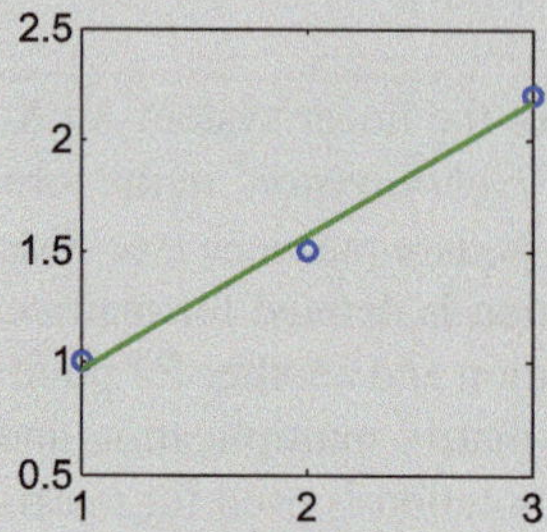

The solution of the overdetermined system $\mathbf{A} \cdot \mathbf{x} = \mathbf{b}$ is given in the MATLAB® answer. The second part of the answer gives the deviation in each of the three elements. The example can be interpreted as linear curve fitting on three given points (see graphic). For such a task the user may also use the polynomial curve fitting tool of MATLAB®, which is introduced in Chap. 10. As an exercise, the novice may confirm the obtained result for `x` by using the graphical user interface tool.

The division operator can also be applied for underdetermined systems $\mathbf{A} \cdot \mathbf{X} = \mathbf{B}$, where the number of columns of $\mathbf{A}$ exceeds the number of rows. In that case the solution is usually not unique, and MATLAB® delivers just one solution.

Moreover the well-known function sin, cos, exp can be performed on matrices. The exponential function of a matrix is based on the formula

$$\exp(A) = \sum_{i=0}^{\infty} \frac{1}{i!} A^i \tag{1.10}$$

in which the powers of A are performed with matrix multiplication. Try:

```
exp(A)

ans =
    2.7183    7.3891
   20.0855   54.5982
```

Exercises.

1. Find an example to show that the commutation-property $\mathbf{A} \cdot \mathbf{B} = \mathbf{B} \cdot \mathbf{A}$, which is valid for single numbers, is not valid for matrix multiplication! Take care that $\mathbf{A}$ and $\mathbf{B}$ are square matrices of the same size. Is it valid for pointwise multiplication?
2. Confirm by random matrices the validity of the following identities:

$$
\begin{aligned}
(\mathbf{A}+\mathbf{B}) \cdot \mathbf{C} &= \mathbf{A} \cdot \mathbf{C} + \mathbf{B} \cdot \mathbf{C} \\
\mathbf{C} \cdot (\mathbf{A}+\mathbf{B}) &= \mathbf{C} \cdot \mathbf{A} + \mathbf{C} \cdot \mathbf{B} \\
(\mathbf{A} \cdot \mathbf{B}) \cdot \mathbf{C} &= \mathbf{A} \cdot (\mathbf{B} \cdot \mathbf{C}) \\
(\mathbf{A} \cdot \mathbf{B})^{-1} &= \mathbf{B}^{-1} \cdot \mathbf{A}^{-1} \\
\left(\mathbf{A}^{-1}\right)^{-1} &= \mathbf{A} \\
\mathbf{A} \cdot (c\mathbf{B}) &= c(\mathbf{A} \cdot \mathbf{B}) = (c\mathbf{A}) \cdot \mathbf{B}
\end{aligned}
$$

3. Find the inverse of the following matrices:

$$\begin{pmatrix} 1 & 3 \\ 4 & 5 \end{pmatrix} \begin{pmatrix} 1 & 2 \\ 4 & 8 \end{pmatrix}$$

4. Confirm by random matrices and some integer values of p and q the validity of the following identity:

$$\mathbf{A}^{p+q} = \mathbf{A}^{p} \bullet \mathbf{A}^{q}$$

1.3 A Simple Environmental Model

A simple example may illustrate the methodology used in this book. As an introductory simple situation consider the population of a biological species, which is denoted by c. As a first approach it seems reasonable that the number of children increases with the population. The number of children stands for the reproduction rate of the species, denoted as $\partial c/\partial t$, the temporal change of the population at each time instant. For the sake of simplicity one may consider that the reproduction rate is proportional to c:

$$\frac{\partial c}{\partial t} \propto c \tag{1.11}$$

When the proportionality factor is denoted by α, the same relation is expressed by the equation

$$\frac{\partial c}{\partial t} = \alpha c \tag{1.12}$$

which is a differential equation for the population c as a function of time t. With (1.12) the first task in modeling is already performed. The conceptual model, the proportionality relationship, is expressed as a differential equation. In this book differential equations are derived for various different processes in different environmental compartments. The user is led from a conceptual model concerning processes to the mathematical formulation of one or more differential equations.

This task is completed with the formulation of the initial condition: at time $t = 0$ the population has the value c_0, or:

$$c(t = 0) = c_0 \tag{1.13}$$

The second step of modeling is the solution of the differential equation under consideration of the boundary condition. There are several different means to do that. For simple equations the solution can be written explicitly in a formula, here:

$$c(t) = c_0 \exp(\alpha t) \tag{1.14}$$

The given exponential function fulfils both requirements. In MATLAB® the formula can be evaluated and plotted directly. The following commands need to be given in the command window (Fig. 1.6):

```
alpha = 1;
c0 = 1;
t = [0:0.1:1]
f = c0*exp(alpha*t)
plot (t,f);
```

```
t =
              0    0.1000    0.2000    0.3000    0.4000    0.5000    0.6000
0.7000    0.8000    0.9000    1.0000
f =
         1.0000    1.1052    1.2214    1.3499    1.4918    1.6487    1.8221
2.0138    2.2255    2.4596    2.7183
```

The concentration obviously increases by a factor of 2.8 during the time period of length 1. Before continuing let us have a short view on the commands given above. The first two commands specify the parameters α and c_0. The third command defines the vector `t`, containing 11 elements: 0, 0.1, 0.2, ..., 1. Note that MATLAB® prints the vector into the command window when the semicolon at the end of the line is omitted. The fourth command initiates several tasks. At first the vector `t` is multiplied by the parameter value of `alpha`. The result of this scalar multiplication is again a vector (see Sect. 1.2). As a next task, the exponential function is calculated for that vector. The command for the exponential function is:

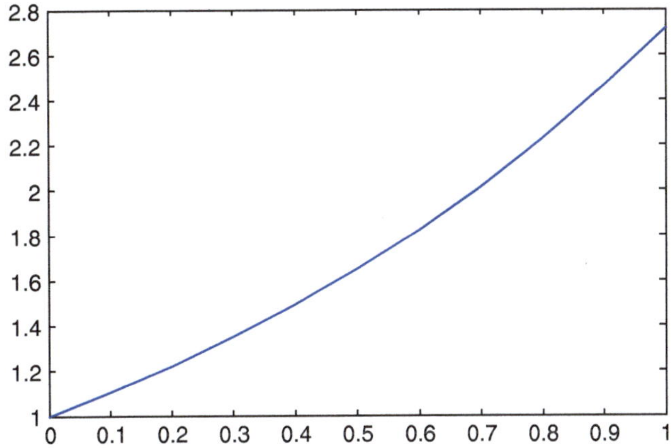

Fig. 1.6 MATLAB® figure; first example

`exp`. This again is a vector. It is the strength of MATLAB® that functions are defined on matrices. The result of the exponential operation is multiplied by `c0` - again by scalar multiplication – and stored in the vector `f`.

Finally, the `plot` command yields the graphical representation: the `t` vector is used on the x-axis, and the vector `f` on the y-axis. The plot is depicted in a new window on the display, the figure editor. More details of the figure editor are given in the next Section.

In many cases the solution can not be expressed by an explicit formula like in (1.14). For that situation MATLAB® offers a command for the numerical solution, which is the approximate solution derived by a computational algorithm. For ordinary differential equations there are several `ode`-commands, for partial differential equations it is the `pdepe`-command. Both situations will be explained in detail below. In addition, it is possible for the modeler to construct a numerical solver oneself. For that more challenging strategy examples will be given, too.

Usually the modeling is not complete with the second step. The third step of modeling is the evaluation of the results, which one may also call post-processing. Examples are simple calculations of derivative variables. In the given example one may be interested in the growth rate at the 10 time periods between the time instances, given by the vector `f`. This can simply be evaluated by using the `diff` – command

```
diff(f)

ans =
    0.1052    0.1162    0.1285    0.1420    0.1569    0.1734    0.1916
0.2118    0.2341    0.2587
```

Looking at the development of real species in the real world, the simple expression (1.12) turns out to be too simplistic to describe the observed behavior.

However, it may be sufficient for certain populations for certain time periods. The most obvious flaw of the model is that it allows the population to increase beyond any arbitrary margin, provided the time period is long enough. Of course this is impossible: as soon as a certain high population density will be reached, conditions will turn to become increasingly unfavourable and the reproduction rate will become smaller than assumed by the linear approach. Extended model approaches, which take a *carrying capacity* into account, will be presented in Chap. 19.

Let's examine the situation in which the proportionality constant in the example given above is lower than 1:

```
lambda = 1; c0 = 10;
t = [0:0.1:1];
f = c0*exp(-lambda*t);
plot (t,f);
grid;
```

Note that it is allowed to write several commands in a single line, as demonstrated in the first line. In such a case it is necessary to finish the writing of commands with ; (the last must not have it). Instead of using a negative parameter, we choose to specify a positive value but write the formula with a minus sign (Fig. 1.7).

Obviously the population is decreasing. This model is particularly interesting for biogeochemical species in the environment. In many situations the concentration of a chemical or biochemical species is declining according to the simple linear model, as presented. The shown development of concentration is well known as exponential decay. Exponential decay depends on the linear decay law (1.12). λ is called the decay constant or degradation constant, depending on the nature of the real process.

The proportionality constant can be related to a characteristic half-life $T_{1/2}$. The relationship is obtained from the condition:

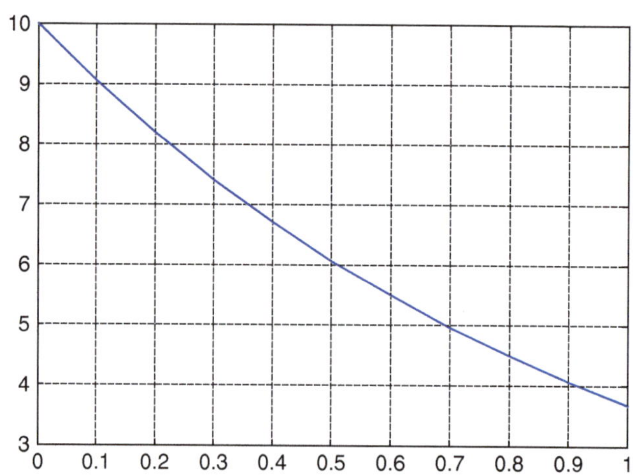

Fig. 1.7 MATLAB® figure; second example

$$\exp(-\lambda T_{1/2}) = \frac{1}{2} \tag{1.15}$$

which can be rearranged to:

$$T_{1/2} = \ln(2)/\lambda \tag{1.16}$$

'ln' denotes the natural logarithm. In MATLAB® the natural logarithm is called by `log`:

```
log(2)
ans =
    0.6931
```

According to formula (1.16) half-life and decay constant are of inverse proportionality. With increasing $T_{1/2}$ the decay rate decreases and vice versa. When the decay rate is 1 with the physical dimension [time^{-1}] in a given time unit, half-life is given by 0.6931 in the same time unit. Vice versa holds: for $T_{1/2} = 1$ the decay constant is $\lambda = 0.6931$.

It is important to realize that condition (1.15) delivers an universal half-life. In fact, it is a unique characteristic of the model for exponential decay that the concentration is halved after a universally fixed time period.

1.4 MATLAB® Graphics – The Figure Editor

As demonstrated above, using the `plot` command leads to a special graphical user interface for the creation and manipulation of graphics: the figure editor. The figure editor is reached directly by the `figure` command. Headline, main menu entries and buttons for the most important commands are shown in Fig. 1.8.

The figure editor has a manifold functionality out of which only few important elements can be mentioned here. Maximum and minimum on both axes are determined automatically, also the grid spacing on the axes. All these settings can be changed by using the sub-menu commands of the figure editor. The sub-menu entries of 'Edit' are depicted in Fig. 1.9.

The axes are changed using the 'Axes Properties…' option. All properties of the graphic can be changed under the 'Figure Properties…' option. The appearing input select box has changed between versions 6 and 7. Figure 1.10 depicts the outlook of the recent version.

Fig. 1.8 Headline, main menu entries and buttons of the build-in MATLAB® figure editor

Fig. 1.9 Edit command of the MATLAB® figure editor

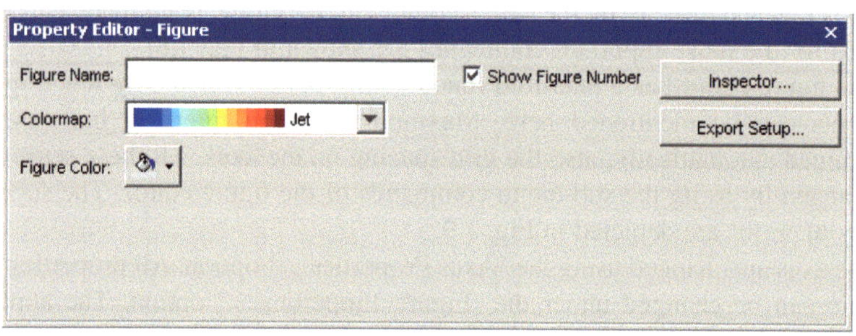

Fig. 1.10 Property editor of the MATLAB® figure editor

The elements of the graphics can be selected by mouse click. The Property Editor Box changes its outlook again, showing relevant properties of the chosen element. If the standard properties are still not sufficient, the 'Inspector…' button opens another input box for more properties to be checked and changed. See an example for a line element in Fig. 1.11.

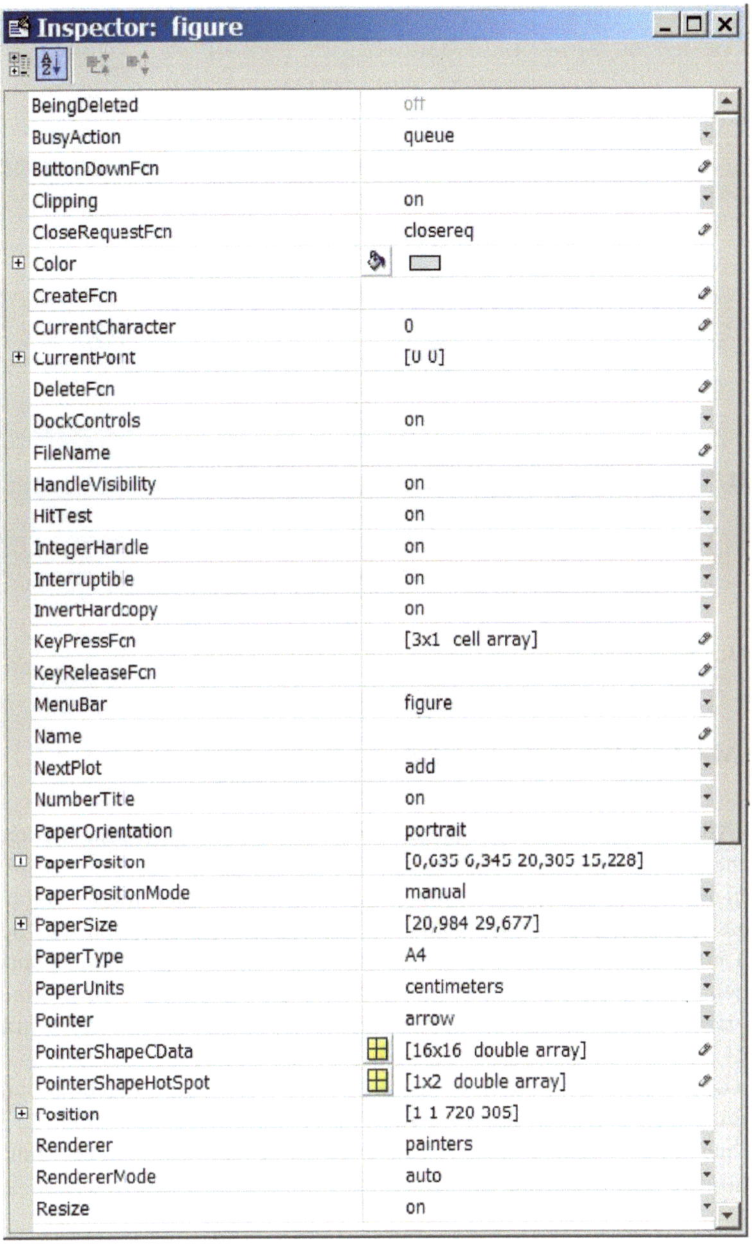

Fig. 1.11 Inspector input box of the MATLAB® figure editor

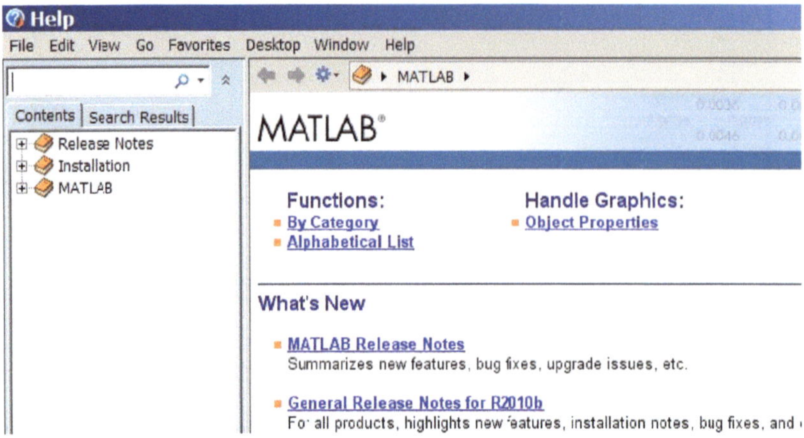

Fig. 1.12 The MATLAB help system

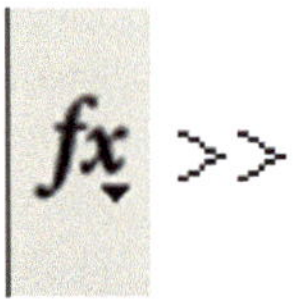

Fig. 1.13 The function browser symbol

1.5 MATLAB® Help System

MATLAB® has a simple to handle and very effective help system. It is reached under the main menu entry 'Help' under 'MATLAB Help.' Particularly useful is the detailed description of key words, which can be obtained under 'Index.' If the exact notation of a command or keyword is not known, one should use the 'Search' section and enter related terms. The term 'rectangle' for example does not appear in the index; but search leads directly to the corresponding MATLAB® command, which is `rectangle`. It must be used when a rectangle is added to a graphics (Fig. 1.12).

In recent versions of MATLA® there is a faster shortcut to functions and help concerning their use. A click on the function browser symbol (see Fig. 1.13) left actual line in the command window delivers a list of lots of MATLAB® functions and instructions to use them.

References

Cantrell RS, Cosner C (2003) Spatial ecology via reaction-diffusion equations. Wiley & Sons, Chichester, p 428

Christakos G, Bogaert P, Serre M (2002) Temporal GIS. Springer, Berlin, p 217

Chwif L, Barretto MRP (2000) On simulation model complexity. In: Joines JA, Barton RR, Kang K, Fishwick PA (eds) Proc. 2000 Winter Simulation Conf., Orlando, pp 449–455

Constanza R, Sklar FH (1985) Articulation, accuracy and effectiveness of mathematical models: a review of freshwater wetland applications. Ecol Model 27:45–69

Deaton ML, Winebrake JI (1999) Dynamic modeling of environmental systems. Springer, New York, p 194

Finlayson BA (2006) Introduction to chemical engineering computing. Wiley & Sons, Hoboken, p 339

Gander W, Hrebicek J (1997) Solving problems in scientific computing using MAPLE and MATLAB. Springer, Heidelberg, p 412

Harper M (2000) Sediment nitrogen and phosphorus processes model (SNAPP): User guide, p 10

Hornberger G, Wiberg P (2005) Numerical methods in hydrological sciences. Am. Geophys. Union, Washington DC(e-book)

Jørgensen SE (1994) Fundamentals of ecological modeling. Elsevier, Dordrecht, p 628

Kiusalaas J (2005) Numerical methods in engineering with MATLAB. Cambridge University Press, Cambridge, p 426

Kumblad L, Gilek M, Naeslund B, Kautsky U (2003) An ecosystem model of the environmental transport and fate of carbon-14 in a bay of the Baltic Sea, Sweden. Ecol Model 166(3):193–201

Luff R, Haeckel M, Wallmann K (2001) Robust and fast FORTRAN and MATLAB libraries to calculate pH distributions in marine systems. Comput Geosci 27(2):157–169

Lynch DR (2005) Numerical partial differential equations for environmental scientists and engineers. Springer, New York, p 388

Martin N, Gorelick SM (2005) MOD_FreeSurf2D: a MATLAB surface fluid flow model for rivers and streams. Comput Geosci 31(7):929–946

McCuen RH (2002) Modeling hydrologic change: statistical methods. Lewis, Boca Raton, p 433

Moler C (2004) Numerical computing with MATLAB. SIAM Books, Philadelphia, p 336

Richter O (1985) Simulation des Verhaltens ökologischer Systeme. VCH Verlagsgesellschaft, Weinheim, p 219 (in German)

Robbin JW (1995) Matrix algebra. A.K. Peters Ltd., Wellesley, p 554

Shampine LF, Gladwell I, Thompson S (2003) Solving ODEs with MATLAB. Cambridge University Press, Cambridge, p 263

Trauth MH (2010) MATLAB recipes for earth sciences, 3rd edn. Springer, Berlin, p 336

Zimmerman WBJ (2004) Process modeling and simulation with finite element methods. World Scientific, New Jersey, p 382

Chapter 2
Fundamentals of Modeling, Principles and MATLAB®

2.1 Model Types

As the term of 'model' has a wide variety of meanings, clearly a very narrow type of models is addressed in this book. Even within the special field of 'environmental models' only the sub-class of *deterministic models* is treated; statistical models do not appear, although statistics plays a significant role in environmental sciences and technology.

In deterministic models all variables and parameters are functions of independent space and time variables. The *independent variables* are referred to by the usual notation x, y, z and t. In most models, especially in the relatively simple examples presented here, it is sufficient to formulate the problem considering only a subset of these four variables.

Depending on the number of space dimensions, one speaks of 0D, 1D, 2D or 3D models. 0D models have no space dependency, only a time dependency. Ecological models concerning populations of biological species in an environmental compartment are in their majority of that type. As t is the only independent variable, the analytical formulation leads to *ordinary differential equations*. These are differential equations, which depend on one variable only; in contrast to *partial differential equations*, where there are at least two independent variables.

Models with no time dependency are denoted as steady, *steady state* or stationary. The corresponding terms for time dependent simulations are: unsteady or *transient*. A steady state is approached in real systems, if the internal processes have time enough to adjust to constant outer conditions. It is a necessary condition for steady state that exterior processes or parameters do not change in time. Otherwise steady conditions cannot be reached.

1D models include one space dimension only. Models for the soil compartment are mostly 1D, as the changes in vertical direction are of concern: seepage to the groundwater table or evaporation to the ground surface. Processes in rivers (image a water level peak or a pollutant plume moving downstream) can be regarded in 1D under certain conditions. Water from surface water bodies infiltrating into aquifers

DOI 10.1007/978-3-642-22042-5_2, © Springer-Verlag Berlin Heidelberg 2012

may be described by a 1D approach, if relevant conditions do not change substantially in the vertical direction and along the shoreline.

1D steady state models lead to ordinary differential equations. Transient models, including at least one space direction, lead to partial differential equations. It is important to know about these differences, as mathematical solution techniques for both types of equations are different and different MATLAB® commands need to be used. Here we use MATLAB® for steady and unsteady modeling in 1D.

2D models include two space variables. One may distinguish between 2D horizontal and 2D vertical models. Terrestrial ecology is a typical field, where this type of model is suitable, describing the distribution or population of species on the land surface. In streams or estuaries or in shallow water models are often set up for vertically averaged variables, for which a 2D horizontal description results. Models for 2D vertical cross-sections are obtained,

- In groundwater flow, where several geological formations are to be included, but no variations of hydraulic conditions in one horizontal direction
- In cross-sections of streams
- In air pollution modeling, if no space direction is preferential around a source; in that case the single radial coordinate r replaces two horizontal space variables x and y

3D models are quite complex in most cases and will marginally appear in this book, as the focus is on simple explanatory examples. Numerical algorithms using the methods of Finite Differences, Finite Volumes or Finite Elements are the methods of choice for modeling in higher space dimensions, steady and unsteady. The MATLAB@ 'Partial Differential Toolbox' can be recommended for the application of these algorithms.[1] As the focus here is on core MATLAB®, we leave numerical methods out. Instead it is outlined, how core MATLAB® can be applied for steady state modeling in higher dimensions, based on computing of analytical solutions.

2.2 Modeling Steps

The task of modeling can be sub-divided in several steps. The way from a real system to the working model contains different tasks, where every step depends on good performance of the previous step. The major steps are to build a conceptual model, to describe it by mathematical analysis, to solve the differential equations by computational methods and finally to perform post-processing tasks. A schematic overview of the procedure is given in Fig. 2.1 (see also: Holzbecher 1998).

[1] In the partial differential equations toolbox, the modeling of advection processes (see below) is difficult and requires numerical skills. All other processes can be simulated using the appropriate commands.

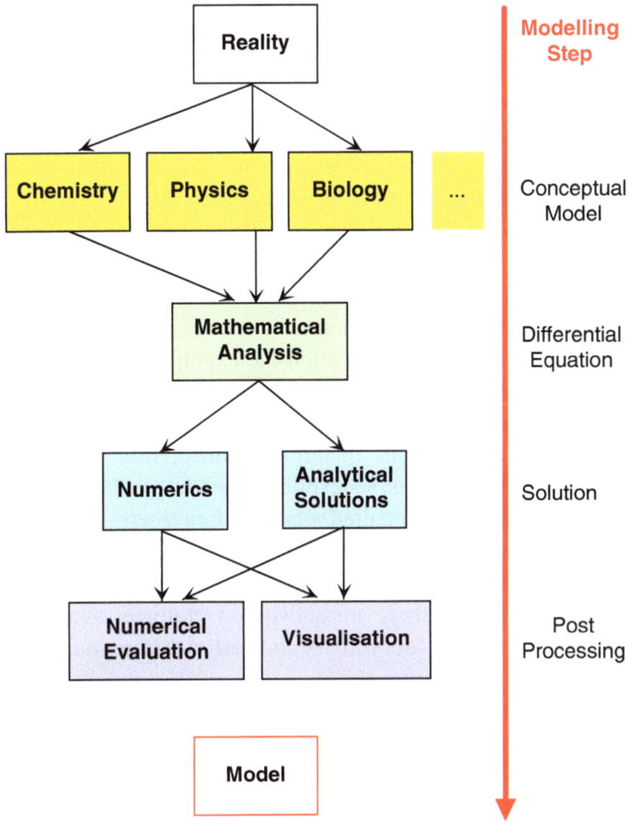

Fig. 2.1 Modeling steps

At first a conceptual model has to be formulated. From the available scientific and technical expertise and knowledge, as well as the experience with observations of the system in question, a concept has to be set up. The concept involves all processes, which could be relevant for the studied system. This first step is a qualitative one, i.e. there are no numbers involved yet. Scientific or technical expertise from the involved branches has to be included. In case of environmental problems advice has to be obtained from several disciplines mostly: chemistry, physics, biology, biochemistry, geology, biogeochemistry, ecology, hydrology, hydraulics, or hydrogeology.

The next step is the formulation of the model in mathematical terms. Variables and parameters as functions of time and space are related to each other by mathematical expressions. Rearrangements and transformations of the expressions usually lead to the formulation of differential equations. Fundamental theoretical or empirical laws and principles are combined to finally yield differential equations. In the simplest case there is a single equation only, in general a system of equations emerges. It can be *ordinary differential equations*, which have a single independent variably. More general *partial differential equations* depend on more than one independent variables. As will be shown in numerous examples the differential equations need to be accompanied by boundary and initial conditions, in order to complete the mathematical formulation.

With the next main step, according to the list in Fig. 2.1, we come to the computer. The solutions of the differential equations, under consideration of initial and boundary conditions, have to be calculated, which is always done on a computer. There is not a single strategy, which delivers these solutions and thus different paths have to be followed. Sometimes the solutions can be expressed by explicit formulae, which are quite easy to implement using any mathematical software. An example for such a situation with an *analytical solution* was presented in Chap. 1.3. With the exponential function the example formula is much simpler than in more general cases. However, in most cases even sophisticated analytical formulae do not suffice.

For most problem formulations in terms of differential equations numerical methods are required. As there is no explicit formula available, the solution has to be found approximately by so called *numerical methods*. It is sufficient to find such an approximate solution, as there are tolerance parameters, which mostly ensure to reach a good accuracy. Fortunately these methods need not to be implemented by the modeler: there are outworked strategies available in software packages. MATLAB® solvers for ordinary and partial differential equations will be presented.

In order to solve a problem usually some data need to be made available to the software program. Thus the computer very often comes into play even before the calculation of the solution is at stake. These tasks are referred to as *pre-processing*. A simple example is to transfer a parameter value from one physical unit into another. The computation of one parameter from another or from several others may be more complex. A more challenging task is the determination of parameter distributions within a model region based on some measured values. It will be shown how such tasks can be performed easily using MATLAB®.

After the approximate solution is computed (the computer always delivers approximate solutions, also when analytical solutions are evaluated!), there are usually several *post-processing* steps. Almost always the modeler and her/his client appreciate to have a graphical representation. Another task is compute fluxes for some key variables, may be in order to set up an entire balance for a model entity. For such a purpose numerical integration is a useful tool, which is also possible using MATLAB®. As another example the user may like to compare calculated and measure values. These few examples demonstrate that post-processing is a problem-specific task.

Finally the steps lead from a real system to a computer model. The step concept, sketched in Fig. 2.1, needs not to be followed strictly. In practise work will be on different steps at the same time. *Feedback loops* within the task list are necessary to improve earlier approaches, to correct errors and to adjust the model in view of measured data. Often new data come in after the start of modeling, which make adjustments necessary.

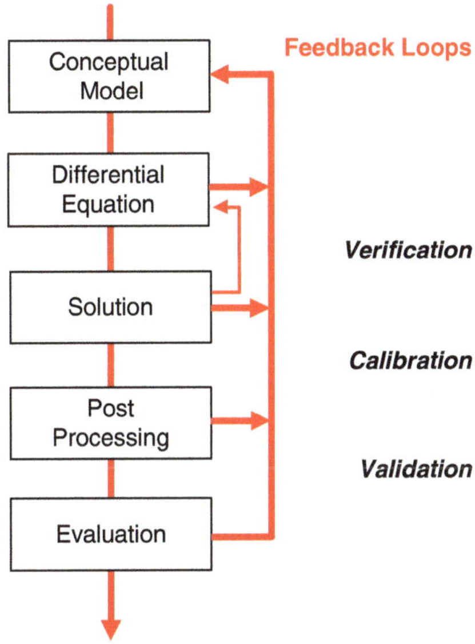

Fig. 2.2 Verification, calibration and validation as feedback loops between different model levels

The feedback loops within the system of tasks can be related to terms as verification, calibration and validation, which is visualized in Fig. 2.2. All these terms do not have a uniquely definition and thus may be found in slightly different contexts in the scientific literature. The term *verification* is mostly used in connection with software testing, which is a crucial step of software development. When a software code is entered and runs regularly without error messages from the computer, it is not sure that there are no 'bugs' in the code. In order to test the correct performance of the computer program, test cases are set up, to check, if the program delivers the correct answer. Test cases can be based on simple post-processing, on analytical solutions, on theoretical derivations and on inter-comparison with results from other codes.

Comparison with test cases, which are generally accepted in the concerned scientific community (so called benchmarks), is called *benchmarking*. Within the outlined step concept verification is thus a feedback loop in which it is checked, whether the computed solution delivers the solution of the differential equation. Unless it is a simple straight forward computation, the MATLAB® modeler also has to verify her/his implemented m-code. In case of errors the code *debugging* becomes necessary. How to 'debug' in MATLAB® is explained in Sect. 2.7.

The term *calibration* is used for the procedure of adjusting the model parameters for a specific application of the code. The term is almost identical to *parameter estimation* and strongly related to the term *inverse modeling*. Within the step concept the term means a feedback loop, in which the solution or some entity that is determined by post-processing is compared with values, obtained from the real site. In case the check is negative, usually some parameter values have to be adjusted in order to obtain a good fit. If that does not help, it may become necessary to make adjustments in the mathematical formulation or even in the conceptual model.

The work of *validation* is the most challenging. As the result of such work it is proven that a model is valid, i.e. behaves like the real system. The term usually is restricted to a field site application, but it sometimes also refers to the fact that a code can be applied for a certain type of applications. In the later case for each application site-specific parameters have to be included. Connected to software tools the term 'valid' remains slightly obscure as its concrete meaning is to be specified for each application field. It has to be clarified, which aspects of the real model should be represented by the model. As a model is not identical to the represented real system, there are always real world aspects for which the model is not sufficient.

The step concept, visualized in Fig. 2.1 and Fig. 2.2, is less a work schedule than a priority list. The mathematical formulation has to be based on a good conceptual model. If the mathematical formulation is insufficient, good solution techniques will not improve the model. One has to be sure that the solvers deliver accurate results, before putting extensive efforts into post-processing.

2.3 Fundamental Laws

The mathematical analytical formulation is based on fundamental principles and on empirical laws. From the former most important are the principles of conservation:

- Mass conservation
- Momentum conservation
- Energy conservation

Total mass, momentum and energy are preserved. If there are losses or gains, these are introduced in the conservation formulation as sources or sinks.

2.3.1 Conservation of Mass

The most nearby and most common application of the continuity equation is that for mass. There are two types of mass conservation. One type is the mass conservation of the medium, which can be solid, aqueous or gaseous. The mass is expressed in terms of a density ρ with a physical unit [M/L^3]. 'M' represents a mass unit and 'L' a length unit. For example the density of fresh water at a temperature of 4°C and the pressure of 101,325 Pa (1 atm) is 1,000 kg/m^3.

The second formulation of mass conservation in a fluid concerns biogeochemical species within a fluid. In that case the mass is expressed in terms of the concentration c. The continuity equation then is formulated in terms of the concentration c of the species. The concentration also has the unit [M/L^3] and measures the mass within a volume of fluid. The content of chloride Cl^- in seawater amounts to 19 g/l.

One has to extend the mass definition in situations, where the fluid does not fill the entire volume. Then the total mass per space volume is expressed by θc with porosity θ as additional factor. Porosity is dimensionless and measures the volumetric share of the fluid phase on the total volume. The physical dimension of the product is thus further [M/L^3]. In aquifers groundwater usually fills approximately 25% of the volume; thus holds: θ=0.25.

The described concept can be extended to situations with several phases, where each phase has its own θ-value. In the unsaturated soil zone below the ground surface, above the groundwater table there are three phases present: the soil as solid phase, seepage water as liquid phase and soil air as gaseous phase.

The mass of a gas component is expressed in terms of partial pressure. According to the ideal gas law the product of pressure and volume is a constant, which changes only with temperature. Thus mass conservation can be formulated in terms of pressures instead of volumes. According to Dalton's Law[2] the pressures of a gas mixture have to be summed up to yield the total pressure.

2.3.2 Conservation of Momentum

The momentum of a fluid is expressed as the product $\rho\mathbf{v}$, where $\mathbf{v}$ denotes the velocity. The physical unit of momentum is [M/(L^{2}T)], where the letter 'T' represents a time unit. As velocity is a vector, the momentum also is a vector, with one vector component for each space dimension of the model.

2.3.3 Conservation of Energy

The kinetic energy of a fluid is expressed as $\frac{1}{2}\rho v^2$ with the physical unit [M/(LT^2)]. If energy is measured in Joule, the given expression measures Joule per volume.

In problems, which include heat transfer, thermal energy is expressed in terms of temperature T. The energy content per volume is given by the product ρCT, where the new factor C is the specific heat capacity. The physical unit of C is [L^2/(T^{2}K)], where 'K' represents the temperature measure unit, mostly °Celsius or °Kelvin. In many tables values for the product ρC can be found, which is addressed simply as heat capacity. Heat capacity has the unit [M/(LT^{2}K)]. If energy is measured in Joule ρC has the physical unit of J per volume and °Kelvin, while C is measured in the unit Joule per mass and °Kelvin.

[2] John Dalton (1766–1844), English chemist and physicist.

2.4 Continuity Equation for Mass

For the mathematical formulation of mass conservation consider the change of mass during the small time Δt within a control volume with spacing Δx, Δy and Δz, each for one direction in the three-dimensional space. There are two ways to calculate changes of mass. One method is to consider the mass within the control volume at the beginning and at the end of the time period and calculate the difference. The other method is to balance all fluxes across the boundaries of the volume. Balancing means that fluxes into the volume have to be taken as positive, while those leaving the volume are negative. In three-dimensional space six faces of the control volume have to be taken into account.

A simpler set-up for the one-dimensional space is depicted in Fig. 2.3. A box contains a certain amount of mass at the start of the time period, and a different amount at the end. During the time period there was influx on one face and outflux at the other. The graphical symbols in the equation at the bottom of the figure will be replaced by mathematical formulae in the following derivation.

The mass at the beginning and the end of the period t and $t+\Delta t$ is given by:

$$\theta \cdot c(x,t) \cdot \Delta x \Delta y \Delta z \quad \text{and} \quad \theta \cdot c(x,t+\Delta t) \cdot \Delta x \Delta y \Delta z$$

where θ denotes the share on the total volume. In case of a saturated porous medium θ denotes porosity. In the unsaturated zone, within soil for example, θ is the volumetric water saturation, when the aqueous phase is concerned. In the situation in which two fluids occupy the space (for example water and oil) the share of each

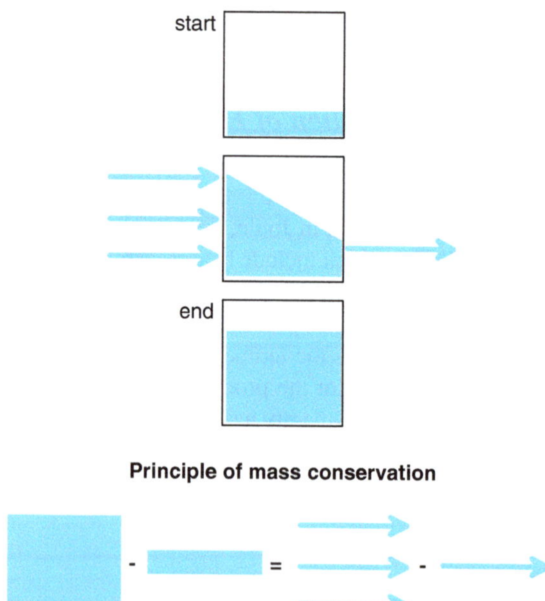

Fig. 2.3 Illustration for the derivation of the mass conservation equation

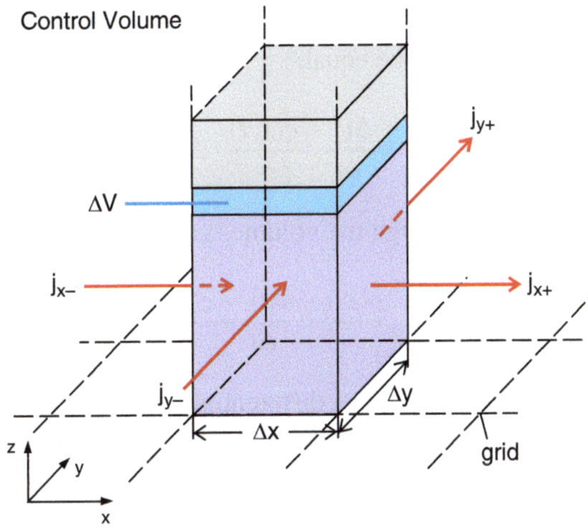

Fig. 2.4 Illustration of a control volume in two space dimensions x and y

phase has to be taken into account, too. $\Delta x \Delta y \Delta z$ is the volume. c denotes the concentration, measured as [mass/volume]. The change of mass per time is given by:

$$\theta \cdot \frac{c(x, t + \Delta t) - c(x, t)}{\Delta t} \cdot \Delta x \Delta y \Delta z$$

Fluxes in x-direction are given across faces of the control volume:

$$\theta j_{x-}(x,t)\Delta y \Delta z \quad \text{and} \quad \theta j_{x+}(x,t)\Delta y \Delta z$$

where j_{x-} denotes mass flux per area across the left face of the volume, in negative x-direction. Analogously j_{x+} denotes the mass flux in x-direction across the right face, in positive x-direction (see Fig. 2.4). The fluxes may change spatially and temporally which do the brackets indicate. Both fluxes are positive, if they add mass to the control volume, and negative otherwise. The physical unit of mass flux is [M/(L^2·T)]. The term $\theta \Delta y \Delta z$ denotes the area, through which flow takes place.[3]

The balance between both flux terms is thus given by:

$$\theta(j_{x-}(x,t) - j_{x+}(x,t))\Delta y \Delta z$$

For simplicity the fluxes across the four other faces are neglected for the derivation at this point. One may assume here that the flux components in y- and

[3] It is generally assumed that the volumetric share and the area share compared to the entire volume or area respectively are both quantified by the same number, here θ. This is not necessarily true. Especially in technical systems such as filters both ratios may vary significantly. The practioner in the field is usually happy, if there is one value at all.

z-direction are zero. As stated above, both formulations measure the change of mass and thus need to be equal:

$$\theta\frac{c(x,t+\Delta t)-c(x,t)}{\Delta t}\cdot\Delta x\Delta y\Delta z=\theta(j_{x-}(x,t)-j_{x+}(x,t))\Delta y\Delta z \qquad (2.1)$$

Division through the volume $\Delta x\Delta y\Delta z$ and porosity θ yields:

$$\frac{c(x,t+\Delta t)-c(x,t)}{\Delta t}=-\frac{j_{x+}(x,t)-j_{x-}(x,t)}{\Delta x} \qquad (2.2)$$

From this equation a differential equation can be derived by the transition of the finite grid spacing Δx and time step Δt to infinitesimal expressions, e.g. by the limits $\Delta x\to 0$ and $\Delta t\to 0$. It follows:

$$\frac{\partial c}{\partial t}=-\frac{\partial}{\partial x}\quad j_x \qquad (2.3)$$

which is a differential formulation for the principle of mass conservation. The presumption for the differentiation procedure is that the functions c and j_x, are sufficiently smooth, mathematically speaking differentiable, which is usually taken for granted. Equation 2.3 is valid for one-dimensional transport and is the basis for the mathematical analysis of transport processes. The unit of the equation is $[M/(L^3\cdot T)]$.

Formulation (2.3) is valid if there are no internal sources or sinks for the concerned biogeochemical species. Sources and sinks are understood here in the most general sense: each process, which creates or destroys some species mass, can contribute to such a source or sink. In the remainder of this volume we will see examples, where chemical reactions and inter-phase exchange of species can be included in that way.

Easily the given mathematical formulation can be extended to consider sources and sinks additionally. If these are described by a source- or sinkrate $q(x,t)$ $[M/(L^3\cdot T)]$, which may vary spatially and temporally, one simply has to add a corresponding integral term

$$\int\limits_{\Delta x}\int\limits_{\Delta t} q(x,t)dtdx$$

on the right side of (2.1) and (2.2). The term is positive, if mass is added (source) and negative, if mass is removed (sink). In the derivation of (2.3) the integral term had to be differentiated, which leads to the general transport equation in one space dimension:

$$\theta\frac{\partial c}{\partial t}=-\frac{\partial}{\partial x}\theta j_x+q \qquad (2.4)$$

Flux components in y- and z-direction can also be taken into account, based on formulae analogous to formula for the x-direction. The fluxes j_{y-}, j_{y+}, j_{z-} and j_{z+} have to be introduced, balanced and the balances added on the right side of (2.1) and (2.2). Taking the limits $\Delta y \to 0$ and $\Delta z \to 0$ one obtains:

$$\theta \frac{\partial c}{\partial t} = -\left(\frac{\partial}{\partial x}\theta j_x + \frac{\partial}{\partial y}\theta j_y + \frac{\partial}{\partial z}\theta j_z\right) + q \tag{2.5}$$

which is the generalized formulation of the mass conservation for three space dimensions. Using the formal ∇-operator (speak: 'nabla'),

$$\nabla = \begin{pmatrix} \frac{\partial}{\partial x} \\ \frac{\partial}{\partial y} \\ \frac{\partial}{\partial z} \end{pmatrix} \text{ in 3D, } = \begin{pmatrix} \frac{\partial}{\partial x} \\ \frac{\partial}{\partial y} \end{pmatrix} \text{ in 2D, } = \frac{\partial}{\partial x} \text{ in 1D} \tag{2.6}$$

the equation can be written more compactly:

$$\theta \frac{\partial c}{\partial t} = -\nabla \bullet \theta \mathbf{j} + q \tag{2.7}$$

With the different forms of the ∇-operator the short notation of the continuity (2.7) is valid in one-, two- or three-dimensional space. On the right side the ∇-operator is multiplied by the flux vector $\theta \mathbf{j} = \theta \begin{pmatrix} j_x \\ j_y \\ j_z \end{pmatrix}$ as a vector product. In the formulae, here and in the following, the $\bullet$ denotes the standard *vector product*,[4] which in MATLAB® is applied by using the * multiplication and the transpose of one column vector. Examine with the following command:

```
[1;2]'*[1;2];
```

An advantage of the formulation (2.7) is that it is valid for one-, two- and three-dimensional situations. The number of components in the flux-vector and ∇-operator is equal to the number of space dimensions. In two dimensions, as illustrated in Fig. 2.4, the flux vector has two components. The illustration is concerned with a fluid, for which the mass conservation principle can also be applied, as for any other chemical species. When the fluid density is not changing,

[4] In three dimensions for vectors arbitrary vectors **u** and **v**:
$\mathbf{u} \bullet \mathbf{v} = \begin{pmatrix} u_x \\ u_y \\ u_z \end{pmatrix} \bullet \begin{pmatrix} v_x \\ v_y \\ v_z \end{pmatrix} = u_x v_x + u_y v_y + u_z v_z$, not to be confused with the cross-product $\mathbf{u} \times \mathbf{v}$; another formulation, found in the literature is: $\frac{\partial c}{\partial t} = -div\mathbf{j}$; divergence 'div' is another expression for a vector product with the nabla-operator.

one may express mass change by volume change (ΔV), which itself is represented by a change of the 'water'-table.

The derived equation for mass conservation alone is not yet sufficient for a complete mathematical formulation. There are too many unknown variables, namely concentration c, and the components of the flux vector **j**. In order to reduce the number of unknowns, one has to utilize a formulation, in which the flux terms are connected with the concentrations. Finally an equation will result, in which c is the only unknown variable.

The advective flux is simply given by the product of the concentration and the velocity. In the three-dimensional case the three flux components are determined by the three velocity components:

$$j_x = v_x c \quad j_y = v_y c \quad j_z = v_z c \tag{2.8}$$

Independent of the dimension, using the scalar multiplication one may write in vector notation:

$$\mathbf{j} = \mathbf{v}c \quad \text{or} \quad \mathbf{j} = c\mathbf{v} \tag{2.9}$$

In formulae we mostly omit the multiplication sign. Note that on the right side of (2.9) there is scalar multiplication: the scalar variable c is multiplied with the vector variable **v**, as already outlined in Chap. 1.

2.5 MATLAB® M-files

The command window is good for an introduction into MATLAB®. It demonstrates that operations, which someone used to compute on a scientific calculator, can be better performed using MATLAB®. However, MATLAB® can do much more.

In MATLAB® a high level programming language is included. It is often called 'M'. Files, containing 'M' source code, are stored in files with extension '.M'. The programming work with m-files replaces extensive operating in the command window. Only for certain tasks, the command window will remain the most direct and simple way to compute with MATLAB®. Program writing and editing, not only with MATLAB@, is done by use of a text editor.

MATLAB has a built-in editor. The editor can be called in different ways. The easiest way is to use

```
edit
```

in the command window. Other convenient ways of calling the editor are given in the next section. Options for the editor are set in the 'Preferences' sub-menu of 'File' (see Fig. 2.5). It is also possible to use other than the built-in editors. With the

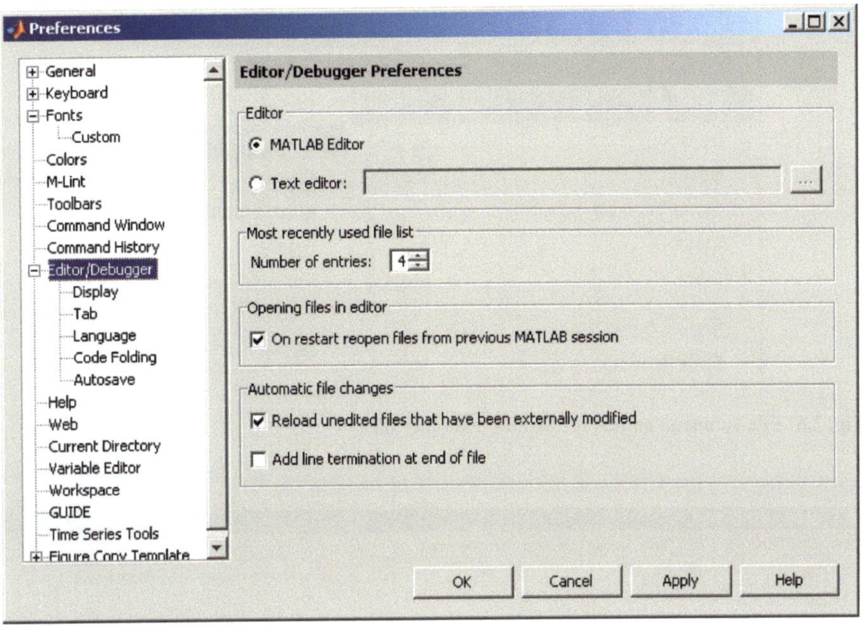

Fig. 2.5 Preferences, relevant for the editor

editor the programmer works on m-files. In the first chapter it was shown how MATLAB® operations are stated in the command window. Even after few exercises the user will recognize that it is often necessary to give a former command again or to give it in a slightly different form; maybe just with a parameter name altered. It was already shown that the command history view of the MATLAB® graphical user interface is an appropriate tool to go back to former commands. As already mentioned an alternative way is to use the up- and down-arrow buttons of the keyboard.

There is another alternative, which for most purposes turns out to be so powerful that most users prefer it for their normal work in comparison to the command window. MATLAB® command sequences can be gathered and stored as files. The extension of these files is simply *.m*; for that reason they are called *M-files*. What the MATLAB® user does with M-files, is what programmers do with other programming languages, as FORTRAN, JAVA or some C variant, just to mention some names.

In order to create a new M-file we use the 'New→M-file' entries of the 'File' main menu, as shown in Fig. 2.6. In the same menu there is an entry for opening an already existing M-file.

A simple example demonstrates the procedure. As just described, create a new M-file. The MATLAB® editor appears. The main menu of the editor is depicted in Fig. 2.7. Type the following commands:

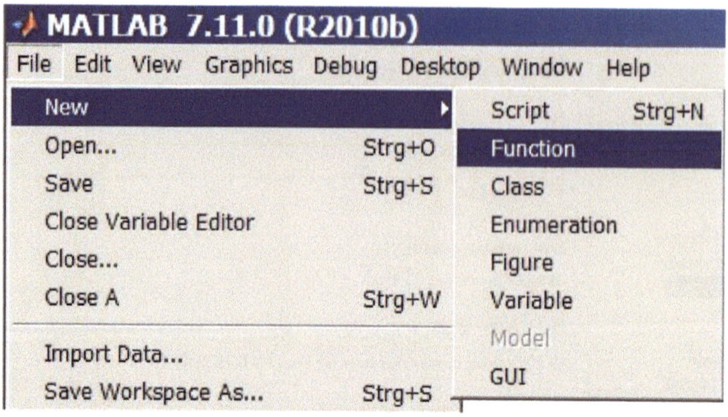

Fig. 2.6 File submenu entries for M-files and other files

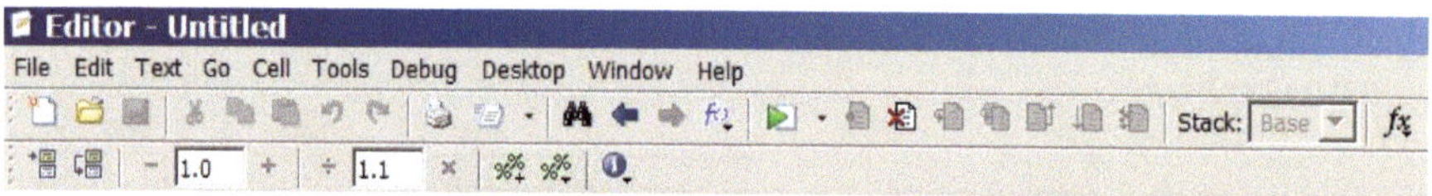

Fig. 2.7 MATLAB® editor; main menu entries of the graphical user interface

```
c0 = 1;
t = [0:0.1:1];
f = c0*exp(-lambda*t);
plot (t,f);
```

Use the 'Save' or 'Save as...' entries in the 'File' menu of the editor, or the corresponding button, to save the file under the name *'example.m'*. Return to the command window and type:

```
lambda = 1;
example;
```

A graph showing concentrations decreasing with time appears.[5] All commands of the *'example.m'* file are executed, when `example` is used as command. A bundle of graphs can now be plotted in the same figure without much typing work:

[5] Depending on the specific installation and organization on the computer, the user could have a problem here, if MATLAB® does not find the *'example.m'* file. The user should store the file in a directory, which is included in the MATLAB® path list. Use the command `path`, to see the current path list of the installation. In case of problems store the file in another directory or use the `addpath` command, to add another directory in the path list. There may also be a problem, if the user chooses an M-file name that already exists within the directories of the path list. Use the `which` command in order to check, if your stored version of the file is the most nearby version in the MATLAB® system on your machine!

Fig. 2.8 Graphical output of '*example.m*' file

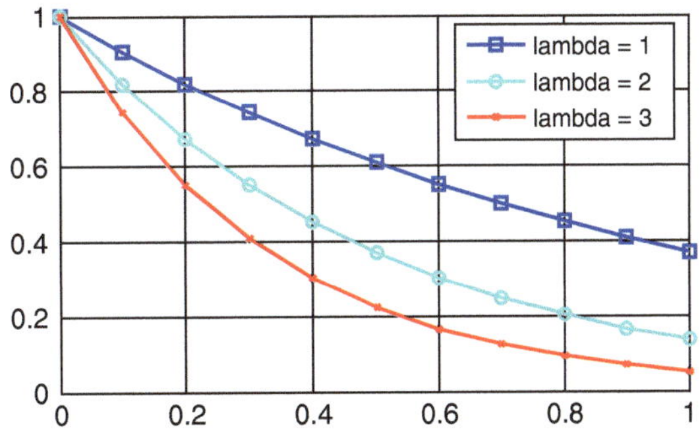


```
hold on;
lambda = 2;
example;
```

A second graph appears. Proceed further:

```
lambda = 3;
example;
legend ('lambda = 1','lambda = 2','lambda = 3');
```

A graph, showing stronger degradation with increasing `lambda` – value, appears. It should look similar to Fig. 2.8

The example shows that it is convenient to call repeatedly appearing command sequences in an M-file. The demonstration application is a mixture of commands on the command window, and of editing a file. Alternatively the entire command sequence can be started in an M-file. There is also an alternative way to start the current M-file: the button  in the main menu of the editor.[6] The procedure is exemplified in the next section. There is an alternative way to create an M-file directly from the command history. One or several command in the command history can be highlighted. By using the right mouse button one gets several options. One option is to create a M-file directly from the selected commands.

There are two types of M-files: *scripts* and *functions*. Scripts are simply files containing a sequence of commands. By typing the filename, the commands are executed subsequently. Variables, which are not specified or initialized in the script have to be available in the workspace, when the script is executed. Variables, specified in a script, are available in the workspace after execution, if they are not cleared explicitly.

Like scripts functions contain sequences of commands, and are stored with the extension '.m'. Formally they differ from scripts as the first no-comment line starts

[6] The button does not appear, when the editor is called outside of MATLAB®, for example by double-click on a M-file from the operating system.

with the `function` keyword. What follows is a list of output parameters, the equality sign, the function name and a list of input parameters:

```
function [out1, out2,…] = myfun (in1, in2, …)
```

Also alike scripts functions are called by using the function name and *formal parameter* sets for input and output.

```
[out1, out2,…] = myfun (in1, in2, …)
```

either from the command window, from a script or from another function. The filename must be identical with the function name, i.e. in the example: 'myfun.m'. Formal parameters in the calling command and the function are identified on basis of the parameter lists. They need **not** be identical! For example the call:

```
myfun (5, A)
```

results in a continuation of the execution with the function commands, where the parameter `in1` obtains the value 5, and `in2` gets the value of variable `A`. No output is taken from the function in this case. By using

```
[B, A] = myfun (5, A)
```

`B` and `A` obtain values calculated during execution of the function. Then the function commands are executed, the execution returns to the calling statement and proceeds with the following command.

Functions may contain subfunctions. Subfunctions are not visible outside the file where they are defined. Normally functions return when the end of the function is reached. The

```
return
```

statement can be used to force an early return. If there are functions with identical names, the ones defined as subfunctions have priority. In general the programmer may use the

```
which
```

command to find the location of function with highest priority. The

```
nargin
nargout
```

commands deliver the numbers for input resp. output parameters. This can be used to deal with varying formal parameters, for example to allow the user to call the file with a reduced number of parameters. This is feasible if for missing parameters default values are specified.

2.6 Ifs and Loops in MATLAB®

In M-files quite often commands or sequences of commands are to be executed under certain conditions only. This can be easily programmed using the `if`, a keyword that is available in all programming languages. The following command sequence in pseudo-programming language explains its application:

```
if condition
    commands1
else
    commands2
end
```

If the `condition` is fulfilled, `commands1` are executed. Otherwise `commands2` are executed. The `else` block can be omitted if there is nothing to be done, if the `condition` is not fulfilled. Conditions can be formulated differently. Instead of a rigorous definition here four mainly self-explaining examples:

```
a==b    a>1    a<b    a
```

In order to be valid conditions, `a` and `b` must be of same data type. There are two equality signs in the first example in order to distinguish a condition from an assignment. In the second example the condition is clear, if `a` is a number. But the condition is also valid if `a` is of different data type, which the reader may explore. The last of the four conditions is fulfilled if `a` is a positive non-zero number, otherwise not. The letter represents the value of a condition, and could also be defined like this in MATLAB®:

```
c = (a>=0);
if c a=1; end
```

By the example sequence the value of `a` is changed to 1, if the former value was positive or equal to zero. Variables as `c` of the example are so-called *logicals*. These are variables that can take only two values: 'true' or 'false.' As other programming languages MATLAB® treats logicals as binary, i.e. they have values 1 for 'true' and 0 for 'false.'

```
2>1
ans =
     1
```

Logical variables can be combined by logical operators (and, or, not and exclusive or), as given in Table 2.1. Example:

```
if A>0 & F==1
```

Table 2.1 Defining table for logical operators

Input parameter		and	or[a]	not	xor
`A`	`B`	`A & B`	`A | B`	`~A`	`xor(A,B)`
0	0	0	0	1	0
0	1	0	1	1	1
1	0	0	1	0	1
1	1	1	1	0	0

[a] The | symbol is reached by Alt-Strg->

It is also possible to check the type or any other characteristic of a variable. For that purpose there are several `is...` keywords (which have at least one variable as formal parameter):

`isfloat` checks if a variable is a float, i.e. a single or double
`islogical` checks if a variable is a logical
`isscalar` checks if variable is a scalar
`isvector` checks if variable is a vector
`isfinite` checks if variable is finite
`isinf` checks if variable is infinite
`isnan` checks if variable is not a number
`isnan` checks if variable is not a number
`isempty` checks whether an array is empty
`isnan` checks if variable is not a number

Look for `is*` in the help system (see Chap. 1.5) to obtain a complete list of all such commands.

The `if`, `else` and `end` keywords are reserved and should not be used by the MATLAB® user as function or variable names. The `if`, `else` and `end` keywords are reserved and should not be used by the MATLAB® user as function or variable names. An `if` without `else` is just a condition. Using the `if- else` construction, one has the option to bifurcate into one of two branches. A bifurcation into several branches can be realized by using nested if-else commands:

```
if condition1
     if condition2
     commands1
     else
     commands2
     end
  else
     if condition3
     commands3
     else
     commands4
     end
end
```

There is also the possibility for a direct multiple branching. For this purpose MATLAB@ has the `switch- case` construction. It is exemplified in the following examples:

```
switch i
case 0
    commands1
case 1
    commands2
case 2
    commands3
end
```

In order to use the functionality of a single M-file for the example from the preceding section, the programming technique of *loops* has to be applied. How loops work, is best explained by an example. The most common `for`-loop is used in the command sequence:

```
c0 = 1;
t = [0:0.1:1];
figure; hold on;
for lambda = 1:1:5
   f = c0*exp(-lambda*t);
   plot (t,f);
end
legend ('1','2','3','4','5');
```

`for` is a MATLAB® keyword, which is indicated by a different color. MATLAB® keywords may not be used as variable names. The statement, starting with the `for` keyword, starts a loop, which ends at the line with the `end` keyword. `lambda` is the loop variable in this example (the user is free to choose the name of the variable). In the first run through the loop `lambda` has the value 1. The commands within the loop are executed with that parameter value; here the function vector `f` is evaluated and a graph is plotted in the figure. Then the commands are executed again with the variable parameter taking the next value. In the example `lambda` gets the values 2, 3, 4 and finally 5. After the last run through the loop the execution continues with the next statement after `end`. In the example after the loop, the final command adds the legend in the figure. As described in Sect. 2.5 store the entire M-file under a new name and call that name in the command window; or use the  button (old version:  (Fig. 2.9)).

Note that commands within loops are indented, as shown in the example above. This is not required by MATLAB®, but it is highly recommended, as it enhances the readability of the M-files significantly. Loops may be nested, i.e. inner loops may appear within outer loops. In that case it is important to recognize the corresponding `end` statements. It is good practise to indent commands with every inner loop even further. In that way the programmer visually obtains a connection between the start of a loop (with either a `for` or a `while` keyword) and the corresponding `end`.

Fig. 2.9 Illustration of a loop

Like in any other programming language, also in MATLAB®, there are several other types of loops. Another main type is the `while`-loop. Here a condition is checked at the start of the loop, and the commands of the loop (between `while` and `end` statements) are executed as long as the condition is fulfilled. Here the use has to take care that some of the variables appearing in the condition change during the loop; otherwise an infinite loop in produced.

If in larger loops it is necessary to have another outlet from the loop aside from the default at the `end`, one may use the `break` command. In nested loops the command only concerns that one loop, in which the statement is placed, i.e. execution proceeds with the command after the next end of loop statement.

`continue` is a related command: the execution is skipped not for the entire concerned loop (as with `break`), but only for the remaining statements in the current run through the loop.

Loops can also be used in the MATLAB® command mode. A trivial example is:

```
for i = 1:4
```

Note that no prompt appears after pressing the return key. The command is not yet executed. Enter:

```
a(i) = i
end
```

When the `end` command is entered the entire loop is executed and the prompt appears in the command window again. Note that in the specification of the loop variable `i` only two values are given. The increment 1 is used as default in such a situation.

In MATLAB® the direct notation of loops can often be avoided by using vector- or vector–matrix notations. For an example examine the 'elegant' solution implementing the Horner scheme, as shown above. By matrix–vector multiplication loops are implicitly performed, as there is a sum over the products of matrix elements. Utilizing this type of looping not only is much shorter and more clearly arranged, it also is usually much more efficient, as special vectorized implementations of the basic operations on a near-machine level are utilized.

2.7 Debugging of M-files

It was already said that the M-file, currently shown in the editor, is executed by pressing the  button in the editor menu list. Alternative ways to do the same thing are (1) pressing the F5 button, or (2) select the sub-menu entry 'Run' from the 'Debug' main entry. For the exploration of M-files some other debugging commands are very convenient, which are explained here.

Debugging is a term, which came up with computer programming. Debugging is the task to find and correct errors, so called bugs. If a program or an M-file does not behave as intended, if there is still at least one erroneous statement, it is convenient to stop the program execution at a certain point and watch the execution of the following steps in detail. For such tasks several tools are available in the MATLAB® editor in the 'Debug' submenu.

A *breakpoint*, at which execution stops, can be set by the user easily by using the column of '-'signs left from the line number counter in the editor window. A mouse click initiates the bar to change into a red circle, indicating the location of a breakpoint, as shown in Fig. 2.10. When the program is run, execution stops at the first breakpoint encountered. A green arrow indicates the current position of the command execution (see Fig. 2.10).

Now the user may check the current values in variables. Moving the curser through the editor window will make the contents of variables pop-up in small boxes. Figure 2.11 depicts an example: variable T, which was touched by the cursor, is a 1×1 double variable and currently contains the value 4. Another method for checking variables at a breakpoint is more convenient, if the variable is a huge array, and its contents can not be shown appropriately in a small box. The user can change into the MATLAB® command window and examine the workspace, as described in Chap. 1.

The  button in the editor's main menu initiates the execution of the next line only. Alternatives are the keyboard F10 button or the 'Step' entry in the 'Debug' sub-menu. Now each command can be checked step by step, examining the effect of the command on the variables involved. The green arrow on the left side of the text window moves further with every step, always indicating the next command, which is not jet executed. Although the 'Debug' commands and options were designed for the programmer to find bugs in programs, these are also convenient tools for novices to understand M-files, written by others. Novices are urged to try the debug functionality on the loop of the M-file of the previous section.

There are further debugging tools, for which the reader may view the MATLAB® help. It is important to know that the  button (or key F5) always

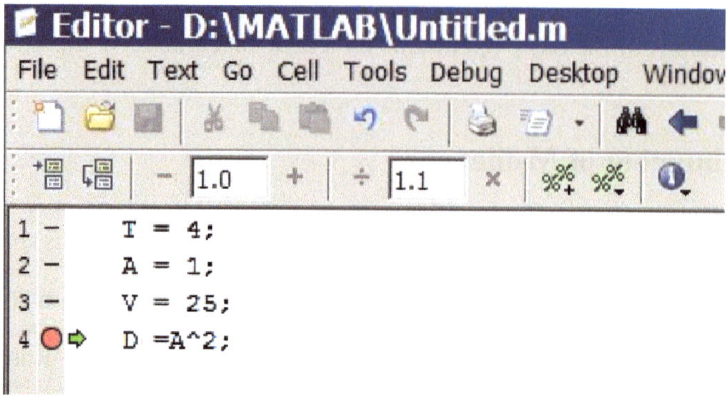

Fig. 2.10 Illustration of debugging; set of breakpoint and stop of execution

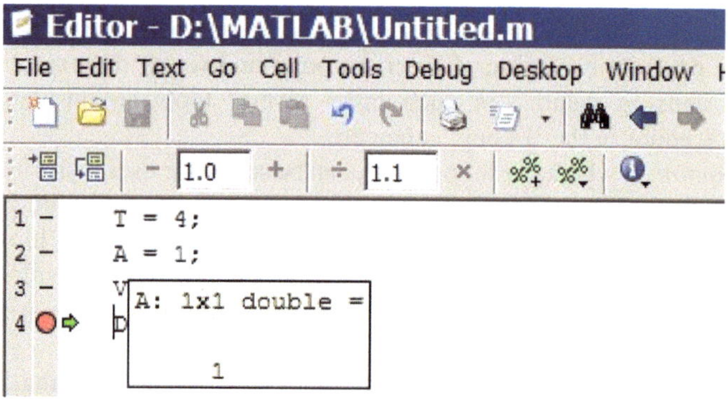

Fig. 2.11 Illustration of debugging; variable check

stands for continuation: starting from the current position the M-file will be executed until it reaches the next breakpoint (it is possible to set several breakpoints!) or the end of the M-file. The  button stops execution; alternatively select sub-menu entry 'Exit Debug Mode.' Breakpoints are deleted by a click, which changes the red circle back to a bar. All breakpoints are cleared by using the  button, or by selection of the corresponding submenu entry.

Reference

Holzbecher E (1998) Modeling density-driven flow in porous media. Springer, Berlin, p 286

Chapter 3
Transport

3.1 The Conservation Principle

Transport is a general term, denoting processes which determine the distribution of biogeochemical species or heat in an environmental compartment. In this chapter transport is understood in a narrower sense as interaction of physical processes with an effect on species or components, or on heat. Other processes, which also may be relevant for the environment, like sorption, degradation, decay and reactions of various types, are not conceived as pure transport processes and are treated in chapters below.

These transport processes are relevant in almost all environmental systems. The term is not restricted to a specific compartment of the environment. Heat and mass transport are a common phenomenon which can be found almost everywhere, in the hydrosphere, in the pedosphere as well as in the atmosphere, in surface water bodies – rivers, lakes and oceans, in sediments, in groundwater, in the soil, in multi-phase systems as well as in single phases.

There are two different types of transport processes in the narrower sense: advection and diffusion/dispersion. Advection denotes transport in the narrowest sense: a particle is purely shifted from one place to another by the flow field. Diffusion and dispersion are processes which originate from concentration differences. Within all systems there is a tendency to equalize concentration gradients. If the species are mobile, e.g. if they have the possibility to move from one place to the other, there will be net diffusive or dispersive flux from those locations with high concentrations to locations with low concentrations.

Transport can be described by differential equations as will be shown below. In fact it is one differential equation for each species or component. The differential equation, the so called transport equation can be derived from the principle of mass conservation and Fick's Law.

Concerning heat transport a differential equation for temperature T as dependent variable results. The equation is derived from the principle of energy conservation and from Fourier's Law. From the mathematical point of view it is the same

E. Holzbecher, *Environmental Modeling*,
DOI 10.1007/978-3-642-22042-5_3,

differential equation, with different meanings of the coefficients only. Thus we refer to the temperature equation as transport equation as well.

The fundamental formulation of a conservation principle is expressed by the general continuity equation. For mass the continuity equation was already derived in Chap. 2. What follows next is a generalization of the mass conservation equation, derived in Chap. 2. The conservation of variable A, which may represent mass, momentum or energy, and which depends on time t and the three space directions x, y and z, is quantitatively expressed by the differential equation:

$$\frac{\partial}{\partial t}A = \frac{\partial}{\partial x}j_{Ax} + \frac{\partial}{\partial y}j_{Ay} + \frac{\partial}{\partial z}j_{Az} + Q \tag{3.1}$$

where j_{Ax}, j_{Ay} and j_{Az} represent fluxes in the three space directions. The three flux terms are components of the flux vector $\mathbf{j}_A$ corresponding to the three space directions. Fluxes, as all other terms in the continuity equation, depend on the independent variables x,y,z and t too. In the term Q all sources and sinks are gathered. If $Q(x,y,z,t)$ is positive, there is a source at time t at position $\mathbf{r} = (x,y,z)$; if Q is negative, there is a sink.

The continuity equation states that the amount of change of variable A in time is equal to the local flux budget. The continuity equation is derived from the budget of a control volume, i.e. a volume of finite small extensions Δx, Δy and Δz(in 3D). Figure 2.3 shows a control volume in 2D for a fluid filling the entire space, where only two finite extensions Δx and $\Delta y$are sufficient.

In the small but finite time interval Δt, the amount of A per volume unit changes from $A(x, y, z, t)$ to $A(x, y, z, t + \Delta t)$. The total amount of change in the control volume is thus given by $(A(x,y,z,t+\Delta t) - A(x,y,z,t))\Delta x\Delta y\Delta z$. In Fig. 2.3 this corresponds to the volume change ΔV. On the other hand, the total budget can be expressed by the fluxes, the sources and the sinks. In each space dimension there are two surfaces, across which mass, momentum or energy may enter or leave according to the corresponding flux component. In x-direction the fluxes across the two faces are given by $\left(j_{Ax}(x+\frac{\Delta x}{2},y,z,t) - j_{Ax}(x-\frac{\Delta x}{2},y,z,t)\right)\Delta y\Delta z\Delta t$; the difference in flux terms of j_{Ax} from one side of the control volume to the opposite side has to be multiplied by the face area of the control volume, which is here given by $\Delta y\Delta z$. In the notation of the fluxes, visualized in Fig. 2.3, the A in the subscript is omitted and the + or – sign denotes the direction. Note the assumption that the time step Δt is small, so that the change of the flux terms and also of the sinks and sources during that time can be neglected.

Both expressions of the change within the control volume with a time step have to be equal, which is expressed in the detailed equation:

$$\begin{aligned}
&(A(x,y,z,t+\Delta t) - A(x,y,z,t))\Delta x\Delta y\Delta z \\
&= \left(j_{Ax}(x+\frac{\Delta x}{2},y,z,t) - j_{Ax}(x-\frac{\Delta x}{2},y,z,t)\right)\Delta y\Delta z\Delta t \\
&+ \left(j_{Ay}(x,y+\frac{\Delta y}{2},z,t) - j_{Ay}(x,y-\frac{\Delta y}{2},z,t)\right)\Delta x\Delta z\Delta t \\
&+ \left(j_{Az}(x,y,z+\frac{\Delta z}{2},t) - j_{Az}(x,y,z-\frac{\Delta z}{2},t)\right)\Delta x\Delta y\Delta t + Q\Delta x\Delta y\Delta z\Delta t
\end{aligned} \tag{3.2}$$

Using the notation of the 2D case in (3.2), the corresponding equation without sources or sinks is simply $\Delta V = (j_{x+} - j_{x-})\Delta y \Delta t + (j_{y+} - j_{y-})\Delta x \Delta t$. Equation 3.2 is simplified in two steps. First one divides through the product of all spatial extensions and the finite time step $\Delta x \Delta y \Delta z \Delta t$ and obtains:

$$\begin{aligned}\frac{A(x,y,z,t+\Delta t) - A(x,y,z,t)}{\Delta t} = & \frac{j_{Ax}(x+\frac{\Delta x}{2},y,z,t) - j_{Ax}(x-\frac{\Delta x}{2},y,z,t)}{\Delta x} \\ & + \frac{j_{Ay}(x,y+\frac{\Delta y}{2},z,t) - j_{Ay}(x,y-\frac{\Delta y}{2},z,t)}{\Delta y} \\ & + \frac{j_{Az}(x,y,z+\frac{\Delta z}{2},t) - j_{Az}(x,y,z-\frac{\Delta z}{2},t)}{\Delta z} + Q\end{aligned} \tag{3.3}$$

The second step is the transition from finite steps to infinitesimal steps, $\Delta x \rightarrow \partial x, \Delta y \rightarrow \partial y, \Delta z \rightarrow \partial z, \Delta t \rightarrow \partial t$, according to the differential calculus in order to get the continuity equation in the formulation given in (3.1). Using the vector notation, the same equation can be expressed briefly as:

$$\frac{\partial A}{\partial t} = \nabla \cdot \mathbf{j}_A + Q \tag{3.4}$$

Thus the aim to express the flux as a function of the concentration is simple for advective transport. In order to achieve this for diffusive/dispersive flux, an empirical relationship has to be introduced, e.g. Fick's Law.

3.2 Fick's Law and Generalizations

3.2.1 Diffusion

There is a natural tendency in natural systems to level out concentration differences. The process which causes this tendency is called diffusion. When in a system there is a high concentration at one place and a smaller concentration at another place, there will be a net diffusive flux of the component from the location with higher concentration to the one with lower concentration. In the molecular scale, diffusion is a random motion of molecules in all directions. In systems without concentration differences all random walks together maintain the same concentration level. But if concentration is not constant, there is a net flux in one direction, from the high to the low concentrations.

A system with initial concentration differences will finally reach a constant concentration level if no other processes are present. Other processes can stabilise the concentration gradient. Then the diffusive flux may be balanced by processes which maintain a permanent out- and inflow. Still the concentration gradient is accompanied by diffusive flux.

The empirical (first) Fick's Law[1] is a quantification of diffusive flux (physical unit: mass/(area·time)), here stated for the fluid phase:

$$\mathbf{j} = -D\nabla c \tag{3.5}$$

In words: the diffusive flux is proportional to the negative concentration gradient. The minus sign guarantees that the direction of the net flux is from high concentrations to low concentrations. The factor of proportionality is the diffusion constant or diffusivity D with the physical unit [area/time]. Note that here too the ∇-operator is used. Here it is not working in connection with a vector product, because the following variable is a scalar.[2] The result of such an operation is a vector. In three dimensions it can be written as:

$$\nabla c = \begin{pmatrix} \partial c/\partial x \\ \partial c/\partial y \\ \partial c/\partial z \end{pmatrix} \tag{3.6}$$

In general, the diffusivity D depends on the fluid and on the transported component; it depends on temperature and pressure and on the geochemical environment. For all substances there is a diffusivity in gases, which differs from the diffusivity in liquids, and it even depends on the type of gas or liquid. The diffusivity in saltwater usually is different from the diffusivity in freshwater.

The diffusivity, as defined by (3.5), is defined in single phase systems, i.e. in liquids or in gases, and is a characteristic of the molecules involved, i.e. of the component and of the medium. For that reason it is common to speak of D as molecular diffusivity. In the following this will be taken into account by writing D_{mol}, while D remains the notation for diffusivity in general.

The order of magnitude of D_{mol} in water at common temperatures of 20°C for most chemical components is around 10^{-9} m²/s. In air it is in the range of 10^{-5} m²/s. For ammonium gas NH_3 for example, D_{mol} in water is $1.46 \cdot 10^{-9}$ m²/s at 20°C and

[1] Adolf Eugen Fick (1829–1901), German physiologist; Fick's second Law is valid for mass conservation in a single phase environment:

$$\frac{\partial}{\partial t}c = D\frac{\partial^2}{\partial x^2}c$$

The formula is obtained when Fick's Law, as given in (3.5), is used as replacement for the flux in (3.4).

[2] In contrast to a vector a scalar has a single value only. A scalar function has a single value, depending on space and/or time. Velocity is a typical vector variable, which in a one-dimensional case is reduced to a scalar variable.

$1.23 \cdot 10^{-9}$ m²/s at 4°C (Häfner et al. 1992). For the same gas the same reference gives a value of $1.98 \cdot 10^{-5}$ m²/s at 0°C for the molecular diffusivity in air.

In order to formulate Fick's Law in multi-phase systems, such as in porous media, two modifications have to be made. At first it has to be taken into account that the area through which diffusive flux may occur is only a part of the total area (see control volume in Fig. 2.4). Usually it is assumed that the area is reduced by the same factor as the volume. The volumetric share of the pore space, porosity, is thus taken as the factor that measures the share of the active area. For that reason a factor θ appears on the right side of (3.5) if applied in porous media. In unsaturated porous media θ represents volumetric water content.

The second correction is necessary to take into account that diffusion pathlengths are necessarily longer if several phases are present. The situation is illustrated in Fig. 3.1. While in single phase systems the shortest path is available for diffusive fluxes of particles, in a multi-phase environment such direct connection is impeded by obstacles. As pathlengths are longer in multi-phases, the diffusive flux in those systems is smaller than in a single phase case. One may also say that pathlenghts are prolonged, which yields a factor ϑ greater than 1 in the denominator of the concentration gradient.

Pathlength prolongation also has to be considered in the calculation of flux $\mathbf{j}$. The flux in normal direction is smaller than the flux following the generally non-normal pathline. The combined effect of both corrections with the *length prolongation factor* ϑ leads to the equation:

$$\mathbf{j} = -\frac{1}{\vartheta^2} D_{mol} \nabla c \tag{3.7}$$

In sedimentological and geochemical science (Boudreau 1996; Drewer 1997) Fick's Law is formulated with the correction factor as given in (3.7), which goes back to Carman (1937). Later Carman (1956) used the term tortuosity factor for ϑ, which is misleading in (3.7), where it appears in the denominator. We prefer here to adopt a notation that is often found in groundwater literature:

$$\mathbf{j} = -\tau D_{mol} \nabla c \tag{3.8}$$

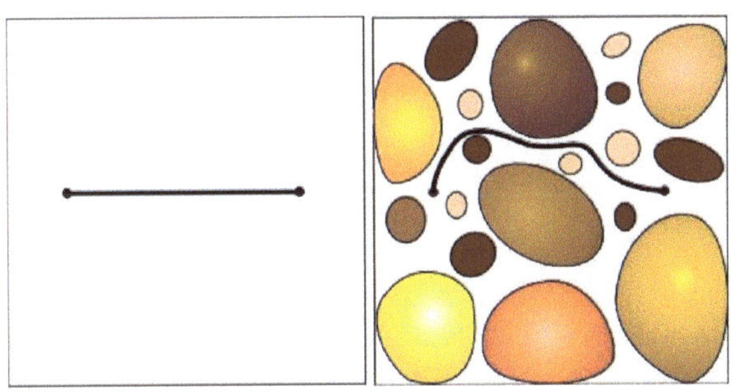

Fig. 3.1 Comparison of diffusion pathlengths in single and multi-phase systems

where the *tortuosity* τ is defined as another factor, aside from factor θ, with values between 0 and 1 in Fick's Law for multi-phase systems, a formulation which was already proposed by Bear (1972).[3] The factor τ and the prolongation factor are thus connected by the formula $\tau = 1/\vartheta^2$. The coefficient of the gradient, consisting of three factors, can be termed *effective diffusivity.*

$$D_{eff} = \theta\tau D_{mol} \tag{3.9}$$

Care is advisable with the term 'effective,' because it is not used in the same way in scientific literature. Sometimes the product of diffusivity and tortuosity, without porosity, gets the predicate 'effective.' Sometimes the term effective is omitted at all. In contrast to effective diffusivity the single phase diffusivity is often referred to as *molecular diffusivity.*

Length prolongation and tortuosity are connected to the *formation factor*, which is determined by electrical resistivity measurements. Using this technique, Archie (1942) found a power law relationship between porosity and formation factor, which in terms introduced above can be noted as

$$D_{eff} = \theta^m D_{mol} \; or \; \tau = \theta^{m-1} \tag{3.10}$$

and can be found as *Archie's Law* in several publications (Sahimi 1993; Boudreau 1996).

Archie (1942) reports values for m (Eq. 3.10) of 1.8–2 for consolidated sandstones, 1.3 for unconsolidated sand in a laboratory experiment, and 1.3–2 for partly consolidated sand. For theoretical or conceptual work the value $m = 2$ is considered, which may be justified if there is no further information. From (3.10) then follows: $\tau = \theta$, and from (3.9):$D_{eff} = \theta^2 D_{mol}$ with formation factor θ^2.

Boudreau (1996) provides an extensive overview of papers about tortuosity and porosity and their relationship. Several fixed relationships between τ and φ have been proposed. The relations given by Archie (1942), Weissberg (1963) and Iversen and Jørgensen (1993) contain parameters that can be estimated based on measured data. The latter propose the relation $\vartheta^2 = 1 + n(1 - \theta)$ with a typical value of $n = 3$ for clay-silt sediments and of $n = 2$ for sandy sediments. As Boudreau (1996) already noted, the resulting parameter curves are identical to those given by the Burger-Frieke equation $\vartheta^2 = \theta + a(1 - \theta)$ with parameter a. Figure 3.2 depicts the curves for Archie's Law and the Iversen-Jørgensen equation for the main parameter range.

[3] ϑ, which is here called the length prolongation factor, is often introduced as tortuosity (Boudreau 1996; Drewer 1997).

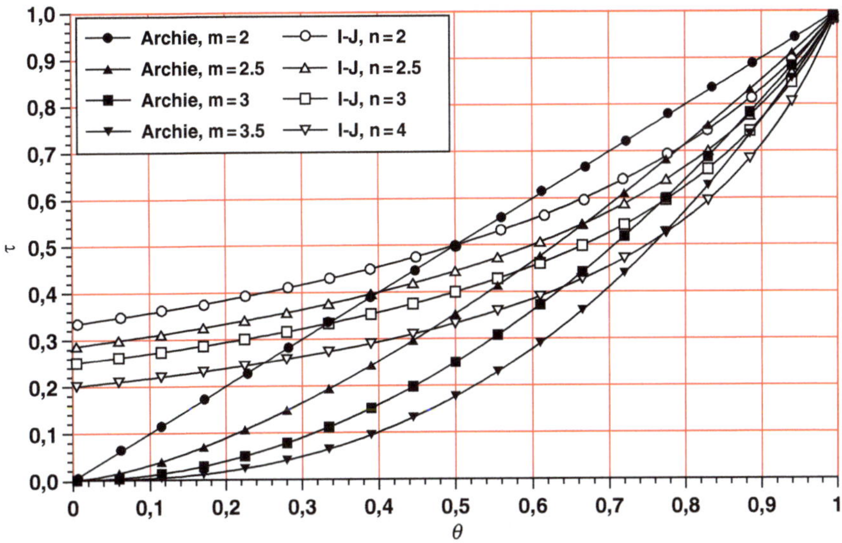

Fig. 3.2 Tortuosity τ as function of porosity θ, according to Archie's law for different values of parameter m, and according to Iversen & Jørgensen (I-J) for parameter values of n

3.2.2 Dispersion

Another generalization of Fick's Law is necessary if advection is also present. It can be observed that in a fluid flowing through a homogeneous porous medium the diffusivity as proportionality gradient in Fick's Law (3.5) is not constant, but shows itself a strong dependency on the flow velocity. In the scientific literature on groundwater this effect is referred to as *dispersion*. Dispersion is thus a general phenomenon that includes diffusion as special case. For the 1D situation one may write:

$$D = \tau D_{mol} + \alpha_L v \tag{3.11}$$

The effective dispersivity, which is used in Fick's Law, consists of two parts. One stems from the molecular diffusion and the other from porous media flow. For high velocities the second part dominates, which is the common situation in groundwater, although flow in aquifers is still rather slow compared to fluxes in other hydrological compartments. The proportionality factor between dispersion and velocity along a flow pathline is given by the parameter α_L, which has the physical dimension of [length]. One may also use the term dispersion length or *longitudinal dispersivity*. The subscript '$_L$' refers to longitudinal, as it is valid only in the direction of the flow.

In the general two- and three-dimensional situation the concept of dispersion has to be generalized. Transverse to the flow direction, the factor α_L as proportionality factor between effective dispersivity and seepage velocity is not valid any more.

Another parameter has to be introduced, the *transversal dispersivity* α_T. The analytical formulation becomes much more complex, because the scalar factor D_{eff} has to be replaced by the *dispersion tensor* **D**:

$$\mathbf{D} = (\tau D_{mol} + \alpha_T v)\mathbf{I} + \frac{\alpha_L - \alpha_T}{v}\mathbf{v}\mathbf{v}^T \tag{3.12}$$

with unity matrix **I**. The elements of the matrix $\mathbf{v}\mathbf{v}^T$ contain the products of the velocity components. Here the usual matrix product of a column vector and a row vector yields a matrix. The formulation may seem a little complex at first sight. It takes into account that the mixing constant in the direction of velocity is different from the mixing constant transverse to the velocity direction and is valid for arbitrary vectors **v**. Note that **v** may change spatially and temporally. With the dispersion tensor the dispersive flux term becomes:

$$\mathbf{j} = -\mathbf{D}\nabla c \tag{3.13}$$

where the product on the right side is performed as matrix–vector multiplication. Transversal dispersivity is smaller than longitudinal dispersivity. Even a factor of one or two orders of magnitude is possible.

An important feature is the scale dependency of longitudinal and transversal dispersivities, which has been observed in groundwater studies. Figure 3.3 shows the scale dependency of longitudinal dispersivity in porous media. Data for that figure were taken from several studies on dispersion in groundwater.

It is also interesting to compare the effective diffusivity with the velocity dependent dispersion, i.e. the two terms which contribute to the effective dispersivity in (3.11). For a length scale of 1 m, a velocity in the range of some mm/a, and a longitudinal dispersivity of 0.1 m, the value of 10^{-4} m^2/a results. Molecular diffusivity in water for most components is around 10^{-9} m^2/s or $3{\cdot}10^{-2}$ m^2/a. Even though the diffusivity has to be reduced by the factor τ, it can be concluded that for the given scale the diffusive flux exceeds dispersive flux. The values are characteristic for lacrustine sediments. Only for very high sedimentation burial rates and for very long mixing pathlengths a non-negligible contribution of dispersion can be expected.

3.3 The Transport Equation

3.3.1 Mass Transport

When both advection and diffusion/dispersion are taken into account, the flux vector in x-direction results as the sum of both contributions:

$$j_x = -D\frac{\partial c}{\partial x} + vc \tag{3.14}$$

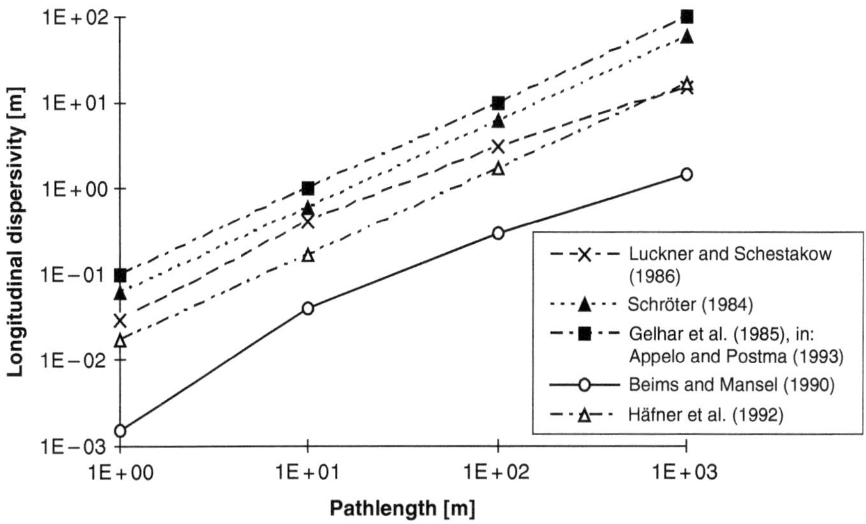

Fig. 3.3 Scale dependency of longitudinal dispersivity in porous media, as observed by different authors

where in the diffusivity D different contributions have to be considered: from the molecular scale, from tortuosity or from dispersion at the regional scale. Similar formulae can be stated for the flux components in y- and z-direction. In vector notation results:

$$\mathbf{j} = -\mathbf{D}\nabla c + \mathbf{v}c \tag{3.15}$$

Here, the coefficient is written as a matrix in order to account for the general case, as described above.

Now it is time to use this result to replace the flux terms in the mass conservation (2.4). The result in one dimension is:

$$\theta \frac{\partial c}{\partial t} = \frac{\partial}{\partial x} \theta \left(D \frac{\partial c}{\partial x} - vc \right) + q \tag{3.16}$$

In case of constant velocity one obtains the most common formulation of the transport equation:

$$\theta \frac{\partial c}{\partial t} = \frac{\partial}{\partial x} \left(\theta D \frac{\partial c}{\partial x} \right) - \theta v \frac{\partial c}{\partial x} + q \tag{3.17}$$

Sidebar 3.1: Mass Conservation in Streams

For 1D models in streams, the mass conservation leads to a slightly different formulation. The changing cross-section along the stream needs to be considered. Instead of (2.1) one obtains:

$$\frac{c(x,t+\Delta t)-c(x,t)}{\Delta t}\cdot A\Delta x = j_{x-}(x,t)A_{x-} - j_{x+}(x,t)A_{x+}$$

where A denotes the cross-section in the y-z- and j fluxes through cross-sections. As the riverbed is usually changing, different upstream and downstream cross-sections A_{x-} and A_{x-}have to be taken into account (Fig. 3.4).

Division by Δx and transition to infinitesimal scale leads to the formulation

$$A\frac{\partial c}{\partial t} = \frac{\partial}{\partial x}(j_x A)$$

After application of Fick's Law for the cross-section the equation becomes:

$$A\frac{\partial c}{\partial t} = \frac{\partial}{\partial x}\left(D_{turb}A\frac{\partial c}{\partial x}\right) - \frac{\partial}{\partial x}(v_{mean}Ac)$$

where D_{turb} stands for mean turbulent diffusivity across a cross-section, and v_{mean} for mean velocity across the cross-section.

Of course, the 1D approach is a simplification of the flow regime in a stream or channel. However, it is justified in order to capture the dominant downstream behavior in the main flow channel. Additional features, such as counterflow in groyne fields or along the bankline and flow in floodplains, can be accounted for by the introduction of additional source and sink-terms in the equation.

In case of constant D the equivalent formulation is:

$$\theta\frac{\partial c}{\partial t} = \theta D\frac{\partial^2 c}{\partial x^2} - \theta v\frac{\partial c}{\partial x} + q \tag{3.18}$$

For higher dimensional problems one may use the ∇-operator to obtain an equally short formulation for the general situation:

$$\theta\frac{\partial c}{\partial t} = \nabla\cdot\theta(\mathbf{D}\nabla c - \mathbf{v}c) + q \tag{3.19}$$

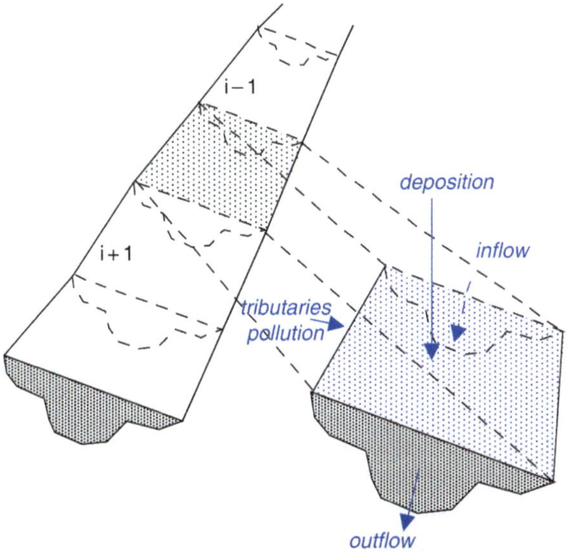

Fig. 3.4 Illustration for the derivation of 1D transport equation in streams

This is the so-called *mass transport equation*, which is valid for biogeochemical species of all kinds. The mathematical characterization is as follows: it is of second order in space, as there appear second derivatives in x, y and z but not third derivatives. It is of first order in time. It is parabolic concerning the mathematical classification of partial differential equations. In case of constant coefficients it is a linear equation.

The simplifications, performed for the 1D equation, can be made for the multi-dimensional situation as well. For the equation

$$\theta \frac{\partial c}{\partial t} = \nabla \cdot (\theta \mathbf{D} \nabla c) - \theta \mathbf{v} \cdot \nabla c + q \tag{3.20}$$

the generalized condition is that the flow field is divergence-free, or in mathematical formulation: $\nabla \cdot \mathbf{v} = 0$.[4] Formulation (3.20) is then obtained from formulation (3.19) due to:

$$\nabla \cdot \mathbf{v}c = \mathbf{v}\nabla c + c(\nabla \cdot \mathbf{v}) = \mathbf{v}\nabla c$$

[4] For an incompressible fluid the condition $\nabla \cdot \mathbf{v} = 0$ means that there are no internal sources or sinks for the fluid (see chap. 12).

3.3.2 Fourier's Law and Heat Transport

In this chapter energy is understood as heat energy throughout. The main aim is to derive the fundamental equations for heat transport, which are based on the energy conservation principle. Other forms of energy, for example the energy consumed or produced in reactions or by phase transitions, may be relevant in certain cases, but will not be treated here.

In analogy to mass conservation, the conservation of thermal energy can be stated in the following form:

$$(\rho C)\frac{\partial}{\partial t}T = -\nabla \mathbf{j}_e + q_e \tag{3.21}$$

where (ρC) denotes heat capacity in [energy/(volume·°temperature)](often: [J (°K)$^{-1}$ m^{-3}]) and $\mathbf{j}_e$ the heat flux in [energy/(area·time)] (often [Watt/area]). The energy sink or source q_e, as well as the entire differential equation, measure the volumetric energy rate in the physical unit [energy/(volume·time)].[5] Some values of heat capacities are listed in Table 3.1.

The left side describes the storage of heat, while in the first term on the right side mass differences are expressed through the spatial change of heat fluxes. The coefficient on the left side relates energy storage in form of heat due to temporal temperature change to the mass. The heat capacity C is the expression of the energy – mass relationship, while specific heat capacity ρC is the expression for energy – volume relation. At first instance, (3.21) is a formulation for pure phases. In porous media as a two-phase system, either storage and fluxes can be added weighted by their relative volumetric share, expressed in terms of porosity:

$$\theta(\rho C)_f\frac{\partial}{\partial t}T_f + (1-\theta)(\rho C)_s\frac{\partial}{\partial t}T_s = -\nabla \mathbf{j}_e + q_e \tag{3.22}$$

In (3.22) both phases may have different temperatures: T_f in the pore water, T_s in the sediment. Usually heat transport is slow in relation to interphase heat transfer, i.e. the heat exchange between fluid and solid phase is fast and as a result temperatures in the two phases are the same: $T_s = T_f = T$. Concerning the long-term development of non-oscillating thermal regimes, as it is mostly met in field situations, the assumption of one temperature level is true. Then holds:

$$\left[\theta(\rho C)_f + (1-\theta)(\rho C)_s\right]\frac{\partial}{\partial t}T = -\nabla \mathbf{j}_e + q_e \tag{3.23}$$

[5] The given formulation is a simplified version already. The left side can be derived by using the following formula for internal energy: $\varepsilon = \varepsilon_0 + \int c_v dT$ with specific heat c_v. The given formulation is also valid for incompressible media under nearly constant pressure (Häfner et al. 1992).

Table 3.1 Thermal conductivities and diffusivities for selected fluid, solid and mixed phases (Sources: Häfner et al. 1992; Lide 1995)

Phase	Density [kg/dm³]	Thermal conductivity [W/m°C]	Specific heat capacity [kJ/kg°C]	Thermal diffusivity [10^{-6} m²/s]
Water at 5°C	1.0	0.5724	4.202	0.13622
Water at 10°C	0.9998	0.5820	4.192	0.13886
Water at 20°C	0.9982	0.5984	4.182	0.14335
Water at 30°C	0.9957	0.6154	4.178	0.14793
Water at 40°C	0.9922	0.6305	4.179	0.15206
Water at 50°C	0.988	0.6405	4.181	0.15505
Water at 90°C	0.962	0.6753	4.210	0.16674
Seawater at 10°C	1.0269	0.5781	3.911	0.1439
Calcite	2.6–2.8	2.2	0.91	0.92984
Sand (dry)	1.2–1.6	0.6 (0.33)	0.8	0.62500
Fine sand (dry)	1.635	0.627	0.76	0.50459
Fine sand (saturated)	2.02	2.75	1.419	0.95940
Gravel (dry)	1.745	0.557	0.766	0.41671
Gravel (saturated)	2.08	3.07	1.319	1.11900

When the coefficient on the left side of the equality is regarded as the heat capacity of the solid/fluid sediment system:

$$(\rho C) = \theta(\rho C)_f + (1-\theta)(\rho C)_s \tag{3.24}$$

one can stay with the formulation of the energy conservation as noted above in (3.21). In a multi-phase environment (ρC) is a property of the system including all phases.

The heat transport will be derived from this equation by specifying the flux $\mathbf{j}_e$ with respect to the relevant processes. In order to reach an equation like that for mass transport, one has to express the heat flux vector in terms of temperature. For advective heat flux this can be achieved in analogy to (2.9):

$$\mathbf{j}_e = \theta(\rho c)_f \mathbf{v} T \tag{3.25}$$

For diffusive heat flux there is an analogue to Fick's Law, which is Fourier's Law.[6] In 1822 J.B. Fourier stated a linear relation between heat flux on one side and the temperature gradient on the other:

$$\mathbf{j}_e = -\lambda_T \nabla T \tag{3.26}$$

The proportionality factor is the thermal conductivity λ_T that depends on the medium through which heat transfer is taking place. λ_T is thus a material property

[6] Jean Baptiste Fourier (1786–1830), French mathematician.

which can be compared with the diffusivity in Fick's Law (Eq. 3.5). In the literature one can also find the term *conductive heat flux*. The physical unit of λ_T is [energy/(length·time·°temperature)] (mostly: [Watt (°Km)$^{-1}$]). Some values of thermal conductivities can be found in Table 3.1.

Like Fick's Law, Fourier's Law was at first stated for a single phase situation, but the formulation (3.26) does not have to be changed for multi-phases. The proportionality factor usually is not the same: in the saturated porous medium, λ_T is a weighted mean of the phases involved, i.e. the fluid and the solid phase:

$$\lambda_T = (1-\theta)\lambda_s + \theta\lambda_f \tag{3.27}$$

where the subscripts denote the thermal conductivities for the pure phases.

Table 3.1 provides a list of 'material properties,' which are relevant for heat transport. Some are related to pure phases (water and calcite), some to mixed phases (gravel, sand).

A parameter with the unit [area/time] results when thermal conductivity is divided by the specific heat capacity. In analogy to mass diffusion, this parameter is termed *thermal diffusivity,* and Fourier's Law can be understood as the law of heat diffusion. Note that the transformation from single phase to multi-phase is not the same for heat and mass diffusion. The important difference is that heat diffusion takes place in all phases, whereas mass diffusion is relevant only in the fluid phase.

Replacing the energy flux vector in (3.21) by the terms for advective and diffusive fluxes, as formulated in (3.25) and (3.26), one obtains the heat transport equation:

$$(\rho C)\frac{\partial}{\partial t}T = \nabla\cdot\left(\lambda\nabla T - \theta(\rho C)_f\mathbf{v}T\right) + q_e \tag{3.28}$$

which for a divergence-free flow field simplifies to:

$$(\rho C)\frac{\partial}{\partial t}T = \nabla\cdot\lambda\nabla T - \theta(\rho C)_f\mathbf{v}\cdot\nabla T + q_e \tag{3.29}$$

Division by ρC delivers:

$$\frac{\partial}{\partial t}T = \nabla\cdot D_T\nabla T - \theta\kappa\mathbf{v}\cdot\nabla T + \frac{q_e}{\rho C} \tag{3.30}$$

with thermal diffusivity $D_T = \frac{\lambda}{\rho C}$ and the ratio of heat capacities $\kappa = \frac{(\rho C)_f}{\rho C}$. Some values of thermal diffusivities are listed in Table 3.1. Note that thermal diffusivities are more than two orders of magnitude higher than molecular diffusivities for species. In systems, in which heat and mass diffusion act simultaneously, heat diffuses much faster than any species.

Aside from the $\theta\kappa$ term, the advection term (3.30) and the different representation of the sources, the heat transport formulation is identical to the mass transport

(3.19). Concerning the mathematical type both differences do not change the type of the differential equation: it remains second order in space and first order in time, and is thus of parabolic type. There is a first time derivative on the left side of the equation, and on the right side there are three space derivative terms: the second order term, representing diffusion and dispersion, a first order term, representing advection, and a source term which usually does not contain derivatives and is thus of zero order.

3.4 Dimensionless Formulation

Without the source term one often reads of the *advection-diffusion-equation*, in connection with the differential equations derived in the preceding sub-chapter. Depending on how the processes are understood, one may also speak of advection-dispersion, convection-diffusion or convection–dispersion-equation. These terms can be found for the transport equations for mass (Eq. 3.19) or for heat (Eq. 3.30).

For a better illustration of the sensitivity of the solutions, it is often appropriate to use the dimensionless formulation of these transport equations. In the sequel, we consider the situation with constant coefficients and no sources or sinks. The transport equation in dimensionless form is:

$$\frac{\partial \omega}{\partial \tau} = \frac{\partial}{\partial \xi} \frac{1}{Pe} \frac{\partial \omega}{\partial \xi} - \frac{\partial \omega}{\partial \xi} \tag{3.31}$$

with dimensionless variables $\omega = \frac{c-c_0}{c_{in}-c_0}, \xi = \frac{x}{L}, \tau = \frac{vt}{L}$ and dimensionless Péclet[7]-number $Pe = \frac{vL}{D}$ for the mass transport equation,and with dimensionless variables $\omega = \frac{T-T_0}{T_{in}-T_0}, \xi = \frac{x}{L}, \tau = \frac{\theta \kappa v t}{L}$ and dimensionless Péclet-number $Pe = \frac{\theta \kappa v L}{D_T}$ for the heat transport equation. The advantage of formulation (3.31) is obvious. There is only one parameter left, which is the Péclet number. The behavior of the solutions can thus be explored by the variation of that single parameter, and can often be visualized nicely in a single diagram (see Chap. 5.3 for an example).

3.5 Boundary and Initial Conditions

In the preceding part of the book, the fundamental theoretical and empirical laws are presented and it is shown how these are combined to yield differential equations. For most differential equations several functions can be found which fulfil the equation. The equation $\partial u/\partial s = -u(s)$ for example is fulfilled by the

[7] Jean Claude Eugène Péclet (1793–1857), French physicist.

functions $u(s) = C\exp(-s)$ for all values of C. Such solutions are called *general solutions* and contain one or more *integration constants*, like C in the example. In order to restrict the solution space, additional conditions have to be specified. The additional requirements are usually formulated as *boundary and initial conditions*. The mathematical formulation based on differential equations is completed by these conditions.

The number of conditions, required to deliver a unique solution, is mainly determined by the order of the differential equation. For first order equations (which contain only first derivatives) one condition is needed, while second order equations require two conditions. The term initial condition usually refers to the variable time t and a condition at $t = 0$. The term boundary condition refers to a space variable x, y or z and a condition at the boundary of the model region. In the just mentioned example $s = t$, the initial condition $u(t = 0) = u_0$ leads to the unique solution $u(t) = u_0 \exp(-t)$. Typical for 1D steady states in sediment layers is $s = z$ and the boundary condition $u(z = 0) = u_0$ at the sediment-water interface. The unique solution $u(z) = u_0 \exp(-z)$is then valid, representing an exponentially declining profile of the unknown variable.

A fundamental classification distinguishes three types of boundary conditions. A first type boundary condition or *Dirichlet* type[8] condition specifies the value of the unknown dependent variable at the boundary. There is a concentration value to be given in a mass transport problem and a temperature value in a heat transport problem.

In a second type boundary condition or *Neumann*[9] type condition, the derivative of the variable is specified. As this gradient is proportional to diffusive flux, one can interpret these conditions best as a specified diffusive flux. In mass transport the concentration gradient is to be given, while in heat transport the temperature gradient is prescribed.

A prominent role plays the condition with a vanishing gradient. According to Fick's Law or Fourier's Law there is no diffusive flux then. Often the condition is simply referred to as '*no-flow*' condition; but it should be kept in mind that a vanishing gradient still allows advective flux. If there is a non-zero velocity component across the boundary, then there usually is heat or mass flux across that boundary even when the so called 'no-flow' condition is declared. Thus, it is more precise to use the characterization 'no-diffusive flow' instead of 'no-flow.' Only a zero velocity normal to the boundary and a vanishing gradient together guarantee no flux.

The third type, *Cauchy*[10] - or *Robin* boundary condition, is the general condition as it requires a relationship between the gradient and a given value of the variable:

---

[8] Peter Gustav Lejeune Dirichlet (1805–1859), German mathematician.

[9] Carl Gottfried Neumann (1832–1925), German mathematician.

[10] Augustin Louis Cauchy (1789–1857), French mathematician.

Table 3.2 Classification of boundary conditions, overview

Type	Name	Condition for variable u(s)
First	Dirichlet	$u = u_1$ specified
Second	Neumann	$\partial u/\partial s$ specified
Third	Cauchy or Robin	$\alpha_0 u + \alpha_1 \partial u/\partial s = j$

$$\alpha_0 c + \alpha_1 \frac{\partial c}{\partial n} = j \text{ for mass transport or}$$
$$\alpha_0 T + \alpha_1 \frac{\partial T}{\partial n} = j_e \text{ for heat transport} \tag{3.32}$$

with given coefficients α_0 and α_1 and given mass flux j or heat flux j_e. In flow problems third type boundary conditions are formulated analogously in terms of hydraulic head, pressure, pressure head or streamfunction. The third type condition includes first and second type conditions as special cases. Third type boundary conditions are connecting advective and diffusive fluxes (Table 3.2).

Values of boundary conditions may change with time. There are applications where even the type of the boundary condition changes with time.

In transient problems, another form of conditions appears in addition to boundary conditions: the *initial conditions*. As the name tells, an initial condition concerns the knowledge of a variable at the beginning of the simulation, usually at time $t = 0$. It is necessary to know the starting position if the temporal development for $t > 0$ is to be simulated.

References

Appelo CA, Postma D (1993) Geochemistry, groundwater and pollution. Balkema, Rotterdam, p 536

Archie GE (1942) The electrical resistivity log as an aid in determining some reservoir characteristics. Trans AIME 46:45–61

Bear J (1972) Flow through porous media. Elsevier, New York, p 764

Beims U, Mansel H (1990) Assessment of groundwater by computer model design and pumping tests. In Groundwater Monitoring and Management, IAHS Publ., No. 173, 11–22

Boudreau BP (1996) The diffusive tortuosity of fine-grained unlithified sediments. Geochim Cosmochim Acta 60(16):3139–3142

Carman PC (1937) Fluid flow through porous rock. Trans Inst Chem Eng London 15:150–157

Carman PC (1956) Flow of gases through porous media. Butterworths Scient. Publ, London, p 180

Drewer J (1997) The geochemistry of natural waters. Prentice-Hall, Upper-Saddle River, p 436

Häfner F, Sames D, Voigt H-D (1992) Wärme- und Stofftransport. Springer, Berlin, p 626

Iversen N, Jorgensen BB (1993) Diffusion coefficients of sulfate and methane in marine sediments: influence of porosity. Geochim Cosmochim Acta 57:571–578

Lide DR (ed) (1995) Handbook of chemistry and physics, 76th edn. CRC Press, Boca Raton

Luckner L, Schestakow WM (1986) Migrationsprozesse im Boden- und Grundwasserbereich. Deutscher Verlag für Grundstoffindustrie, Leipzig
Sahimi M (1993) Flow phenomena in rocks: from continuum models to fractals, percolation, cellular automata, and simulated annealing. Rev Mod Phys 65(4):1394–1534
Schröter J (1984) Mikro- und Makrodispersivität poröser Grundwasserleiter. Meyniana 36:1–34
Weissberg H (1963) Effective diffusion coefficients in porous media. J Appl Phys 34:2636–2639

Chapter 4
Transport Solutions

4.1 1D Transient Solution for the Infinite Domain

An analytical solution for the transport (3.18) with constant coefficients and $q = 0$ was given by Ogata and Banks (1961). It reads:

$$c(x,t) = \frac{c_{in}}{2}\left(\operatorname{erfc}\left(\frac{x - vt}{2\sqrt{Dt}}\right) + \exp\left(\frac{v}{D}x\right)\operatorname{erfc}\left(\frac{x + vt}{2\sqrt{Dt}}\right)\right) \tag{4.1}$$

'erfc' denotes the complementary error-function, which is defined as follows:

$$\operatorname{erfc}(\xi) := 1 - \frac{2}{\sqrt{\pi}}\int_0^{\xi}\exp(-\varsigma^2)d\varsigma$$

'erfc' is related to the error-function 'erf':

$$\operatorname{erfc}(\xi) := 1 - \operatorname{erf}(\xi) \text{ mit } \operatorname{erf}(\xi) := \frac{2}{\sqrt{\pi}}\int_0^{\xi}\exp(-\varsigma^2)d\varsigma$$

Both functions can directly by called by a MATLAB® command, equal to the abbreviation used above. The sequence of commands

```
x = [-4:0.04:4]; plot(x,erf(x),'--r',x,erfc(x),'b');
legend ('erf','erfc',2); grid;
```

illustrates the graphs of the two functions (Fig. 4.1).

In MATLAB® other functions relevant in connection with the error-function are implemented. Details can be obtained by using the keyword 'error function' in the MATLAB® help system.

E. Holzbecher, *Environmental Modeling*,
DOI 10.1007/978-3-642-22042-5_4,

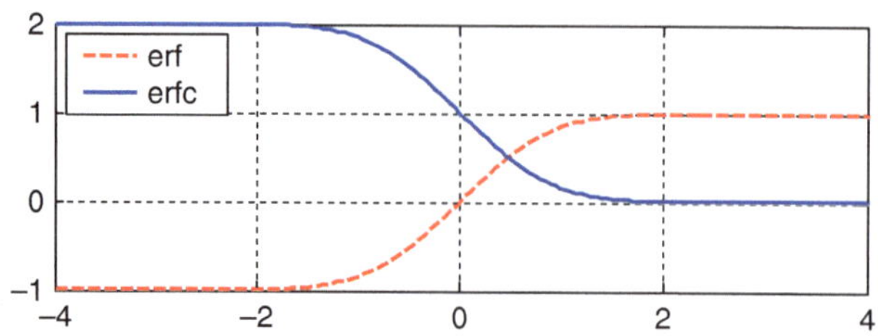

Fig. 4.1 Error-function (erf) and complementary error-function (erfc)

The 1D-solution for transport given in (4.1) is valid in the infinite half-space $x \geq 0$ for the initial condition

$$c(x, t = 0) = 0$$

and the boundary condition:

$$c(x = 0, t) = c_{in} \qquad c(x = \infty, t) = 0$$

Sidebar 4.1: Special Functions (1)

In MATLAB® there are various ways to call functions. There are special functions like the trigonometric functions

```
sin, cos, tan, cot
```

their inverse functions

```
asin, acos, atan, acot
```

their inverse functions with °-output

```
asind, acosd, atand, acotd
```

their hyperbolic counterparts

```
sinh, cosh, tanh, coth
```

for which there are also inverse functions with arguments in radian or degree.

There are also all secant-functions; for example $\csc(x) = 1/\sin(x)$:

```
csc, acsc, ascsd, acsch
```

The logarithm functions are based on different basis (e, 2 or 10):

```
log, log2, log10
```

(continued)

The exponential function is reached simply by `exp`.

To mention is also the *error function* and its complement, the complementary error function, defined by

$$erf(x) = \frac{2}{\sqrt{\pi}} \int_0^x \exp(-t^2)dt$$

$$erfc(x) = 1 - erf(x) = \frac{2}{\sqrt{\pi}} \int_x^\infty \exp(-t^2)dt$$

$$erfcx(x) = e^{x^2} erfc(x) \tag{4.2}$$

which are called using MATLAB® by `erf`, `erfc` and `erfcx`. There are also the inverse error functions, which are reached by the keywords `erfinv` and `erfcinv`.

Another integral of the exponential function, the so-called *exponential integral* or *well function*, is defined as

$$E(x) = \int_x^\infty \frac{\exp(t)}{t} dt \tag{4.3}$$

and is reached by `expint`. Other special functions, which one may meet, are implemented in MATLAB®, like the Bessel functions, and the modified Bessel functions, the Airy functions, the Beta-function, the incomplete Beta-function, the Gamma-Function, the incomplete Gamma-function:

```
besseli, besselj, airy, beta, betainc, gamma, gammainc
```

e.g. when the component is not present at the initial simulation time and is introduced into the system by inflow of concentration c_{in} at position $x = 0$. The second boundary condition concerns the behavior of the system at a point that is infinitely far away from the inflow boundary, where the concentration remains constant. For the application of the formula in practice follows that the outflow boundary is far away during all time instants of interest. When the front approaches the outflow boundary the solution of Ogata-Banks is not valid anymore.

The generalisation of the formula of Ogata & Banks for a non-zero initial condition $c(x = 0, t) = c_0$ is given by:

$$c(x,t) = c_0 + \frac{c_{in} - c_0}{2} \left(\text{erfc}\left(\frac{x - vt}{2\sqrt{Dt}}\right) + \exp\left(\frac{v}{D}x\right) \text{erfc}\left(\frac{x + vt}{2\sqrt{Dt}}\right) \right) \tag{4.4}$$

(Wexler 1992). In MATLAB® the formula is computed in an m-file '*analtrans.m*', which is described in the following. First input data are specified:

```
T = 1;      % maximum time
L = 1;      % maximum length
v = 1;      % velocity
D = 0.1;    % diffusivity / dispersivity
c0 = 0;     % initial value
cin = 1;    % inflow value
M = 50;     % number of timesteps
N = 50;     % number of nodes
```

Before the formula is implemented, some auxiliary variables are introduced:

```
e = ones (1,N);              % ones-line-vector
t = linspace (T/M,T,M);      % time discretization
x = linspace (0,L,N);        % space discretization
c = c0*e;                    % initial distribution
```

`e` is a row vector of length `N`. In another row vector `t` the time instants are gathered, for which the formula is to be evaluated. There are `M` time instants, where the initial time $t = 0$ is not counted. In the row vector `x` there are the positions, at which the formula is evaluated. There are `N` locations, without start- and end-position. In the row vector `c` the initial condition at $t = 0$ of the concentration at the positions `x` is given.

During the execution run of the program the vector `c` becomes a matrix, for which a new line is created for each of the time-instants given by vector `t`. This happens in the following loop, in which the formula is implemented:

```
for i = 1:length(t)
    h = 1/(2.*sqrt(D*t(i)));
    c = [c;c0+(cin-c0)*0.5*(erfc(h*(x-e*v*t(i)))...
      +exp((v/D)*x).*erfc(h*(x+e*v*t(i))))];
end
```

The loop-index is `i`, the current time within the loop is `t(i)`. The `length` function determines the number of elements in a vector[1]. The `h` is the coefficient $1/2\sqrt{Dt}$[2], which appears in the brackets. The lengthy expression in the third and forth line correspond with the formula of Ogata-Banks. Note that `x` is a vector. In order to add or subtract the term vt (see (4.4)) it has to be ensured that the term is also a vector: this is done by multiplication with the ones-vector `e`. Also note that the multiplication of the 'exp'-factor and the 'erfc'-factor in the last term of the expression has to be performed element-wise: for that reason the multiplication-operator `.*` is necessary.

---

[1] Alternatively one may use the `size` function. Here `size(t,2)` replacing `length(t)` works well also. `size` has the advantage that it can equally be applied to matrices: `size(A)` delivers the number of rows and columns of array `A`.

[2] `sqrt` denotes the square root.

When the MATLAB® `plot` command is applied for a matrix, the values in the columns are plotted against line numbers. As an exercise check that the command

```
plot ([1 2; 2 6; 7 8; 8 1])
```

produces two function graphs based on four nodes each. In order to obtain a plot of lines against column numbers one has to use the command for the transpose matrix. The final command

```
plot (c');
```

produces the plot (Fig. 4.2):

In case of pure diffusion the solution of Ogata & Banks simplifies to the form:

$$c(x,t) = c_{in}\mathrm{erfc}\left(\frac{x}{2\sqrt{Dt}}\right) \tag{4.5}$$

The dimensionless solution c/c_{in} is identical to the complementary error function with the dimensionless argument $\xi = x/2\sqrt{Dt}$. The evaluation can be performed easily in the classical manner: if one is interested in the concentration at location x and at time t, it is convenient to calculate ξ and to go with the obtained value into the graphical plot of error function (Fig. 4.1) to get the corresponding functional value. The latter has to be multiplied by c_{in} to receive the wanted c value.

Exercise 4.1: Heat diffusion $D = 10^{-6}$ m^2/s, $T_0 = 5$°C, $T_1 = 15$°C, $L = 1$ m; how long does it take until the temperature on the other side of the wall reaches

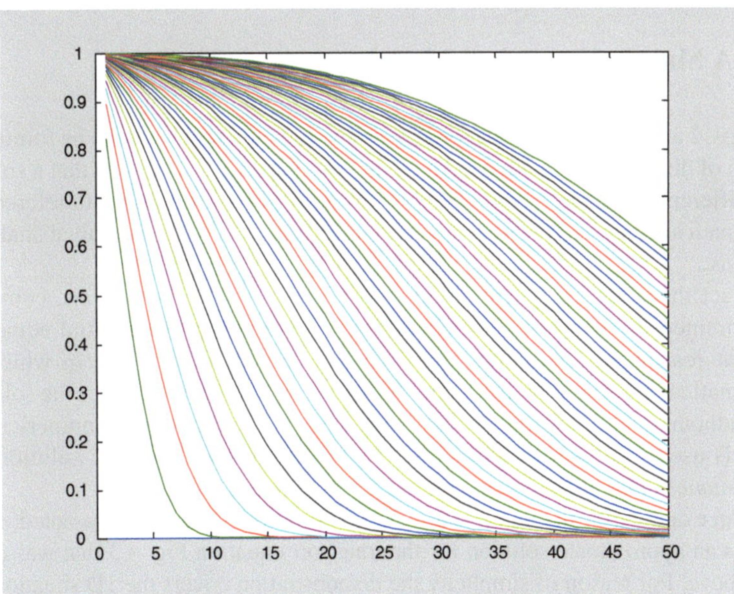

Fig. 4.2 The solution for the transport equation; analytical solution computed with MATLAB®

10°C? Answer: $1.1 \cdot 10^6$ s, to be read in the plot produced by the following sequence of commands:

```
t=[10e5:5e4:20e5]; D=1.e-6;
plot(t,5+10*erfc(ones(1,length(t))./(2*sqrt(D*t))));
```

Also for the constant heat flux boundary condition:

$$-\lambda\frac{\partial T}{\partial x}(x=0,t)=j_0 \tag{4.6}$$

with a constant flux j_0 an analytical solution exists (Baehr and Stephan 1994). It is given by:

$$T(x,t)=T_0+2\frac{j_0}{\lambda}\sqrt{Dt}\cdot\text{ierfc}\left(\frac{x}{2\sqrt{Dt}}\right) \tag{4.7}$$

with the integral error function

$$\text{ierfc}(\xi)=\frac{1}{\sqrt{\pi}}\exp(-\xi^2)-\xi\text{erfc}(\xi) \tag{4.8}$$

The integral error function is not specified in MATLAB®, but can easily be computed by use of (4.8). A corresponding m-file is included in the accompanying software under the name *'ierfc.m'*.

4.2 A Simple Numerical Model

In Chaps. 2 and 3 it is shown that processes and fundamental laws can be formulated in form of differential equations. Above in this chapter it was shown that a solution for a differential equation could be given by an explicit formula. With reference to mathematical analysis, functions given in formulae, as in (4.4), are called analytical solutions.

In fact there are analytical solutions for relatively few situations, compared to the immense complexity, which can be represented in differential equations. For that reason an alternative approach has gained importance, in which the mathematical algorithm on the computer yields an approximation for the solution. The mathematical discipline dealing with these approximations is numerics. The methods used in numerics are called numerical methods, and the solutions are called *numerical solutions* – in contrast to analytical solutions.

As an example for the numerical method a simple procedure is presented, which delivers an approximate solution for the transport equation Fig. 4.3 that was developed above. For reason of simplicity the demonstration covers the 1D situation.

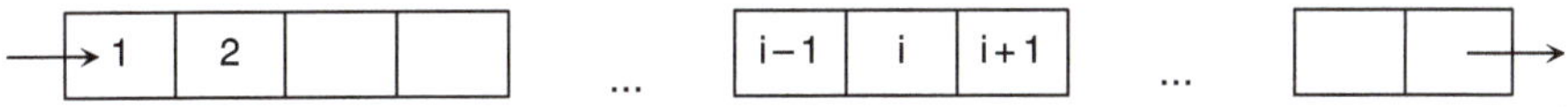

Fig. 4.3 Cells in series

Imagine a series of cells of equal geometry, as depicted in Fig. 4.3, representing an idealized situation in an environmental compartment. It is assumed that there is a constant flow with velocity v through all cells. When an increased concentration c_{in} enters the first cell with the flow, how does that affect the concentration distribution in the entire system? To keep it simple it is assumed that the concentration in each cell has a single value: the concentration in the i-th cell is designated as c_i. The transport problem is to compute the concentrations c_i, depending on time and the transport parameters.

A simple algorithm is developed in the sequel to mimic the major processes, advection and dispersion. Let's start with advection: after an appropriate time step Δt the entire system will be shifted by one cell. The time step depends on the velocity v and the spatial extension of the cells in flow direction Δx. The formula is: $\Delta t = \Delta x / v$. In MATLAB® we start with some settings in the command window:

```
N = 100;
c0 = 0;
cin = 1;
c = c0*ones (1,N);
```

and the following command sequence in the editor:

```
c = circshift(c',1)';        % advection
c(1)=cin;
plot (c); hold on;
```

Store the m-file under the name *'advection.m'*. The first commands specify the number of cells `N` and the initial concentration, which holds for all cells `c0`. In the first row of the m-file the initial concentration is set for all cells. In MATLAB® it is convenient to work with vectors and thus the concentration distribution in the system of cells is represented by a row vector `c`. In the next row the concentration is shifted by one position to the right. For such an operation MATLAB® offers the `circshift` command. Parameters of the command are the vector and the number of positions to be shifted. One has to use the transpose-`'`, because `circshift` operates on column vectors, only. Additionally the circular shift puts the concentration from the last cell into the first cell, which is not the intention here. The concentration in the first cell should be the inflow value c_{in}. This setting is performed in the next command, which overrides the preceding value in that cell.

In the final row the concentration is plotted. The `hold on` command ensures that the new graph is plotted in the same figure-window and that the old graph is not deleted. Run the small m-file several times, and plots similar to Fig. 4.4 will be visible on the display:

A front of increased concentration is passing from the left to the right. This animation is user-controlled, as it is the user who initiates each step with a mouse-

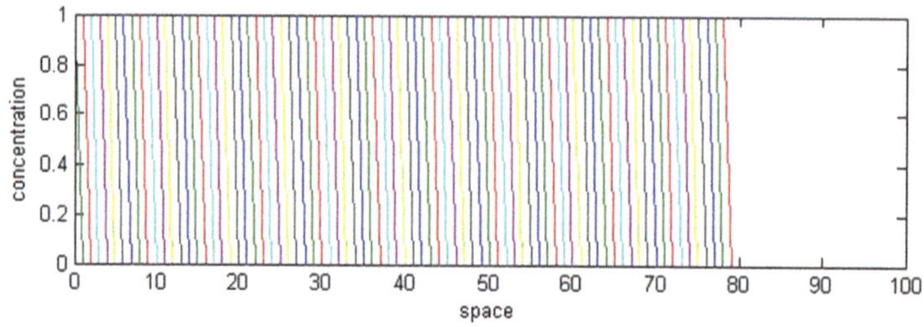

Fig. 4.4 Pure advective passing of a concentration front through a 1D compartment

click. The sequence shows advection, i.e. pure transport with the flow field. What is not correct is that the vertical front-line is slightly tilted. One could think that this represents a tiny transition zone between the low and the high concentration regime, but such a zone would be the result of diffusion or dispersion, which are not yet included in the model. In fact the tilting is an effect of the spatial discretization, e.g. the representation of a 1D-space interval by a finite number of discrete cells. The plotting algorithm, implemented in MATLAB®, connects neighbouring positions by a straight line. The tilting becomes less pronounced when the number of cells is increased.

For further use modify the advection m-file slightly:

```
c1 = circshift(c1',1)';
c1(1) = cin;
```

The concentration variable is re-named to `c1`, as it contains the current value of the concentration in a row vector. The `c` will be reserved for a matrix containing concentrations for all cells at all time-instants. The *'advection.m'* file is called as a sub-module from a main m-file, in which the initialization and the post-processing are performed. The main module should look like this:

```
N = 100;                % number of cells
c0 = 0;                 % initial concentration
cin = 1;                % inflow concentration
c1 = c0*ones (1,N); c = c1;
hold on;
for i = 1:N
   advection;
   plot (c1);
   c = [c;c1];
end
```

In the first three lines, the input part, variables are initialised, tasks which were done in the command window in the previous demonstration. In the forth line the row vector of current concentrations is set to the constant initial value, and also the first row of what is to become the concentration matrix `c` is filled with the initials. A *for*-loop is introduced, in which `N` time steps are simulated. Advection is simulated in *'advection.m'*, called within the loop. Then the concentration is plotted, as already shown in Fig. 4.4. Finally the row of current concentrations is attached to the matrix `c`. Store this file under the name *'simpletrans.m'*.

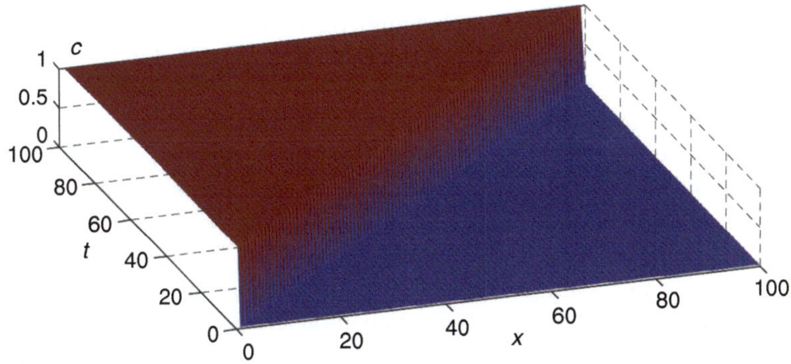

Fig. 4.5 Pure advection in a space (x)-time (t)-diagram

Run the program. After that a surface plot in the x-t-diagram is obtained with the *surf*-command (use MATLAB® command-window) (Fig. 4.5):

```
figure; surf(c)
```

As a next step we introduce diffusion into the model. For that purpose we set up another m-file with the task to mimic diffusive/dispersive processes:

```
for i = 2:N-1
    c2(i) = c1(i) + Neumann*(c1(i-1)-2*c1(i)+c1(i+1));
end
c2(1) = c1(1) + Neumann*(cin-2*c1(1)+c1(2));
c2(N) = c1(N) + Neumann*(c1(N-1)-c1(N));
c1 = c2;
```

Store this m-file under the name *'diffusion.m'*. The Neumann number *Neumann*, which appears here, has to be specified in the main module. The definition of the dimensionless Neumann number is given in Sidebar 4.2; let's take it as simply a parameter. The influence of that number is to examine. An auxiliary concentration row `c2` is computed in the sub-module, where the value in each cell is calculated from the last concentrations `c1` in the same cell and the two neighbouring cells. Why such a procedure mimics diffusion is derived in the following. The corresponding main m-file has a Neumann-number specification additionally and a call of the diffusion sub-module (called by the `diffusion` command):

```
...
Neumann = 0.5;
c1 = c0*ones(1,N); c = c1;
hold on;
for i = 1:N
   diffusion;
   advection;
   plot (c1);
   c = [c;c1];
end
```

Store the m-file under the name *'simpletrans.m'*. Advection and diffusion are simulated in the corresponding sub-modules, which are called within the *for*-loop of the main file. A general term for such a numerical procedure is *operator splitting*. The *for*-loop mimics advancing time. A snapshot of the output of the m-file looks as follows (Fig. 4.6):

The graph was obtained by ending the loop in the main module at a value lower than `N`. In the graph the build-up of a transition zone with concentrations between initial concentration and inflow concentration is obvious (which here is not the result of a graphic routine on discrete data). The transition zone moves with the advancing front, in the figure from left to right. It is also obvious that the transition zone widens with time. In other words: the gradients of the concentration curves become less steep. That is even better visible in the (*x*,*t*)-diagram, produced with the `surf`-command:

```
figure; surf (c )
```

As depicted in Fig. 4.7, the concentration is shown as a surface above the *(x,t)*-plane.

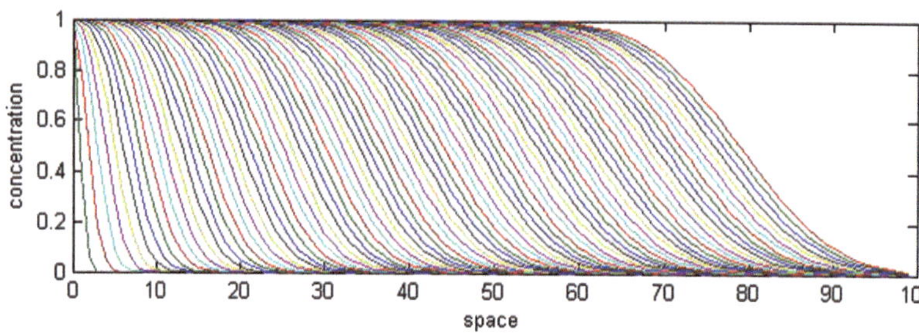

Fig. 4.6 Transport due to advection and dispersion; snapshot from an animation using '*simpletrans.m*'

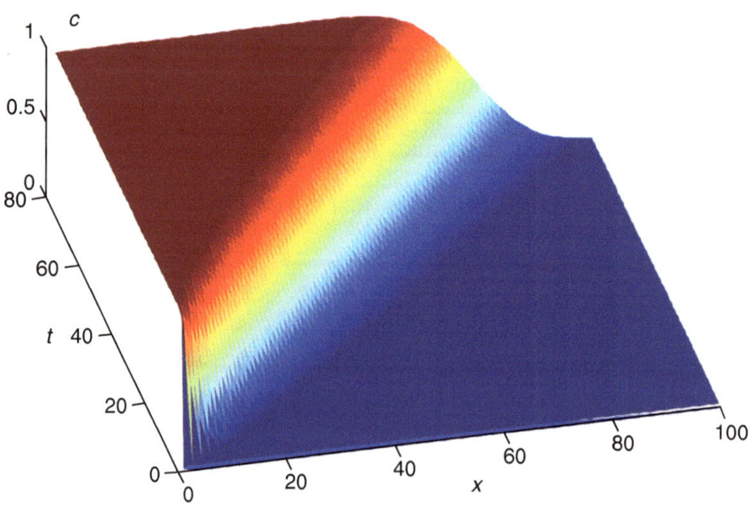

Fig. 4.7 The solution of 1D transport in the space (x) – time (t) diagram, visualized as surface plot

An alternative is the `contourf`-command

```
figure; contourf(c)
```

by which the transition zone can be visualized nicely. The transition zone is the multi-colored region between the plateaus of initial concentration (blue) and the inflow concentration (red) –see Fig. 4.8. For more details concerning 2D graphics see Chap. 14.

Sidebar 4.2: Derivation of the diffusion algorithm
As noted, diffusion is described by *Fick's Law*, which in 1D is stated as:

$$j = -D\frac{\partial c}{\partial x}$$

with diffusivity or dispersivity parameter D (see (3.5)). It was shown that the application of the principle of mass conservation together with Fick's Law leads to the transport equation, which is a differential equation for the concentration c. In the derivation the mass balance was set up for a control volume (Eq. 3.3):

$$\frac{c(x,t+\Delta t) - c(x,t)}{\Delta t} = -\frac{j_{x+}(x,t) - j_{x-}(x,t)}{\Delta x}$$

In order to describe the change of concentration in the finite system of cells, the same equation can be used for each cell. We choose an arbitrary cell at position i of the series with neighbour cells at positions $i + 1$ and $i-1$. The concentrations are designated as c_i, c_{i+1} and c_{i-1}. The fluxes in x-direction across the corresponding faces of the cell can be approximated by a finite version of Fick's Law:

$$j_{x+} = -D\frac{c_{i+1} - c_i}{\Delta x} \quad \text{and} \quad j_{x-} = -D\frac{c_i - c_{i-1}}{\Delta x}$$

Note that these formulae are not valid exactly, but may serve well as approximations. It is the idea that small errors in the approximation may lead to small deviations between analytical and numerical solutions after performing the algorithm. Mathematicians speak of stability at this point, as not all algorithms tend out to be stable. Replacing the finite difference terms in the mass conservation equation above, yields:

$$\frac{c_{i,new} - c_i}{\Delta t} = -\frac{1}{\Delta x}\left(-D\frac{c_{i+1} - c_i}{\Delta x} + D\frac{c_i - c_{i-1}}{\Delta x}\right)$$

and

$$\frac{c_{i,new} - c_i}{\Delta t} = D\frac{c_{i+1} - 2c_i + c_{i-1}}{\Delta x^2}$$

(continued)

where $c_{i,new}$ denotes the concentration in the i'th cell after a time step, e.g. $t + \Delta t$, while all other concentration terms are relevant at the time t. The ratio on the right side is a finite difference representation of the second derivative $\partial^2 c/\partial x^2$. There is an entire class of numerical methods, which is founded on such *Finite Differences*, whereas the simple algorithm here adopts this methodology for the simulation of diffusive fluxes only. In these equations the cell concentrations between neighbouring cells are related. Thus one may use them applying an explicit formula, from which the new concentrations in the system of cells can be computed:

$$c_{i,new} = c_i + \frac{D \cdot \Delta t}{\Delta x^2}(c_{i-1} - 2c_i + c_{i+1})$$

The coefficient, which appears in front of the brackets, the parameter combination $\frac{D \cdot \Delta t}{\Delta x^2}$ is also known as *Neumann-number*[3] (abbreviation: *Neu*), (see Holzbecher and Sorek 2005). It is the given explicit formula for $c_{i,new}$ that is computed in the m-file *'diffusion.m'*.

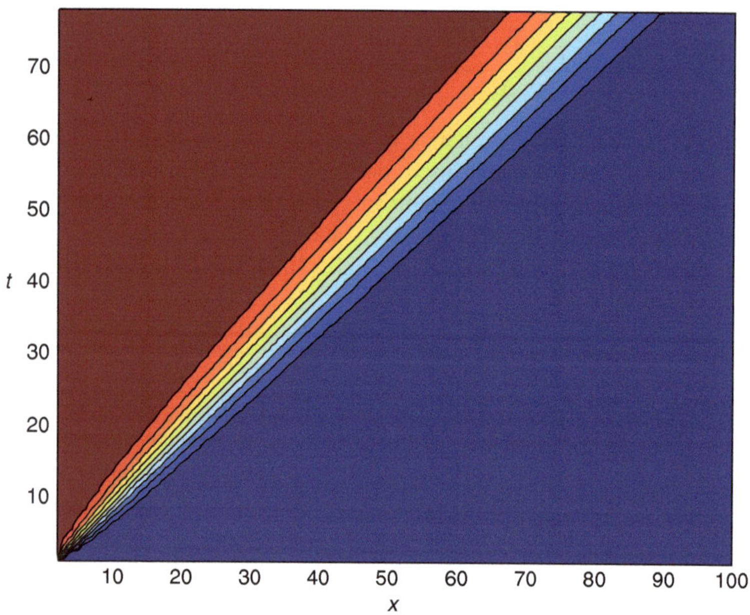

Fig. 4.8 The solution of 1D transport in the space (x) – time (t) diagram, visualized by filled contours

[3] John von Neumann (1903–1957), US-American mathematician, physicist, chemist and computer scientist

Exercise 4.2: Instability of the Explicit Diffusion Simulator Check the simple transport algorithm, introduced above, for Neumann-numbers $Neu = 0.1, 0.2, 0.3,$ 0.4, 0.5 ! Show that the algorithm becomes unstable, when *Neu* takes values greater than ½!

The algorithm can be improved by using several *(M)* diffusion steps within one time step simulation. In other words: a diffusion time step $\Delta t_{diff} = \Delta t/M$ can be introduced in addition. Note that the time step of the splitting algorithm is not arbitrary. It is determined by the velocity v and the cell extension Δx. The following m-file, which is included in the accompanying software under the name '*simpletrans.m*', takes such a refinement into account.

```
T = 30;                         % maximum time [s]
L = 50;                         % length [m]
D = 1;                          % dispersivity [m*m/s]
v = 1;                          % velocity [m/s]
c0 = 0;                         % initial concentration [kg/m*m*m]
cin = 1;                        % inflow concentration [kg/m*m*m]

dtout = 1;                      % output timestep[s]
dxmax = 1;                      % maximum gridlength [m]
%----------------------------
dx = dtout/v;
K = 1; if (dxmax<dx) K = floor(dx/dxmax); end
dx = dx/K; dtadv=dtout/K;
N = ceil(L/dx);
x = linspace(0,N*dx,N);
Neumann = D*dtadv/dx/dx;
M = ceil(2*Neumann);
Neumann = Neumann/M;
dtdiff = dtadv/M;
t = 0;

clear c c1 c2;
c(1:N) = c0; c1 = c;
k = 1;
while (t < T)
    for i=1:M
        diffusion;
    end
    advection;
    c = [c;c1];
    h = plot (x,c1); hold on;
    t = t + dtadv;
end
%----------------------------
plot (x,c','--')            % profiles
xlabel ('space'); ylabel ('concentration');
```

After the specification of the input parameters the grid spacing `dx` is calculated in several steps. First it is derived from the two input parameters time step and velocity. The user may require smaller grid spacing, if the user-defined value for `dxmax` is smaller. In the latter case `K` is the integer factor, by which `dx` is reduced. The time step, corresponding to the reduced `dx` is `dtadv`. `N` is the number of blocks, which is required to reach the user-specified length `L`. `M` denotes the number of diffusion time steps necessary to fulfil the Neumann condition:

$$Neu = \frac{D \cdot \Delta t_{diff}}{\Delta x^2} \leq \frac{1}{2} \tag{4.9}$$

Note that in the m-file the time is specified explicitly, by a maximum simulation time and an output time step. The algorithm is also described by Appelo and Postma (1993).

4.3 Comparison Between Analytical and Numerical Solution

Compare analytical and numerical solutions, as obtained with the m-files *'analtrans.m'* and *'simpletrans.m'*! A typical result is shown in Fig. 4.9, which was obtained for input values $T = 1, L = 1, v = 1, D = 0.1, c_0 = 0, c_{in} = 1, M = 50, N = 50$. There are differences at the start and the end of the simulation, while for intermediate times the two curves coincide.

As was shown above the presented algorithm treats advection exactly to the truncation error of numbers on the computer. The deviances between analytical and numerical solutions are thus due to the discretization of diffusion. Directly after start of the simulation the concentration gradient is very steep, and thus the error

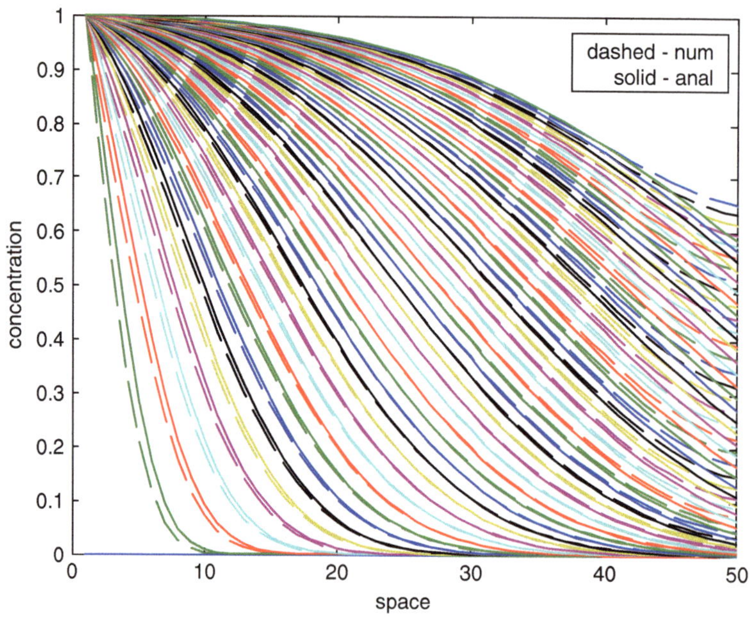

Fig. 4.9 Comparison of analytical and numerical results for the 1D transport equation

due to the approximate representation of the differentials by finite differences is higher than in the later simulation. The error can be reduced by an increase of the number of cells, which is done here by lowering the input value for `dxmax`.

The differences at the end of the outflow boundary are due to a different reason. In fact the boundary condition, which is included in the numerical algorithm, does not coincide with the analytical solution, which is valid for the infinite half space $x \geq 0$. The 'erfc'-solutions (4.1) and (4.4) do not fulfil the $\partial c/\partial x = 0$ Neumann condition at any finite location.

The Neumann boundary condition is in fact approximated by the numerical algorithm, more precisely in the command

```
c2(N) = c1(N) + Neumann*(c1(N-1)-c1(N));
```

appearing in the module *'diffusion.m'*. The formula results from the general finite difference formulation by setting `c1(N + 1) = c1(N)`, where `c1(N + 1)` represents the concentration in the outflow reservoir behind the final cell. The slope of the numerical solution vanishes at the right boundary, which is clearly visible in Fig. 4.9.

The boundary condition is not altered by a finer discretization, and thus the deviation on the outflow side remains. Figure 4.10, obtained for 100 cells, demonstrates that the deviation on the left side is reduced in comparison to Fig. 4.9, but the deviation on the right side remains.

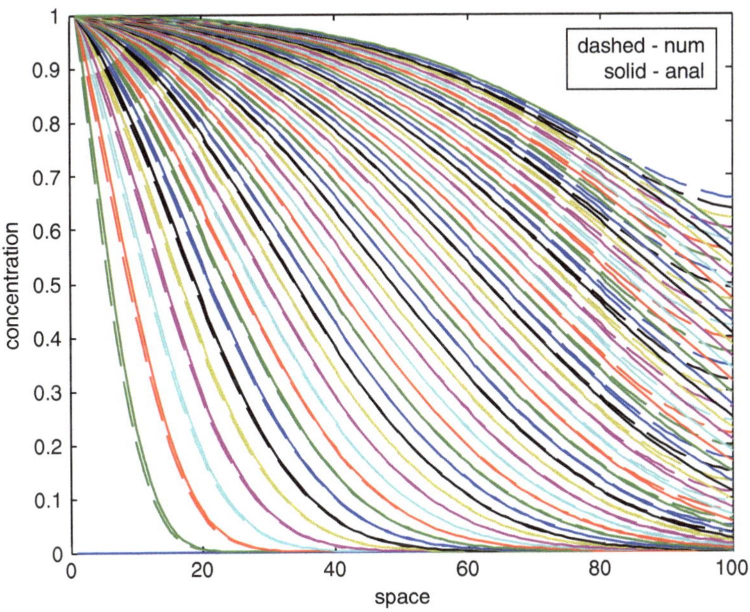

Fig. 4.10 Comparison of analytical and numerical results for the 1D transport equation, for a finer discretization

4.4 Numerical Solution Using MATLAB® pdepe

Here another method for solving the transport equation is presented which is based on the `pdepe` solver for partial differential equations (pde's). It is necessary to introduce this third method, as it offers more possibilities and can be applied for a much broader class of problems. Several species and/or temperature can be treated simultaneously; for that reason the vector variable `u` is used in this sub-chapter to gather all dependent unknown variables. The coefficients may have dependencies, either on time or on space, i.e. on the independent variables x and t, or on the dependent variables `u`. Various forms of additional terms can be taken into account, in order to consider complex sources or sinks. This capability of `pdepe` opens the possibility to simulate networks of reacting biogeochemical species. Moreover, initial conditions are allowed to be space dependent and boundary conditions to be time dependent. The field of possible applications is so wide that only a few examples can be presented within here.

`pdepe` is a MATLAB® command. Like for any other command the help system delivers some information and instructions. In the command window one may also use

```
help pdepe
```

in order to get the basics. The information supplied by the help system is brief and directed to a mathematically skilled audience. Therefore we provide an introduction, which can be understood without being used to mathematical presentations. On the other hand the focus here is on transport equations, which are typical for environmental models, and not on those numerous other partial differential equations, which can also be treated using `pdepe`.

`pdepe` solves 'pde'-systems of partial differential equations which can be written in the following form

$$\mathbf{c} \cdot \frac{\partial \mathbf{u}}{\partial t} = x^{-m} \frac{\partial (x^m \mathbf{f})}{\partial x} + \mathbf{s} \tag{4.10}$$

In the notation of (4.10) the terminology of the MATLAB® help system is adopted. It was already mentioned that the unknown variables which have to be determined are gathered in the vector `u`. The coefficients of the time derivatives are gathered in a diagonal matrix `c` (has nothing to do with concentrations). The functions `f` and `s` on the right side of (4.10) are vector functions too, depending on $x, t, \mathbf{u}$ and $\partial \mathbf{u}/\partial x$. Also for `c` these dependencies can be valid. The integer value m may take the values 0, 1 and 2, representing slab, cylindrical, or spherical symmetry, respectively. In favour of simplicity $m = 0$ will be valid in the introductory examples.

Let's look at an illustrating example. The distributions of species A and B are to be simulated in a system in which advection, dispersion and reaction are the relevant processes. According to the derivations in Chap. 3 such a situation can be described by two transport equations:

$$\begin{aligned}\frac{\partial c_A}{\partial t} &= \frac{\partial}{\partial x}\left(D\frac{\partial}{\partial x}c_A - vc_A\right) - r\\ \frac{\partial c_B}{\partial t} &= \frac{\partial}{\partial x}\left(D\frac{\partial}{\partial x}c_B - vc_B\right) + r\end{aligned} \tag{4.11}$$

c_A and c_B denote the concentrations of species A and B, D the dispersivity, v the velocity, and r the reaction rates. The system (4.11) fulfils the form (4.10) with $m = 1$ and the following functions:

$$\mathbf{c} = \begin{pmatrix}1 & 0\\ 0 & 1\end{pmatrix} \mathbf{u} = \begin{pmatrix}c_A\\ c_A\end{pmatrix} \mathbf{f} = \begin{pmatrix}D\frac{\partial}{\partial x}c_A - vc_A\\ D\frac{\partial}{\partial x}c_B - vc_B\end{pmatrix} \mathbf{s} = \begin{pmatrix}-r\\ +r\end{pmatrix} \tag{4.12}$$

Another possible representation is:

$$\mathbf{c} = \begin{pmatrix}1 & 0\\ 0 & 1\end{pmatrix} \mathbf{u} = \begin{pmatrix}c_A\\ c_A\end{pmatrix} \mathbf{f} = \begin{pmatrix}D\frac{\partial}{\partial x}c_A\\ D\frac{\partial}{\partial x}c_B\end{pmatrix} \mathbf{s} = \begin{pmatrix}-v\frac{\partial}{\partial x}c_A - r\\ -v\frac{\partial}{\partial x}c_B + r\end{pmatrix} \tag{4.13}$$

Note that the latter formulation requires the condition $\partial v/\partial x = 0$. The difference between both formulations is that in formulation (4.12) the flux term f includes dispersive and advective fluxes, while in formulation (4.13) only dispersive fluxes are included. This difference is also important for the boundary conditions, as is shown in the following.

For the complete formulation of the mathematical problem initial and boundary conditions need to be set. Using the terminology introduced above, the initial condition at time t_0 is given by the vector equation:

$$\mathbf{u}(x, t_0) = \mathbf{u}_0(x) \tag{4.14}$$

The boundary condition, valid at locations $x = x_0$ and $x = x_n$, is formulated in MATLAB® by

$$\mathbf{p} + \mathbf{q} \cdot \mathbf{f} = 0 \tag{4.15}$$

The function `p` may to depend on x, t and `u`, the function `q` on x and t. Note that `f` is the flux-vector from the differential (4.10). At first sight the formulation (4.15) seems rather different from the formulations of boundary conditions given above. But it turns out that the MATLAB® formulation offers a lot of flexibility and extended options for all types of conditions. An overview is given in Table 4.1.

Table 4.1 Implementation of transport boundary condition, using MATLAB® 'pdepe'

Type	Name	Representation	Formula	p	q
First	Dirichlet	(4.13), (4.12)	$u = u_1$	$u - u_1$	0
Second	Neumann	(4.13)	$D\frac{\partial u}{\partial x} = -p_1$	p_1	1
Third	Cauchy or Robin	(4.12)	$q_1 D\frac{\partial u}{\partial x} = u_1 - u$	$u - u_1$	q_1
"	"	(4.13)	$D\frac{\partial u}{\partial x} - v \cdot u = 0$	0	1
"	"	(4.12)	$D\frac{\partial u}{\partial x} - v \cdot u = p_1$	p_1	1

In the notation in Table 4.1 a single unknown variable u is used, representing a single unknown component. The boundary conditions, even the type of the boundary conditions may be different for each component of the vector `u`.

The reference to the equation representation indicates, which processes are included in the flux term `f`, either in case of diffusion/dispersion alone, or with advection. In case of several unknown variables the representation can be chosen differently for the different components.

Sidebar 4.3: Syntax of the MATLAB® pdepe-command

Syntax

sol = pdepe(*m,pdefun,icfun,bcfun,xmesh,tspan*)
sol = pdepe(*m,pdefun,icfun,bcfun,xmesh,tspan,options*)
sol = pdepe(*m,pdefun,icfun,bcfun,xmesh,tspan,options,p1,p2...*)

Arguments

m	geometry parameter (slab = 0, cylindrical = 1, spherical = 2).
m	geometry parameter (slab = 0, cylindrical = 1, spherical = 2).
pdefun	function submodule for coefficients of the differential equation (**c**, **f** and **s**).
icfun	function submodule for initial condition ($\mathbf{u}_0$).
bcfun	function submodule for boundary conditions (**p** and **q** at x_0 and x_n).
xmesh	vector [x_0, x_1, ..., x_n], positions, at which the solution vector is calculated; elements of *xmesh* need to fulfill: $x_0 < x_1 < \ldots < x_n$ and there must be at least three entries
tspan	vector [t_0, t_1, ..., t_f], time instants, at the solution vector is calculated; elements of *tspan* need to fulfill: $t_0 < t_1 < \ldots < t_f$ and there must be at least three entries
options	options concerning the numerical algorithm: *RelTol*, *AbsTol*, *NormControl*, *InitialStep*, and *MaxStep*; see, odeset' in help system for details; use default first
p1,p2,...	optional parameters für *pdefun*, *icfun* und *bcfun*
sol	solution matrix, containing values for all elements at all positions of *xmesh* at all time instants of *tspan*.

A complete description can be found in MATLAB®-help. With the `pdepe` command the m-file *'pdepe.m'* is called. The user finds the location of that file by using

```
which pdepe
```

The user may open the file to see that the implemented solution algorithm is also written in m-language. The inexperienced user is not recommended to alter that file, but it is worth to know that alterations of the algorithm are possible in MATLAB®. The syntax of the `pdepe` command is presented in Sidebar 4.3 in brief form.

4.5 Example: 1D Inflow Front

The first test case for the `pdepe` method is the simulation of a situation for which the Ogata-Banks solution holds. At first the functions have to be specified. Because the transport equation is concerned, the names `transfun`, `ictransfun` and `bctransfun` are chosen, for the equation specification, the initial conditions and the boundary conditions, respectively. The entire specification is given by:

```
%----------------------functions------------------------------
function [c,f,s] = transfun(x,t,u,DuDx,D,v,c0,cin)
c = 1;
f = D*DuDx;
s = -v*DuDx;
% -----------------------------------------------------------
function u0 = ictransfun(x,D,v,c0,cin)
u0 = c0;
% -----------------------------------------------------------
function [pl,ql,pr,qr] = bctransfun(xl,ul,xr,ur,t,D,v,c0,cin)
pl = ul-cin;
ql = 0;
pr = 0;
qr = 1;
```

The functions `f` and `s` are specified in accordance with formulation (4.13), as the flux term includes diffusion only. Parameters in this example are diffusivity `D`, velocity `v`, initial concentration `c0` and inflow concentration `cin`. These parameters have to be included in the formal parameter list appearing in the header of each function module. Number and order of these parameters need to be the same. Note that in the boundary module `p` and `q` have to be specified for both boundaries. The subscripts `l` (left) and `r` (right) indicate the boundary. The specifications for the left boundary coincide with the settings in line 1 of Table 4.1; i.e. the Dirichlet boundary condition is specified, while the specifications for the right boundary coincide with line 2, the Neumann condition.

The main module of the m-file looks as follows:

```
function pdepetrans
% transport-solver using 'pdepe'

T = 4.;                    % maximum time [s]
L = 2.5;                   % length [m]
D = 0.01;                  % diffusivity [m*m/s]
v = 1;                     % velocity [m/s]
c0 = 0;                    % initial concentration [kg/m*m*m]
cin = 1;                   % boundary concentration [kg/m*m*m]
M = 100;                   % number of timesteps
N = 100;                   % number of nodes

t = linspace (T/M,T,M);    % time discretization
x = linspace (0,L,N);      % space discretization

%----------------------execution---------------
options = odeset;
c = pdepe (0,@transfun,@ictransfun,@bctransfun,x,...
 [0 t],options,D,v,c0,cin);

%---------------------- output ----------------
plot ([0 t],c)              % breakthrough curves
      xlabel ('time'); ylabel ('concentration');
```

The execution part of the module consists of a single call of `pdepe`. The previous command has to be made in order to specify the `options` structure for the next command. The parameter set for the example appears behind the `options`–parameter in the `pdepe` call. The results are stored on matrix `c`.

The function modules may be located in the same m-file behind the main module. As a result of the `plot` command one obtains *breakthrough curves* shown in Fig. 4.11. Breakthrough curves result from plotting concentration vs

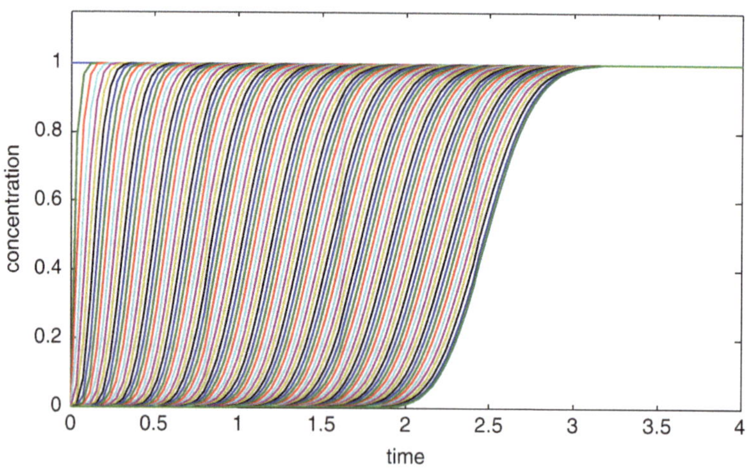

Fig. 4.11 Breakthrough curves as result of the transport solution using MATLAB® pdepe

time at a specified location. The term is common in experimental sciences, where breakthrough curves are recorded in column experiments. The set-up in a column experiment corresponds to the situation described by the Ogata-Banks solution. A front of usually increased concentration enters a 1D system in which at certain positions and certain time instants measurements are taken. A breakthrough curve results if the values measured at one of the locations are plotted against time.

In MATLAB® the `plot (c)` command yields breakthrough curves, if the concentrations for the current time step are added in another row of the matrix, as it is done by MATLAB® `pdepe`. The first formal parameter `[0 t]` determines the x-axis.

Extended versions of the m-files *'analtrans.m'*, *'simpletrans.m'* and *'pdepetrans.m'* are included in the accompanying software. The extensions are explained in the following chapters.

References

Appelo CA, Postma D (1993) Geochemistry, groundwater and pollution. Balkema, Rotterdam, p 536

Baehr HD, Stephan K (1994) Wärme- und Stoffübertragung. Springer, Berlin, p 697

Holzbecher E, Sorek S (2005) Numerical models of groundwater flow and transport, In: Anderson M (ed.) Encyclopaedia of hydrological sciences, Wiley, Vol. 4, Part 13, Art. 155, 2401–2414

Ogata A, Banks RB (1961) A solution of the differential equation of longitudinal dispersion in porous media, US Geological Survey, Professional Paper No. 411-A

Wexler EJ (1992) Analytical solutions for one-, two-, and three-dimensional solute transport in groundwater systems with uniform flow, Techniques of Water-Resources Investigations of the United States Geological Survey, Book 3, Chapter B7, p 190

Chapter 5
Transport with Decay and Degradation

5.1 Decay and Degradation

Organic matter and organic substances are subject of *degradation*. The degradation processes are of biochemical nature as they are mediated by bacteria. The details of these processes are usually quite complex. For their activity, the different bacteria cultures depend strongly not only on the biogeochemical environment but also on temperature and pressure conditions. One crucial condition, for example, is the availability of oxygen. In an aerobic environment bacteria dominate that consume oxygen aside from organic matter. These components are transformed into various products which always include carbon dioxide. In an anaerobic environment, when the available oxygen is consumed, other bacteria take over the role of major contributors to organic matter degradation. Other electron acceptors become more important, as manganese and iron in the solid phase, or nitrogen and sulphate in the fluid phase.

As an example the decay of the herbicide Linuron in soil, taken from Walker and Brown (1983) is given in Fig. 5.1

The term *decay* generally is used for physical or chemical processes that cause a loss of substance. The term is well known in connection with *radioactive decay* for the transformation of radionuclides into daughter products. Uranium U^{238} decays with a half-life of $4.5{\cdot}10^6$ a, i.e. after that time half of the initial mass is still present while the other half is transformed into Th^{234} (if no other processes are involved). As the daughter product Th^{234} has a half-life of 2.1 days only, most of it decays into Pa^{234}, which is even more short-lived with a half-life of only 12 min. The next daughter product U^{234} has a long half-life of $2.5{\cdot}10^5$ a.

There is a basic mathematical formulation that is commonly used to describe decay and degradation. More complex approaches will be presented in the following Chaps. 7 and 9. One general approach recognizes losses q being proportional to a power of the concentration c:

$$q = -\lambda c^n \tag{5.1}$$

E. Holzbecher, *Environmental Modeling*,
DOI 10.1007/978-3-642-22042-5_5, © Springer-Verlag Berlin Heidelberg 2012

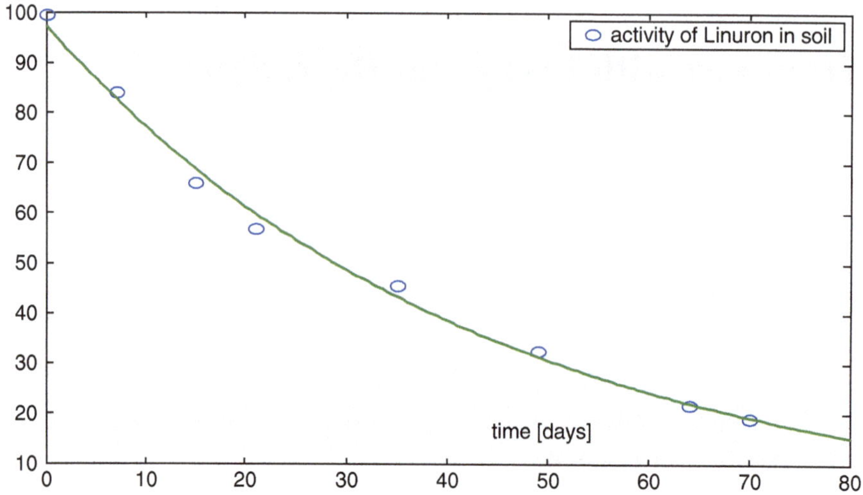

Fig. 5.1 Exponential decay; computed from analytical solution using MATLAB®; circles show measured data, line represents optimized model, as described below

where the integer n is the *order of degradation or decay*. λ is the so called decay or degradation constant, which generally depends on all variables in the environment of the system in question. Second order decay is proportional to the square of the concentration. Note that the physical unit of λ depends on the exponent n. For the important case with $n = 1$, the unit of λ is [1/T].

The sink term, given by (5.1), is substituted in the general mass transport equations as (3.19) and (3.20). Note that both, q and the differential equation have the same physical units (M/T/L^3). Because the porosity θ appears in all terms, it can be omitted. One obtains:

$$\frac{\partial c}{\partial t} = \nabla \cdot \mathbf{D} \nabla c - \nabla \cdot \mathbf{v} c - \lambda c^n \tag{5.2}$$

With this approach, decay and degradation processes are included in the transport equation, and it becomes possible to treat transport, decay and degradation simultaneously. Before we give solutions and solution strategies for the general situation, special cases will be treated first.

If no other processes are relevant remains:

$$\frac{\partial c}{\partial t} = -\lambda c^n \tag{5.3}$$

This is an ordinary differential equation for the independent variable t. Most important is first order decay or degradation, i.e. the case $n = 1$, where losses are proportional to the concentration. Then the solution of the differential (5.3) can be noted directly:

$$c = c_0 \exp(-\lambda t) \tag{5.4}$$

which holds for the initial condition $c(t = 0) = c_0$. The exponential function obviously is the solution for a component with first order decay – that explains the notation *exponential decay*.

The *half-life* $t_{½}$ is the time period in which the component concentration declines to half of the initial value. Thus according to (5.4) the $t_{½}$ is characterized by the condition

$$1/2 = \exp(-\lambda t_{1/2}) \tag{5.5}$$

which is equivalent to the condition $t_{1/2} = \ln(2)/\lambda$. This is the reciprocate relation between decay constant and half-life. With $t_{½}$ exponential decay can be noted in dimensionless form as:

$$\frac{c}{c_0} = \exp(-\ln(2)\frac{t}{t_{1/2}}) \tag{5.6}$$

for the dimensionless variables c/c_0 and $t/t_{½}$. For the time period of five half-lifes the function is depicted by the following MATLAB® commands in Fig. 5.2.

```
plot ([0:0.1:5],exp(-log(2)*[0:0.1:5])); grid;
xlabel('time t/t_{1/2}'); ylabel('c/c_0');
```

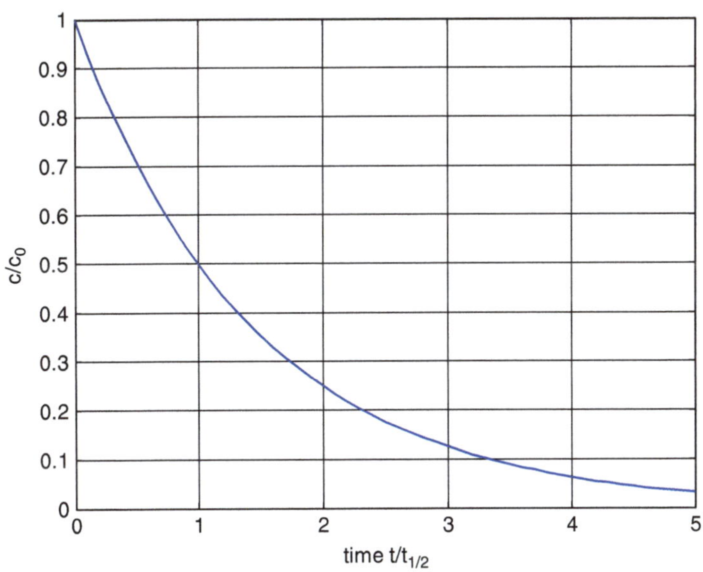

Fig. 5.2 Exponential decay as represented by dimensionless variables

Table 5.1 Half-lifes of selected radionuclides

Radionuclide	Half-life	Radionuclide	Half-life
U-238	$4.5 \cdot 10^9$ years	H-3	12.35 years
U-235	32,500 years	Ra-228	5.8 years
Ra-226	1,600 years	Th-228	1.91 years
Am-241	432.2 years	Gd-153	242 days
Pu-238	87.74 years	Po-210	138 days
Cs-137	30 years	Sr-89	50.5 days
Pb-210	22.3 years	Th-234	24.1 days

From mathematical point of view, the major difference between radioactive decay and other forms of first order losses lies in the decay constant. The half-life of radionuclides is constant under all conditions known. The rate of exponential decay of a nuclide is not influenced by any environmental condition, neither by temperature, nor by pressure, nor by the biogeochemical surrounding. Table 5.1 lists some radionuclides and their half-lifes. The simulation of a chain of radionuclides is described in Chap. 18.2.

In contrast, chemical and biochemical rates are strongly affected by environmental variables. In fact, the decay law can often be understood as a most simplified rule, in which the interaction of several complex processes are gathered and where λ is a lumped parameter. Clearly, in a changed environment the parameter is different. More complex degradation rules, using the Michaelis–Menten or Monod kinetics, are treated below (see Chap. 7).

5.2 1D Steady State Solution

As mentioned above, the differential equation for the steady state is obtained by setting the time derivatives in the transport (5.2) to zero. The right hand side of the 1D transport equation has thus been set to zero:

$$\frac{\partial}{\partial x} D \frac{\partial c}{\partial x} - v \frac{\partial c}{\partial x} - \lambda c = 0 \tag{5.7}$$

This is an ordinary differential equation for the independent variable x. With MATLAB® ordinary differential equations can be solved numerically (see Chap. 9). Here, an analytical solution, which provides the solution in an explicit formula, is an alternative if the coefficients are constants, i.e. independent of x and t. In order to solve differential (5.7) analytically, it is appropriate to note it in a different form:

$$\left(\frac{\partial}{\partial x} - \mu_1\right)\left(\frac{\partial}{\partial x} - \mu_2\right) c = 0 \tag{5.8}$$

The parameters μ_1 and μ_2 can be obtained by comparison of coefficients in (5.7) and (5.8):

$$\mu_1 + \mu_2 = v/D \quad \mu_1 \cdot \mu_2 = -\lambda/D \tag{5.9}$$

A quadratic equation results, which has the solutions:

$$\mu_{1,2} = \frac{1}{2D}\left(v \pm \sqrt{v^2 + 4\lambda D}\right) \tag{5.10}$$

Equation 5.8 can now be solved in two steps. First the solution $\bar{c}$ of the equation

$$\left(\frac{\partial}{\partial x} - \mu_1\right)\bar{c} = 0 \tag{5.11}$$

is determined, which is given by $\bar{c} = C_0 \exp(\mu_1 x)$. C_0 is an integration constant that is determined below in order to fulfil the boundary conditions. In a second step, c is found as solution of the differential equation

$$\left(\frac{\partial}{\partial x} - \mu_2\right)c = \bar{c} \tag{5.12}$$

One obtains a formula for the general solution:

$$\begin{aligned} c(x) &= \exp(\mu_2 x)\left(C_1 + C_0 \int \exp(\mu_1 x)\exp(-\mu_2 x)dx\right) \\ &= C_1 \exp(\mu_2 x) + \frac{C_0}{\mu_1 - \mu_2}\exp(\mu_1 x) \end{aligned} \tag{5.13}$$

where C_1 is the second integration constant. Both C_0 and C_1 are determined by the boundary conditions. The usual condition at the inlet $c(x=0) = c_{in}$ yields the condition: $C_1 + C_0/(\mu_1 - \mu_2) = c_{in}$ or $C_1 = c_{in} - C_0/(\mu_1 - \mu_2)$. Note that μ_1 and μ_2, as given by (5.10), have opposite signs. As μ_1 is positive, the first term in (5.13) is decreasing with depth, while the second is increasing with depth. For that reason, the solutions approach infinity for all values $C_0 \neq 0$. Vice versa, the function with $C_0 = 0$ is the only solution which guarantees finite concentration for arbitrary high values of x. It is this property which makes the choice $C_0 = 0$ favourable in studies, where there is no information concerning the downstream boundary condition (Anderson et al. 1988; Henderson et al. 1999). Then the solution simply reads:

$$c(x) = c_{in} \exp(\mu_2 x) \tag{5.14}$$

When the second boundary condition requires a vanishing concentration gradient at depth L, i.e. $(\partial c/\partial x)(L) = 0$, the second equation for C_0 and C_1 is given by:

$$C_1\mu_2\exp(\mu_2 L) + \frac{C_0\mu_1}{\mu_1-\mu_2}\exp(\mu_1 L) = 0 \tag{5.15}$$

which leads to the solution:

$$\begin{aligned} C_0 &= \frac{c_{in}\mu_2(\mu_2-\mu_1)\exp(\mu_2 L)}{\mu_1\exp(\mu_1 L)-\mu_2\exp(\mu_2 L)} \\ C_1 &= \frac{c_{in}\mu_1\exp(\mu_1 L)}{\mu_1\exp(\mu_1 L)-\mu_2\exp(\mu_2 L)} \end{aligned} \tag{5.16}$$

With these formulae the solution is complete to be computed in MATLAB®. This will be done in the next sub-chapter. Here, we want to point out that the given procedure can be applied to obtain the solution for different types of boundary conditions. In all cases the free constants C_0 and C_1 in a formula for the general solution, like in (5.13), have to be determined to fulfil the conditions. For Dirichlet boundary conditions on both sides $c(0) = c_{in}$ and $c(L) = c_0$one obtains:

$$C_0 = \frac{c_{in}(\mu_2-\mu_1)\exp(\mu_2 L)}{\exp(\mu_1 L)-\exp(\mu_2 L)} C_1 = c_{in} - \frac{C_0}{\mu_1-\mu_2} \tag{5.17}$$

Higher order decay usually is much more difficult to handle than decay of first order. In order to tackle more complex formulations MATLAB® offers the possibility to use numerical methods for ordinary differential equations. Such methods are treated in Chap. 9.

5.3 Dimensionless Formulation

In dimensionless form the solution (5.13) can be written as:

$$\frac{c(\tilde{x})}{c_{in}} = \tilde{C}_0\exp(\tilde{\mu}_1\tilde{x}) + \left(1-\tilde{C}_0\right)\exp(\tilde{\mu}_2\tilde{x}) \tag{5.18}$$

with dimensionless depth $\tilde{x} = x/L, \tilde{\mu}_1 = \mu_1 L, \tilde{\mu}_2 = \mu_2 L$ and one integration constant $\tilde{C}_0$.

The parameters $\tilde{\mu}_1$ and $\tilde{\mu}_2$ can be expressed as function of the dimensionless Péclet number $Pe = vL/D$ (see Chap. 3.5) and the dimensionless second Damköhler[1] number $Da_2 = \lambda L^2/D$:

[1] Gerhard Damköhler (1908–1944), German chemist.

$$\tilde{\mu}_{1,2} = \frac{1}{2}Pe \pm \sqrt{\frac{1}{4}Pe^2 + Da_2} \quad (5.19)$$

If the first Damköhler number for the fluid phase is defined by $Da_1 = \lambda L/v$, the two μ-values can also be expressed as function of Pe and Da_1. Using the identity $Da_2 = Pe \cdot Da_1$, one obtains $\tilde{\mu}_1$ and $\tilde{\mu}_2$ as functions of Pe and Da_1:

$$\tilde{\mu}_{1/2} = \frac{1}{2}Pe \pm \sqrt{\frac{1}{4}Pe^2 + Pe \cdot Da_1} \quad (5.20)$$

The solution with vanishing concentration gradient at depth L can be expressed by the formula:

$$\frac{c(\tilde{x})}{c_{in}} = \frac{\tilde{\mu}_2 \exp(\tilde{\mu}_2) \exp(\tilde{\mu}_1 x) - \tilde{\mu}_1 \exp(\tilde{\mu}_1) \exp(\tilde{\mu}_2 \tilde{x})}{\tilde{\mu}_2 \exp(\tilde{\mu}_2) - \tilde{\mu}_1 \exp(\tilde{\mu}_1)} \quad (5.21)$$

A special case of (5.21) is

$$\frac{c(\tilde{x})}{c_{in}} = \frac{\exp(\sqrt{Da_2}) \exp(-\sqrt{Da_2}\tilde{x}) + \exp(-\sqrt{Da_2}) \exp(\sqrt{Da_2}\tilde{x})}{\exp(\sqrt{Da_2}) + \exp(-\sqrt{Da_2})} \quad (5.22)$$

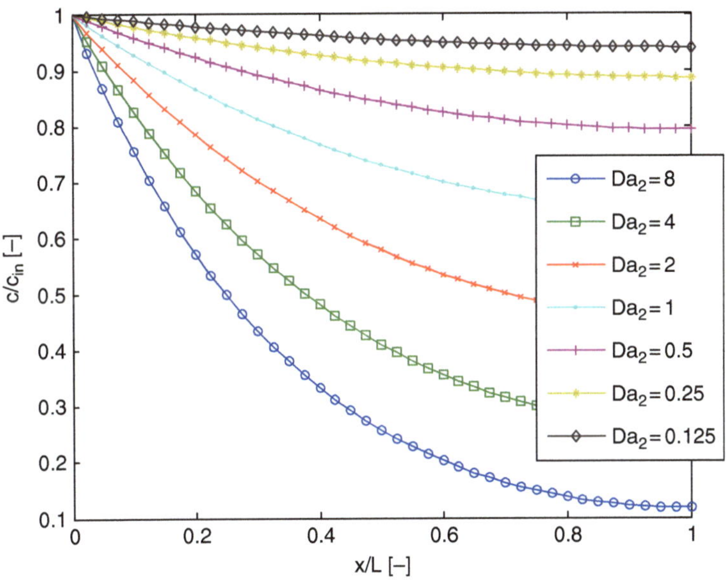

Fig. 5.3 Concentration profiles in the case of diffusion and decay with constant parameters D and λ; as function of dimensionless second Damköhler number Da_2

which is obtained for $Pe = 0$, i.e. for no advection. The graphs of functions (5.22) for different values of the second Damköhler number are presented in Fig. 5.3. The following MATLAB® code is used for the plot. Markers in lines were added by post-processing with the MATLAB® figure-editor.

```
x = [0:0.01:1];
Da2 = 8; mu1 = sqrt(Da2); mu2 = -mu1;
s = mu2*exp(mu2)-mu1*exp(mu1);
c = (mu2*exp(mu2)*exp(mu1*x)-mu1*exp(mu1)*exp(mu2*x))./s
for Da2 = [4 2 1 0.5 0.25 0.125];
    s = sqrt(Da2); mu1 = s; mu2 = -s;
    s = mu2*exp(mu2)-mu1*exp(mu1);
    c = [c;(mu2*exp(mu2)*exp(mu1*x)-mu1*exp(mu1)*exp(mu2*x))./s];
end
plot (x,c);
legend('Da_2=8','Da_2=4','Da_2=2','Da_2=1','Da_2=0.5','Da_2=0.25',
'Da_2 =0.125');
xlabel ('x/L [-]'); ylabel ('c/c_{in} [-]');
```

The corresponding M-file *'analtrans_s1'* is included in the accompanying software.

The result of a varying Péclet number is illustrated in Fig. 5.4. All but one of the depicted curves are calculated for $Da_1 = 1$ and variable Pe. The unit value of Da_1 guarantees equal importance of advection and degradation or decay. The increasing

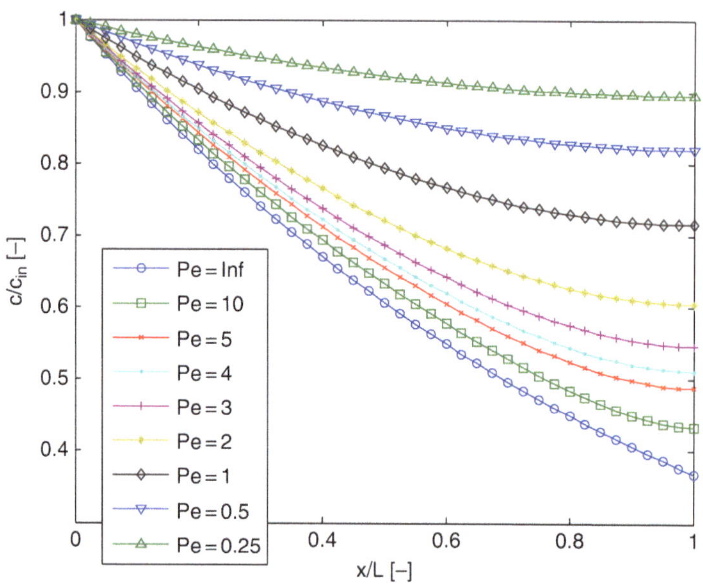

Fig. 5.4 Profiles for fluid phase concentration in case of transport with constant parameters for $Da_1 = 1$ in dependence of the Péclet number Pe

value for Pe then represents a reduced importance of diffusion, which in the profiles is reflected by steeper gradients.

The corresponding M-file *'analtrans_s2'* is included in the accompanying software.

The lower most curve in Fig. 5.4 shows the profile for advection and decay only, i.e. for no diffusion, calculated as solution of the simple first order ordinary differential equation

$$-v\frac{\partial c}{\partial x} - \lambda c = 0 \text{ or } \frac{\partial c}{\partial x} = -\frac{\lambda}{v}c \tag{5.23}$$

The solution is:

$$c(x) = c_{in}\exp(-\frac{v}{\lambda}x) \tag{5.24}$$

or in dimensionless form (for dimensionless concentration, with dimensionless parameter Da_1 and dimensionless independent variable $\tilde{x}$)

$$c/c_{in} = \exp(-Da_1\tilde{x}) \tag{5.25}$$

Formally one can represent the no-diffusion case by $Pe = \infty$. As could be expected for increasing values of Pe, the function of (5.25) is approached as asymptote. But the convergence is quite slow. For the value of $Da_1 = 1$, the value of the Péclet number should be distinctly above 10 in order to obtain a good correspondence between the concentration distribution and the asymptote. For a good correspondence close to the outlet (x close to L), Pe should be 100 or higher.

```
x = [0:0.01:1];
Da1 = 1;
c = exp(-Da1*x);
for Pe = [10 5 4 3 2 1 0.5 0.25];
    s = sqrt(0.25*Pe*Pe+Pe/Da1);
    mu1 = 0.5*Pe+s; mu2 = 0.5*Pe-s;
    s = mu2*exp(mu2)-mu1*exp(mu1);
    c = [c; (mu2*exp(mu2)*exp(mu1*x)-mu1*exp(mu1)*exp(mu2*x))./s];
end
plot (x,c);
 legend('Pe=Inf','Pe=10','Pe=5','Pe=4','Pe=3','Pe=2','Pe=1','Pe=0.5',
   'Pe =0.25');
xlabel ('x/L [-]'); ylabel ('c/c_{in} [-]');
```

Figure 5.5 shows graphs of functions according to (5.21) for a fixed value of the Péclet number $Pe = 1$ and selected values of the first Damköhler number Da_1. The constant value of Pe guarantees a constant relation between diffusion and advection, while with changing values of Da_1 decay or degradation processes change

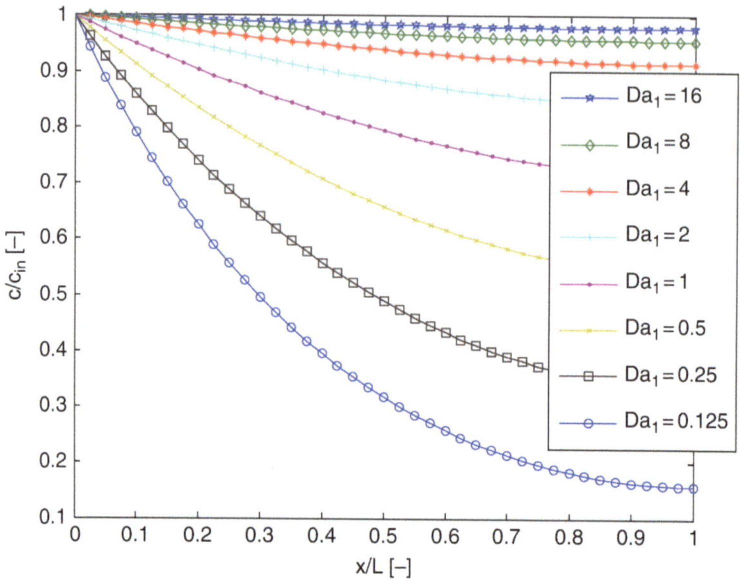

Fig. 5.5 Profiles for fluid phase concentration in case of transport with constant parameters for $Pe = 1$ in dependence of the first Damköhler number Da_1

their relative importance. For high values of the Damköhler number decay is relatively important and thus concentrations are significantly reduced from the inlet ($x = 0$) to the outlet ($x = L$). With decreasing value for Da_1, decay becomes less relevant. Then the profile for advection and diffusion is approached, which is the straight line representing constant concentration $c = c_0$ (for the given boundary conditions). The figure is produced in MATLAB® by the following commands:

```
x = [0:0.01:1];
Pe = 1; Da1 = 16;
s = sqrt(0.25*Pe*Pe+Pe/Da1);
mu1 = 0.5*Pe+s; mu2 = 0.5*Pe-s;
s = mu2*exp(mu2)-mu1*exp(mu1);
c = (mu2*exp(mu2)*exp(mu1*x)-mu1*exp(mu1)*exp(mu2*x))./s;
for Da1 = [8 4 2 1 0.5 0.25 0.125];
    s = sqrt(0.25*Pe*Pe+Pe/Da1);
    mu1 = 0.5*Pe+s; mu2 = 0.5*Pe-s;
    s = mu2*exp(mu2)-mu1*exp(mu1);
    c = [c;(mu2*exp(mu2)*exp(mu1*x)-mu1*exp(mu1)*exp(mu2*x))./s];
end
plot (x,c);
legend('Da_1=16','Da_1=8','Da_1=4','Da_1=2','Da_1=1','Da_1=0.5',
  'Da_1=0.25','Da_1=0.125');
xlabel ('x/L [-]'); ylabel ('c/c_{in} [-]');
```

The corresponding M-file *'analtrans_s3'* is included in the accompanying software.

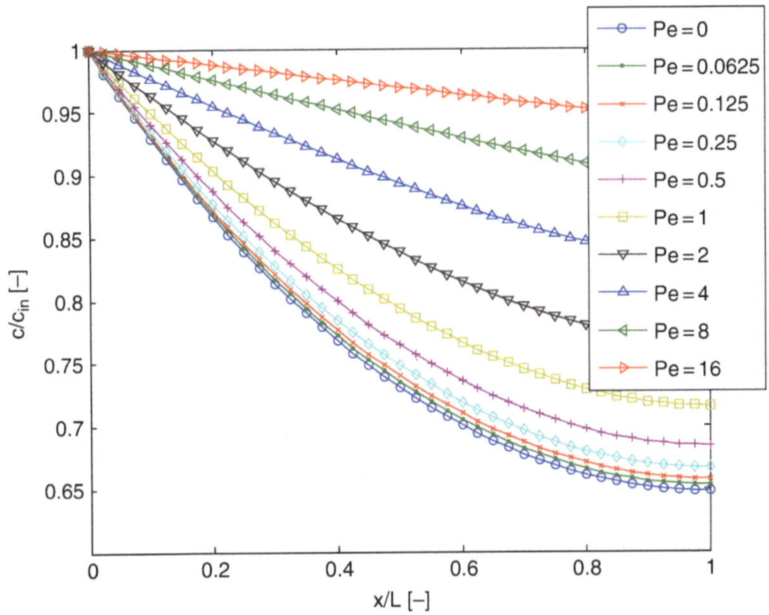

Fig. 5.6 Profiles for fluid phase concentration in case of transport with constant parameters for second Damköhler number $Da_2 = 1$ in dependence of the Pe

Figure 5.6 represents the solutions for increasing advection if the relation between diffusion and decay remains equal. For this plot it is assumed that both processes are equally important, which is expressed by the second Damköhler number $Da_2 = 1$. With increasing Péclet number advection becomes more relevant, the front can penetrate further, and the concentration gradients become less steep.

```
x = [0:0.01:1];
Da2 = 1;
mu1 = sqrt(Da2); mu2 = -mu1;
s = mu2*exp(mu2)-mu1*exp(mu1);
c = (mu2*exp(mu2)*exp(mu1*x)-mu1*exp(mu1)*exp(mu2*x))./s;
for Pe = [0.0625 0.125 0.25 0.5 1 2 4 8 16];
    s = sqrt(0.25*Pe*Pe+Da2);
    mu1 = 0.5*Pe+s; mu2 = 0.5*Pe-s;
    s = mu2*exp(mu2)-mu1*exp(mu1);
    c = [c;(mu2*exp(mu2)*exp(mu1*x)-mu1*exp(mu1)*exp(mu2*x))./s];
end
plot (x,c);
legend('Pe=0','Pe=0.0625','Pe=0.125','Pe=0.25','Pe=0.5','Pe=1','Pe=2','
    Pe=4','Pe=8','Pe=16');
xlabel ('x/L [-]'); ylabel ('c/c_{in} [-]');
```

The corresponding M-file *'analtrans_s4'* is included in the accompanying software.

The limit solution for $Pe = 0$ was obtained from the differential equation for no advection:

$$\frac{\partial}{\partial x}D\frac{\partial c}{\partial x} - \lambda c = 0 \quad \text{or} \quad \frac{\partial^2 c}{\partial x^2} = \frac{\lambda}{D}c \tag{5.26}$$

with the general solution:

$$c(x) = C_0 \exp\left(\sqrt{\frac{v}{D}}x\right) + C_1 \exp\left(-\sqrt{\frac{v}{D}}x\right) \tag{5.27}$$

In terms of dimensionless parameter and variable this is

$$c(x) = C_0 \exp(Da_2\tilde{x}) + C_1 \exp(-Da_2\tilde{x}) \tag{5.28}$$

The free constants C_0 and C_1 are again obtained from the boundary conditions.

5.4 Transient Solutions

The analytical solution for the inflow of a front with concentration c_{in} into a region with concentration c_0 is given by:

$$\begin{aligned} c(x,t) = {} & c_0 \exp(-\lambda t)\left(1 - \frac{1}{2}\operatorname{erfc}\left(\frac{x - vt}{2\sqrt{Dt}}\right) - \frac{1}{2}\exp\left(\frac{vx}{D}\right)\operatorname{erfc}\left(\frac{x + vt}{2\sqrt{Dt}}\right)\right)\ldots \\ & + \frac{c_{in}}{2}\left(\exp\left(\frac{v-u}{2D}x\right)\operatorname{erfc}\left(\frac{x - ut}{2\sqrt{Dt}}\right) + \exp\left(\frac{v+u}{2D}x\right)\operatorname{erfc}\left(\frac{x + ut}{2\sqrt{Dt}}\right)\right) \end{aligned} \tag{5.29}$$

with $u = \sqrt{v^2 + 4\lambda D}$ (Wexler 1992). The solution consists of two parts: the first describes the decline of the original concentration c_0 and the second the change of the inflow concentration c_{in} in the 1D set-up.

In MATLAB® the solution is to be included in the M-file *'analtrans.m'*, which was introduced in the previous chapter. The decay parameter λ is added in the input part of the module:

```
lambda = 0.1;     % decay rate
```

Then the auxiliary parameter `u` is computed by:

```
u = sqrt(v*v+4*lambda*D);
```

and the formula (5.29) is programmed by the lengthy term:

```
c = [c; c0*exp(-lambda*t(i))*(e-0.5*erfc(h*(x-e*v*t(i)))-
0.5*exp((v/D)*x).*erfc(h*(x+e*v*t(i)))) +...
(cin-c0)*0.5*(exp((v-u)/(D+D)*x).*erfc(h*(x-
e*u*t(i)))+exp((v+u)/(D+D)*x).*erfc(h*(x+e*u*t(i))))];
```

Also the *,simpletrans.m'* model can be extended easily by introducing decay as additional process. Write another submodule, named *'kinetics.m'* that includes only one command:

```
c1 = exp(-lambda*dtdiff)*c1; % simple first order kinetic
```

In the main module *,simpletrans.m,'* the command `kinetics` is called before `diffusion`. Appelo and Postma (1993) describe a similar procedure.

In dimensionless formulation the solution is:

$$c(\tilde{x},\tilde{t}) = c_0 \exp(-Da_2\tilde{t})\left(1-\frac{1}{2}\text{erfc}\left(\frac{\tilde{x}-\tilde{t}}{2\sqrt{\tilde{t}/Da_2}}\right)-\frac{1}{2}\exp(Pe\cdot\tilde{x})\text{erfc}\left(\frac{\tilde{x}+\tilde{t}}{2\sqrt{\tilde{t}/Da_2}}\right)\right)\ldots$$
$$+\frac{c_{in}}{2}\left(\exp\left(Pe\frac{1-u}{2}\tilde{x}\right)\text{erfc}\left(\frac{\tilde{x}-u\tilde{t}}{2\sqrt{\tilde{t}/Da_2}}\right)+\exp\left(Pe\frac{1+u}{2}\tilde{x}\right)\text{erfc}\left(\frac{\tilde{x}+u\tilde{t}}{2\sqrt{\tilde{t}/Da_2}}\right)\right) \quad (5.30)$$

with $u = \sqrt{1+4Da_2/Pe}$. Figure 5.7 illustrates the solution for a high Péclet number $Pe = 100$ and a moderate second Damköhler number. The front proceeds in positive x-direction from left to right. The time to proceed from one front line to the next amounts to the 10th part of the mean time which a tracer would need to migrate through the entire system.

With the intruding front the concentration is reduced. Because of the high Péclet number, the front line remains relatively steep, which has the additional effect that

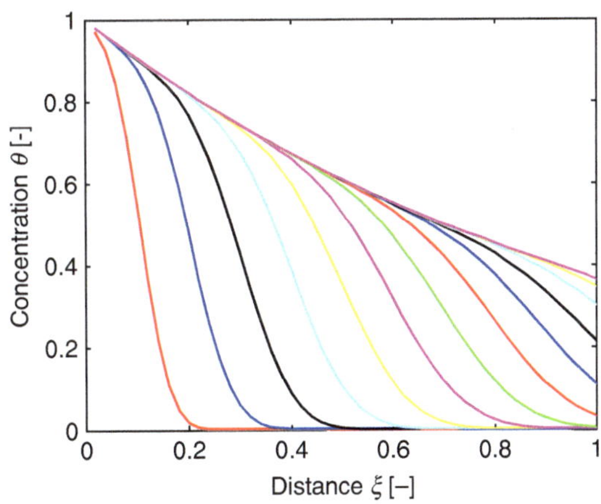

Fig. 5.7 Steady and transient solution for transport equation with degradation; dimensionless space variable ξ and dimensionless concentration θ for $Pe = 100$ and $Da_2 = 1$

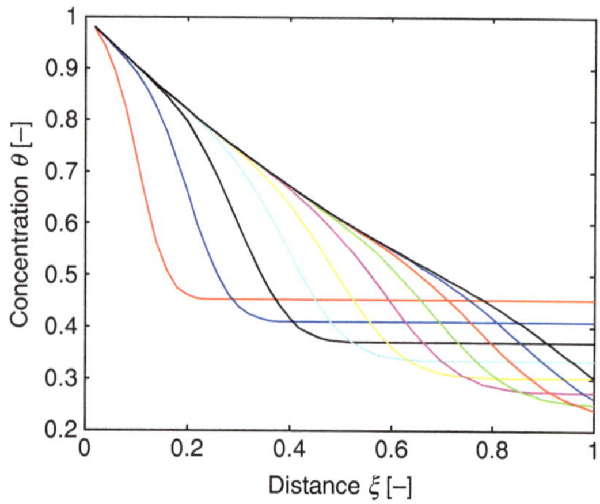

Fig. 5.8 Transient solution for transport equation with degradation; dimensionless space variable ξ and dimensionless concentration θ for $Pe = 100$ and $Da_2 = 1$, for $c_0 = c_{in}/2 = 0.5$

the deviation from the steady state is relatively moderate but increasing. The plot was obtained by using the M-file *'analtransnodim.m'* with appropriate parameters.

Figure 5.8 shows a typical behavior when c_0 is not equal to zero. In that case, both terms in the formulae (5.29) and (5.30) have to be considered. The decline of concentrations from one line to the other on the right side of the figure illustrates the decay of the material, which was initially present in the system. Increasing concentrations on the left side stem from the advancing front.

For analytical solutions of decay chains see chap. 10.4. For multi-species transport of decay-chains see corresponding references in that chapter.

References

Anderson RF, Bopp RF, Buesseler KO, Biscaye PE (1988) Mixing of particles and organic constituents in sediments from the continental shelf and slope off cape cod: SEEP-1 results. Cont Shelf Res 8(5–7):925–946

Appelo CA, Postma D (1993) Geochemistry, groundwater and pollution. Balkema, Rotterdam, p 536

Henderson GM, Lindsay FN, Slowey NC (1999) Variation in bioturbation with water depth on marine slopes: a study on the little Bahamas bank. Mar Geol 160:105–118

Walker A, Brown PA (1983) Measurement and prediction of chlorsulforon persistence in soil. Bull Environ Contam Toxicol 30:365–372

Wexler EJ (1992) Analytical solutions for one-, two-, and three-dimensional solute transport in groundwater systems with uniform flow, Techniques of Water-Resources Investigations of the United States Geological Survey, Book 3, Chapter B7, p 190

Chapter 6
Transport and Sorption

6.1 Interphase Exchange

Aside from advection, diffusion and dispersion, which can be formulated separately for the solid and the fluid phases, the interaction between the phases is another process which is relevant in many environmental systems. Under certain conditions, particles in the pore space are attracted by the surface of the porous medium where some chemical processes tend to bind them in various ways. Different processes like electrical attraction and repulsion, complexation or chemical reaction can be distinguished in a detailed look, gathered under the general term *sorption*.

In multiphase environments, sorption denotes processes that affect an exchange of components between phases. One can speak of interphase exchange. In porous media there is the exchange between solid and fluid phase, i.e. between the water in the pore space and the solid matrix. *Adsorption* denotes fluxes from the fluid to the solid phase, while *desorption* is the opposite process, in which there is a flux from the solid to the fluid phase. In the following the term *mobilization* is used frequently. A pollutant particle which has been fixed at the surface of the solid matrix in a first time period, may be re-mobilized and freed for dispersion and advection processes in a second time period.

A schematic view on adsorption and desorption in the pore space of a porous medium is given in Fig. 6.1. In the remainder of this chapter, we will mostly refer to that multi-phase set-up as it is the standard concept for environmental models concerning groundwater, seepage water and pore water in aquatic sediments.

In a first fundamental distinction, the speed of the processes governing interphase exchange in relation to the transport processes is of concern. One speaks of *fast sorption* if sorption is faster than the transport processes; and of *slow sorption* if sorption is slower than transport. Thus the characterization of sorption depends on the specific situation.

Let c denote the fluid phase concentration and c_s the solid phase concentration. Normally c and c_s are not independent but connected. High or low concentrations in

E. Holzbecher, *Environmental Modeling*,
DOI 10.1007/978-3-642-22042-5_6, © Springer-Verlag Berlin Heidelberg 2012

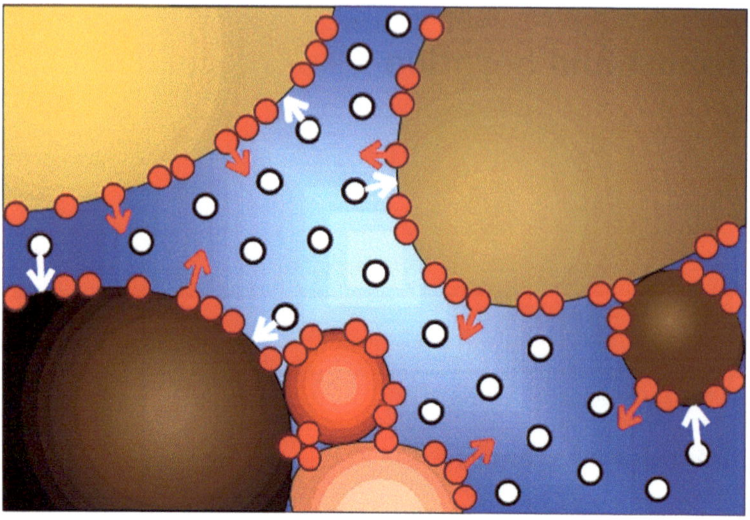

Fig. 6.1 Schematic illustration of ad- and desorption processes

one phase are usually connected with high or low concentrations in the other phase. Such phenomenon can be formulated by a mathematical relationship. In the case of fast sorption, the relationship is mostly stated in the functional form

$$c_s(c) \tag{6.1}$$

in which the solid phase concentration is given in dependency of the fluid phase concentration. This is called a (sorption-) *isotherm*[1] and can be understood as an equilibrium, very much like the equilibrium in chemical reactions. The concentrations in one phase are adjusted if, for whatever reason, the concentration in the other phase is changing.

The simplest example is the *linear isotherm*

$$c_s = K_d c \tag{6.2}$$

where the distribution coefficient K_d determines the ratio between solid phase and fluid phase concentrations. The physical unit [volume/mass] can be attributed to the fact that the concentrations in the fluid and solid phase are usually not measured in the same physical units. Strongly sorbing components have a high K_d, while it is low for weakly sorbing components. Non-sorbing components do not interact

[1] The notation 'isotherm' stems from the fact that such measurements are mostly performed for constant temperatures, i.e. isothermal conditions. In general the isotherm changes with temperature.

with the solid phase and are called *tracers*. Chloride is such a tracer in most environments.

K_d not only depends on the component but also on the solid material. In clays sorption can be expected to be high due to the high surface area per volume and due to the high electric potential. Clay minerals have an excess of imbalanced negative charges, thus favoring the adsorption of cations. Divalent cations are usually more strongly adsorbed than monovalent ions (Fetter 1994).

In the literature, K_d values are treated extensively for different kinds of chemical species, for inorganic and organic components, for chemicals of the natural environment and for contaminants. K_d values extend over several orders of magnitude, from low values as $2 \cdot 10^{-4}$ m^3/kg for sodium (Hölttä et al. 1997) up to high values like 400 m^3/kg for protactinium (Geibert 2001). For tracer-like components even lower values may be found and for strongly fixed components even higher values.

Often ad- and desorption are not taking place at the surfaces of the porous matrix directly but on organic material that itself is fixed at the solid matrix. Especially in aquatic sediments near the sediment-water surface this type of connection may be dominant. Synthetic organic chemicals tend to adsorb on organic carbon. If c_{org} denotes the concentration of organic material, K_{org} denotes the distribution coefficient on organic carbon, and the distribution coefficients are related by the formula:

$$K_d = K_{org} c_{org} / \rho_s \tag{6.3}$$

where the ratio c_{org}/ρ_s represents the weight fraction of organic matter in the solid phase (Karickhoff et al. 1979; Karickhoff 1984). For pure sand, which does not contain any organics, the adsorption is thus zero. As an alternative to K_{org}, the octanol-water-distribution coefficient K_{ow} can be taken. For several chemical components a relation between K_{ow} and K_{org} is given in the form

$$\log(K_{org}) = \alpha \log(K_{ow}) + \beta \tag{6.4}$$

where α and β are empirical constants. A mathematically similar relation often can be stated for K_{org} and solubility S of a component:

$$\log(K_{org}) = -\bar{\alpha} \log(S) + \bar{\beta} \tag{6.5}$$

with empirical constants $\bar{\alpha}$ and $\bar{\beta}$ (Karickhoff et al. 1979). While K_{org} and K_{ow} are correlated positively, the correlation between K_{org} and S is negative. Highly soluble chemicals can be expected to interact only marginally with the porous material. Vica versa, chemicals with low solubility show a strong tendency of interaction with the solid matrix. This is illustrated in Table 6.1 showing a classification concerning *mobility* using K_{ow}. Mobile components have low distribution coefficients, low K_{ow}, and a high solubility. Immobile, strongly sorbing components have high distribution coefficients and low solubility.

Table 6.1 Mobility classes, octanol-water-distribution coefficients K_{ow} according to Fetter (1994) and the range of solubility for organic pollutants

Mobility class	K_{ow} [mL/g]	Solubility S [ppm]
Very high	1–50	$4.4{\cdot}10^3$–$1.4{\cdot}10^5$
High	50–150	850–3,570
Moderate	150–500	110–1,100
Low	500–2,000	30–156
Slight	2,000–$2{\cdot}10^4$	0.275–10
Immobile	$>2{\cdot}10^4$	<0.252

A generalized formulation for fast sorption is the *Freundlich*[2] isotherm

$$c_s = \alpha_{F1} c^{\alpha_{F2}} \tag{6.6}$$

with coefficients α_{F1} and α_{F2}. The exponent α_{F2} usually is smaller than 1, which corresponds to the observation that for low concentrations of c the gradient of the isotherm is higher than for higher concentrations. For $\alpha_{F2} = 1$, the Freundlich isotherm also describes a linear relationship between the concentrations in both phases. The Freundlich isotherm is favoured mostly by experimental scientists, who fit their experimental data using a power law relationship.

The third important formulation is the *Langmuir*[3] isotherm that is written as:

$$c_s = \frac{\alpha_{L1} c}{\alpha_{L2} + c} \tag{6.7}$$

with coefficients α_{L1} and α_{L2}. The Langmuir isotherm also has the property that for low concentrations the gradient of the isotherm is higher than for high concentrations. In contrast to the Freundlich isotherm, the Langmuir isotherm approaches a finite asymptote for $c \to \infty$, given by the parameter α_{L1}. The argument for the relevance of the Langmuir isotherm is that for high concentrations the limited number of sorption sites at the surface of the pore space is occupied, so that no further increase of c_s is possible.

Figure 6.2 depicts examples of the three major isotherm types. Formulae of further isotherms are listed in Table 6.2.

In the soil compartment, sorption and cation exchange are closely connected; see the paper of Johnson et al. (1998), which is concerned with forest eco-systems for an example study. Cations as Ca^{2+}, Na^+, $NH_4{}^+$, Sr^{+2}, Al^{+3} exchange sorption places if the equilibrium is disturbed when water of different composition enters. This is a typical situation for the soil column with water from precipitation or irrigation entering.

[2] Herbert Freundlich(1880–1941), German chemist.

[3] Irving Langmuir (1881–1957), US-American chemist and physicist.

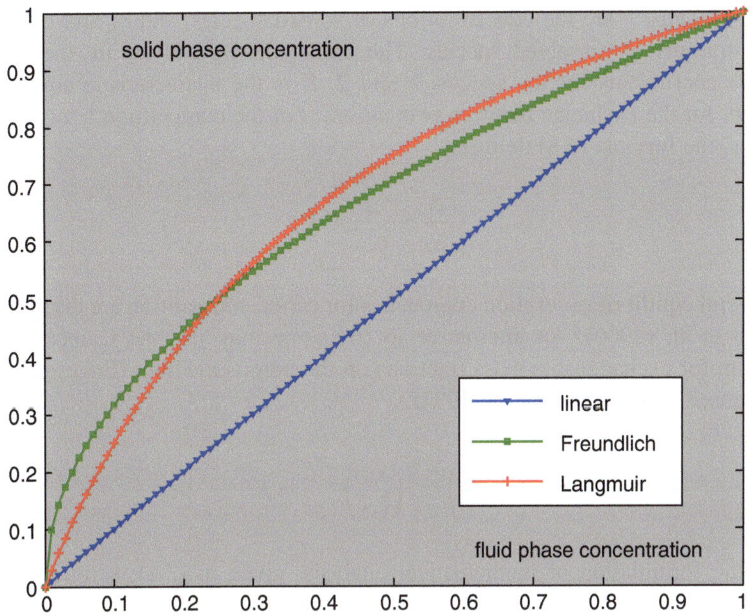

Fig. 6.2 Illustration of sorption isotherms: linear, Freundlich and Langmuir[4]

Table 6.2 Overview on further sorption isotherms

Sorption isotherm	Formula	Number of parameters
Linear	$c_s = K_d c$	1
Freundlich	$c_s = \alpha_{F1} c^{\alpha_{F2}}$	2
Langmuir	$c_s = \frac{\alpha_{L1} c}{\alpha_{L2} + c}$	2
Tempkin	$c_s = \alpha_{T1} + \alpha_{T2} \log(c)$	2
Frumkin	$c_s = \frac{K_d \exp(2\alpha_F c_s) c}{1 + K_d \exp(2\alpha_F c_s) c}$	2
Langmuir-Freundlich	$c_s = \frac{\alpha_1 c^{\alpha_3}}{\alpha_2 + c^{\alpha_3}}$	3
Redlich-Petersen	$c_s = \frac{\alpha_1 c}{\alpha_2 + c^{\alpha_3}}$	3
Toth	$c_s = \frac{\alpha_1 c}{(\alpha_2 + c^{\alpha_3})^{1/\alpha_3}}$	3
Dubinin-Raduskevich	$\log(c_s) = -\alpha_1 \log^2(\alpha_2 c) + \log(\alpha_3)$	3

A very popular description for the equilibrium between the cations is the *Gapon* isotherm:

$$\frac{c_{s1} {c_2}^{1/n_2}}{c_{s2} {c_1}^{1/n_1}} = K \tag{6.8}$$

[4] produced using MATLAB® by: `c = [0:0.01:1]; cs1 = c; cs2 = c.^0.5; cs3 = 3*c./(1 + 2*c); plot (c,cs1,c,cs2,c,cs3);` and some additional design changes from the Figure editor.

where c_{s1} and c_{s2} are the concentrations of sorbed species, and c_1 and c_2 are the concentrations of dissolved species. The exponents n_1 and n_2 are the electric valence coefficients for the species 1 and 2. K is the characteristic equilibrium constant for the exchange between two cations. For the competition between Ca^{2+} and Na^+ the formula (6.8) delivers:

$$\frac{c_{Ca,s}c_{Na}}{c_{Na,s}\sqrt{c_{Ca}}} = K \tag{6.9}$$

Several equilibrium sorption approaches for cation competition are discussed by Vulava et al. (2000). An alternative to (6.8) one may use the Gaines-Thomas isotherm for exchangeable mass fractions on the porous medium (Engesgaard and Christensen 1988; Appelo et al. 1993):

$$\left(\frac{c_{s1}}{c_1}\right)^{1/n_1}\left(\frac{c_2}{c_{s2}}\right)^{1/n_2} = K \tag{6.10}$$

In the formulation of mathematical analysis, given in (2.4), sorption can be included by the introduction of the exchange terms. Considering advective and diffusive fluxes, first order decay or degradation, neglecting additional sinks and sources, the analytical formulation of the mass balances in both phases is:

$$\begin{aligned}\frac{\partial}{\partial t}(\theta c) &= -\nabla\cdot(\theta\mathbf{j}) - \theta\lambda c - e_{fs}\\ \frac{\partial}{\partial t}(\rho_b c_s) &= -\nabla\cdot(\rho_b\mathbf{j}_s) - \rho_b\lambda_s c_s - e_{sf}\end{aligned} \tag{6.11}$$

with concentrations c and c_s, porosity θ and sorption exchange terms e_{fs} and e_{sf}. The exchange term with subscript *fs* denotes the losses from the mobile to the immobile phase and *sf* vice versa. The exchange terms have a positive sign for losses in the first phase and a negative sign if the first phase gains due to exchange. In comparison to the formulation, given in Chap. 3, porosity appears as coefficient in the storage, in the flux and in the decay terms. In these terms the additional factor is relevant in order to take into account that storage, flux and decay occur in the pore space only.

The second (6.11) describes the mass balance for the solid phase, the porous medium. As the species concentration at the solid surface is usually given as a mass fraction, the coefficient ρ_b has to appear in order to obtain the mass balance for the species. ρ_b [kg/m^3] is the *bulk density* of the porous medium that is given by.

$$\rho_b = (1-\theta)\rho_s \tag{6.12}$$

where ρ_s is the density of the solid material without pores. On the right side of the equation, the flux j_s appears in order to denote fluxes in the solid phase.

In groundwater or soil systems, advective or diffusive fluxes in the solid phase can be neglected. But there are exceptions: imagine, the upper soil horizon is turned over due to agricultural practice. That could be described by a diffusion term. In aquatic sediments even more processes contribute to diffusive and advective processes.

As both (6.11) denote a total mass balance, the exchange terms are necessarily equal. In a two-phase environment the sinks of one phase are the sources of the other. Therefore, it is sufficient to introduce one exchange term only and omit the other one (e_{fs}), i.e. $e_{fs} = -e_{sf}$. What is gained in one phase from the sorption process must be lost in the other phase. The describing set of equations then becomes:

$$\begin{aligned} \frac{\partial}{\partial t}(\theta c) &= -\nabla \cdot (\theta \mathbf{j}) - \theta \lambda c - e_{fs} \\ \frac{\partial}{\partial t}(\rho_b c_s) &= -\nabla \cdot (\rho_b \mathbf{j}_s) - \rho_b \lambda_s c_s + e_{fs} \end{aligned} \tag{6.13}$$

6.2 Retardation

In case of fast sorption it turns out that the exchange terms in (6.13) can hardly be quantified. They surely change with time and space. Also the sign changes: in front of an advancing concentration front there is a net gain of the solid phase and losses of the fluid phases. The situation is contrary after a front has passed: there are net gains of the fluid phase and losses of the solid phase. For a quantitative analysis of transport problems it is therefore convenient to find a mathematical formulation in which the exchange term disappears. This is achieved here easily by adding both equations of (6.13). If one neglects decay or degradation, one obtains:

$$\frac{\partial}{\partial t}(\theta c + \rho_b c_s) = -\nabla \cdot (\theta \mathbf{j}) - \nabla \cdot (\rho_b \mathbf{j}_s) \tag{6.14}$$

In order to take advantage of the summation the unknown variable c_s is eliminated. In the case of fast sorption it is possible to reduce the system by utilizing the isotherm relationship (6.1). Equation 6.14 can be re-written as:

$$\frac{\partial}{\partial t}(R\theta c) = -\nabla \cdot (\theta \mathbf{j}) - \nabla \cdot (\rho_b \mathbf{j}_s) \quad \text{with} \quad R = 1 + \frac{\rho_b}{\theta}\frac{c_s}{c} \tag{6.15}$$

where R is the so called *retardation factor*. The formulation (6.15) is frequently used by geochemists (Postma and Appelo 2000). In groundwater studies an alternative formulation often can be found that is valid for the constant porosity situation. Using the chain rule $\partial c_s/\partial t = (\partial c_s/\partial c)(\partial c/\partial t)$ on the left side, the retardation factor appears outside of the time derivative:

$$R\theta \frac{\partial}{\partial t} c = -\nabla \cdot (\theta \mathbf{j}) - \nabla \cdot (\rho_b \mathbf{j}_s) \quad \text{with} \quad R = 1 + \frac{\rho_b}{\theta} \frac{\partial c_s}{\partial c} \tag{6.16}$$

(Kinzelbach 1987). For formulation (6.16) it is assumed that porosity and bulk density are constant in time. One can interpret the role of the retardation factor on the left side of the equation as changing the time scale (see below in this sub-chapter for a more detailed discussion). As the factor is always greater than 1 (all terms appearing in its defining equation are positive), R is responsible for *retardation*.

In the case of a linear isotherm $c_s/c = K_d$, there is no difference between the factors R in (6.15) and (6.16):

$$R = 1 + \frac{\rho_b}{\theta} K_d \tag{6.17}$$

For constant θ there is a constant retardation, for which one often finds the definition (6.17). Retardation factors range from values slightly above 1 up to 10^7, as measured for example by Luo et al. (2000) for Thorium 232.

In general, R depends on the concentrations and on porosity and thus may change with time and space. Both definitions given above differ in the general situation. The left hand side of the differential equations (6.14) and (6.15) is then cause for nonlinearity. For the Freundlich-isotherm holds:

$$R = 1 + \frac{\rho_b}{\theta} \alpha_{F1} \alpha_{F2} (c)^{\alpha_{F2}-1} \tag{6.18}$$

and the Langmuir isotherm:

$$R = 1 + \frac{\rho_b}{\theta} \frac{\alpha_{L1} \alpha_{L2}}{(\alpha_{L2} + c)^2} \tag{6.19}$$

The same procedure can be followed in modeling ion exchange. This will be exemplified for two species, for which the cation exchange capacity CEC can be noted as the sum of the solid phase concentrations of the two species:

$$CEC = c_{s1} + c_{s2} \tag{6.20}$$

Using the Gapon isotherm (6.8) in addition one obtains:

$$\left(1 + K \frac{c_1{}^{1/n_1}}{c_2{}^{1/n_2}}\right) c_{s1} = K \cdot CEC \frac{c_1{}^{1/n_1}}{c_2{}^{1/n_2}} \quad \text{and} \quad \left(1 + \frac{1}{K} \frac{c_2{}^{1/n_2}}{c_1{}^{1/n_1}}\right) c_{s2} = \frac{CEC}{K} \frac{c_2{}^{1/n_2}}{c_1{}^{1/n_1}} \tag{6.21}$$

in order to write the problem setting in two differential equations:

$$\begin{aligned} R_1 \frac{\partial c_1}{\partial t} &= \frac{\partial}{\partial x}\left(D \frac{\partial c_1}{\partial x}\right) - v \frac{\partial c_1}{\partial x} \\ R_2 \frac{\partial c_2}{\partial t} &= \frac{\partial}{\partial x}\left(D \frac{\partial c_2}{\partial x}\right) - v \frac{\partial c_2}{\partial x} \end{aligned} \tag{6.22}$$

with

$$\begin{aligned} R_1 &= 1 + \frac{\rho_b}{\theta} \frac{CEC}{n_1} \frac{K{c_1}^{-1+1/n_1}}{{c_2}^{1/n_2} + K{c_1}^{1/n_1}} \left(1 - \frac{K{c_1}^{1/n_1}}{{c_2}^{1/n_2} + K{c_1}^{1/n_1}}\right) \\ R_2 &= 1 + \frac{\rho_b}{\theta} \frac{CEC}{n_2} \frac{{c_2}^{-1+1/n_2}}{{c_2}^{1/n_2} + K{c_1}^{1/n_1}} \left(1 - \frac{{c_2}^{1/n_2}}{{c_2}^{1/n_2} + K{c_1}^{1/n_1}}\right) \end{aligned} \tag{6.23}$$

Using post-processing the solute phase concentrations can then be obtained by

$$\begin{aligned} c_{s1} &= K \cdot CEC{c_1}^{1/n_1} \Big/ \left({c_2}^{1/n_2} + K{c_1}^{1/n_1}\right) \text{ and} \\ c_{s2} &= CEC{c_2}^{1/n_2} \Big/ \left({c_2}^{1/n_2} + K{c_1}^{1/n_1}\right) \end{aligned} \tag{6.24}$$

In systems with one static phase as it is in groundwater, formulations (6.14) and (6.15) have profound advantage in comparison with the (6.13). If the solid phase is fixed in space (if it is static), the genuine processes of advection and diffusion are not present, or in mathematical formulation: $\mathbf{j}_s = 0$. Then (6.16) is a differential equation for the unknown variable c:

$$R\theta \frac{\partial}{\partial t} c = \nabla \cdot (\theta \mathbf{D} \nabla c) - \cdot \ \mathbf{v} g \nabla c \tag{6.25}$$

On the right side appear terms for diffusion, dispersion and advection in the fluid phase, but there is no contribution from the solid phase. When the differential equation is solved, the other unknown variable c_s and its change in space and time can be computed easily with the help of the isotherm. Division by θ yields:

$$R \frac{\partial}{\partial t} c = \nabla \cdot (\mathbf{D} \nabla c) - \mathbf{v} \cdot \nabla c \tag{6.26}$$

For a further interpretation, (6.25) is compared to (6.26) (the latter without sources, sinks and exchange, as it is valid for tracers). In (6.25) retardation is nothing more than the validity of a prolonged time scale in comparison to the tracer. In mathematical analysis one could formally express that by the notation $R \cdot \partial/\partial t = \partial/\partial(t/R)$. Using the new timescale $\bar{t} = t \cdot R$ one can say that the spatial concentration distribution of the retarded component at time $\bar{t}$ is equal to the distribution of the tracer at time t.

Following the concept of pure retardation sorption has no effect on stationary concentration distributions. This can easily be seen in (6.26) and (6.25): the left side vanishes, and the concentration c is determined as solution of the remaining terms on the right side of the differential equation, in which the retardation factor does not appear.

The concept of retardation can also be maintained if degradation or decay have to be taken into account. The equations above have to be extended by decay terms. Instead of (6.14) one obtains:

$$\frac{\partial}{\partial t}(\theta c + \rho_b c_s) = -\nabla \cdot (\theta \mathbf{j}) - \theta \lambda c - \nabla \cdot (\rho_b \mathbf{j}_s) - \rho_b \lambda_s c_s \tag{6.27}$$

and instead of (6.15):

$$\frac{\partial}{\partial t}(R\theta c) = -\nabla \cdot (\theta \mathbf{j}) - \nabla \cdot (\rho_b \mathbf{j}_s) - \tilde{R}\theta\lambda c \quad \text{with} \quad \tilde{R} = 1 + \frac{\rho_b}{\theta}\frac{\lambda_s}{\lambda}\frac{c_s}{c} \tag{6.28}$$

If there is the same decay constant in both phases (which is surely valid for the radioactive decay of radio-nuclides), both R-factors are identical: $\tilde{R} = R$. For a fixed porous matrix, instead of (6.26) the following differential equation results:

$$R\frac{\partial}{\partial t}c = \nabla \cdot (\mathbf{D}\nabla c) - \mathbf{v} \cdot \nabla c - R\theta\lambda c \tag{6.29}$$

6.3 Analytical Solution

For a homogeneous 1D constant flow field and constant parameters, the differential equation for a retarded species (6.29) has an analytical solution. For the inflow of a front with concentration c_{in} into a region with concentration c_0 holds:

$$c(x,t) = c_0 \exp(-\lambda t)\left(1 - \frac{1}{2}\operatorname{erfc}\left(\frac{Rx - vt}{2\sqrt{DRt}}\right) - \frac{1}{2}\exp\left(\frac{vx}{D}\right)\operatorname{erfc}\left(\frac{Rx + vt}{2\sqrt{DRt}}\right)\right)\ldots$$
$$+\frac{c_{in}}{2}\left(\exp\left(\frac{v-u}{2D}x\right)\operatorname{erfc}\left(\frac{Rx - ut}{2\sqrt{DRt}}\right) + \exp\left(\frac{v+u}{2D}x\right)\operatorname{erfc}\left(\frac{Rx + ut}{2\sqrt{DRt}}\right)\right) \tag{6.30}$$

with $u = \sqrt{v^2 + 4\lambda RD}$ (Kinzelbach 1987). This is an extension of the formula of Ogata and Banks (1961), which was presented in Chaps. 4 and 5. In contrast to the original formula there are two terms, one describing the decline of the original concentration c_0 and the second concerning the change of the inflow concentration c_{in}. If one of these two concentrations is zero, the formula becomes less lengthy as one of the two terms can be omitted.

First the already introduced MATLAB® M-file *'analtrans.m'* is extended to account for fast sorption. The retardation factor R is a new input parameter:

```
R = 2;     % retardation factor
```

The computations for variables `h` and `u` have to be extended:

```
u = sqrt(v*v+4*lambda*R*D);
h = 1/(2*sqrt(D*R*t(i)));
```

Then the explicit formula is written as follows:

```
c = [c; c0*exp(-lambda*t(i))*(e-0.5*erfc(h*(R*x-e*v*t(i)))-...
0.5*exp((v/D)*x).*erfc(h*(R*x+e*v*t(i)))) +...
(cin-c0)*0.5*(exp((v-u)/(D+D)*x).*erfc(h*(R*x-e*u*t(i)))+...
exp((v+u)/(D+D)*x).*erfc(h*(R*x+e*u*t(i))))];
```

That's all. Let's examine some solutions calculated with the extended code.

The complete code is included in the accompanying software under the name *'analtrans.m'*.

Exercise 6.1: Compare results with $R = 1$ and $R = 3$, with parameters: $T = 1$, $v = 1, D = 0.1, L = 1, \lambda = 0$!

Figure 6.3 depicts the results for exercise 6.1. The effect of retardation is nothing but a factor in the time-scale. The concentration distribution at time $t{\cdot}R$ for the retarded species is identical to the curve at time t for the tracer – at least that is the

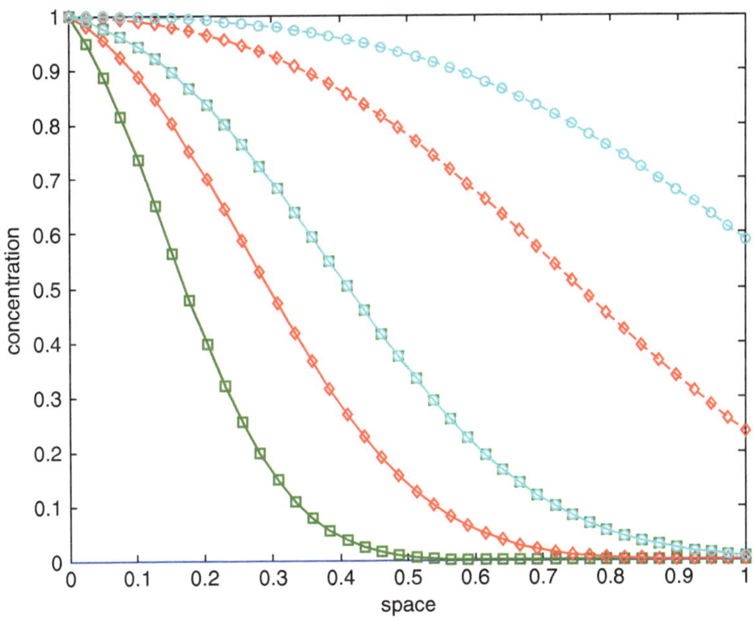

Fig. 6.3 Result of exercise 6.1; both situations are represented by three concentration distributions. Squares mark $t = 1/3$, diamonds $t = 2/3$ and circles $t = 1$. The graph for $R = 1$ and $t = 1/3$ falls together with the graph for $R = 3$ and $t = 1$

expectation due to the mathematical description. Uncertainties in measurement will, of course, provide differences in observed laboratory or field data.

6.4 Numerical Solutions

Fast sorption can also be introduced into the *'simpletrans.m'* code. After the definition of R in the input part of the code, internally only timesteps have to be adjusted. The advection timestep is reduced by the factor R:

```
dtout = dtout/R;
```

Also the Neumann-number:

```
Neumann = Neumann/M/R;
```

The time-end criterion for the loop also depends on the retardation:

```
while (t < T/R)
```

The complete code is included in the accompanying software under the name *'simpletrans.m'*

Exercise 6.2: Use *'simpletrans.m'* to compare results with $R = 1$ and $R = 2$, with parameters: $T = 1, v = 1, D = 0.1, L = 1, \lambda = 1.2$!

Figure 6.4 illustrates the effect of increasing R for a species that is also subject to degradation. Compared are the solutions for no retardation, with parameters as in the example above and with $R = 2$. The two advancing fronts are depicted at 10 time instants, which represent the 10th part of the mean time for a tracer to pass through the entire system. The difference between both solutions is not only due to retardation but also due to higher degradation. This stems from the fact that the retardation factor R appears twice in the differential equations: as coefficient in the storage term on the left side and in the decay/degradation term.

Retardation can also be considered in the *'pdepetrans.m'* code.

```
c = pdepe(0,@transfun,@ictransfun,@bctransfun,x,[0 t],options,...
    D,v,lambda,R,c0,cin);
```

The complete parameter list needs to appear in all function calls. However, the variable R contributes to computations only in the main sub-routine, which reads:

```
function [c,f,s] = transfun(x,t,u,DuDx,D,v,lambda,R,c0,cin)
c = R;
f = D*DuDx;
s = -v*DuDx -lambda*R*u;
```

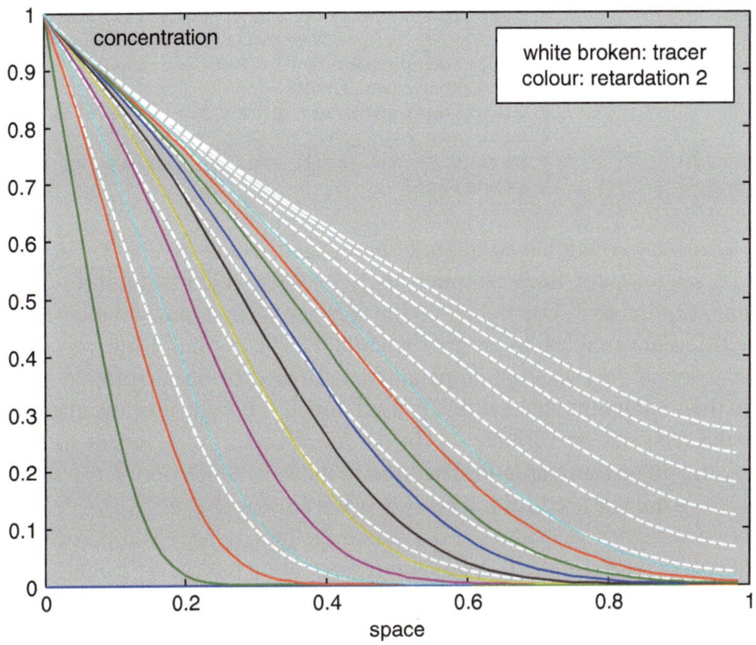

Fig. 6.4 Result of exercise 6.2; both situations are represented by ten concentration curves, which represent the proceeding front. Dotted lines without markers represent the case without retardation; lines with markers represent the front for a sorbing species with the same degradation rate

`R` appears as the coefficient `c` of the time derivative and in the assignment for `s`. With this version of *'pdepetrans.m'* all applications, which were presented above using the *'analtrans.m'* or *'simpletrans.m'* files, can be run. Recall that the advantage of *'analtrans.m'* is that there are no numerical errors as an analytical solution is evaluated explicitly. The advantage of the *'simpletrans.m'* algorithm is that it can be implemented outside of MATLAB® or any other mathematical software package. The advantage of *'pdepetrans.m'* is that it can be easily extended to include other processes which cannot be taken into account by the other methods. This is demonstrated in the following for problems of extended complexity for fast and slow sorption.

First it is shown how the code can be extended to include all types of fast sorption, i.e. linear sorption using formula (6.2), Freundlich sorption using formula (6.18) or Langmuir sorption using formula (6.19). In the extended version of the M-file, the retardation factor is calculated from two sorption parameters, depending on the `sorption` option chosen in the initialization part of the M-file. The new lines in the input part are as follows:

```
sorption = 1;          % sorption-model: no sorption (0), linear (1),
                       %                 Freundlich (2), Langmuir (3)
k1 = 2;                % sorption parameter 1 (R=0 for linear isotherm
                       % with Kd, else k1=R)
k2 = 1;                % sorption parameter 2 (Kd for linear isotherm
                       % with Kd)
rhob = 1300;           % porous medium bulk density [kg/m*m*m]
theta = 0.2;           % porosity [-]
```

The `sorption`-switch has to be set to an integer between 0 and 3. For tracers it should be set to 0, for linear sorption to 1, for Freundlich sorption to 2 and for Langmuir sorption to 3. Depending on the sorption-switch, the variables `k1` and `k2` contain different variables. In case of Freundlich and Langmuir isotherms these two variables contain the two sorption parameters. In case of linear sorption `k1` should be set to the retardation factor if that is used directly. For the user the alternative is to give the K_d-value in variable `k2`. In the latter situation, `k1` needs to be set to 0 (not a valid value for retardation) in order to indicate which option is wanted. The two variables are not used for tracers, i.e. when `sorption = 0`.

The two variables `rhob` and `theta` contain the bulk density and porosity. These are used only if the Freundlich or Langmuir retardation factors are calculated or if R is to be calculated from the K_d-value for the linear isotherm. Some re-calculations have to be performed before the `pdepe`-function is called[5]:

```
if sorption == 1 & k1 <=0
    k1 = 1+k2*rhob/theta;
else
    if sorption > 1 k1 = rhob*k1/theta; end
end
```

For the linear isotherm the retardation factor is calculated in the second line of these commands. In the fourth line, bulk density and porosity are multiplied with the first factor of the Freundlich- or Langmuir-sorption parameters. This is done in order to reduce the number of variables to be transferred to the function sub-routines. The call is lengthy even with that cosmetic:

```
c = pdepe(0,@transfun,@ictransfun,@bctransfun,x,[0 t],options, ...
 D,v,lambda,sorption,k1,k2,c0,cin);
```

The parameter list in the second line needs to appear in all functions of the M-file. The only place where the new introduced variables are needed is the `transfun`-function. The following statements need to be included in the function before the assignment of the other variables:

---

[5] `&` is the logical 'and' operator; for other logical operators see MATLAB® help under 'logical array functions'; the logical 'or' operator is: `|`.

```
switch sorption
    case 0
        R = 1;
    case 1
        R = k1;
    case 2
        R = 1+k1*k2*u^(k2-1);
    case 3
        R = 1+k1*k2*u/(k2+u)/(k2+u);
end
c = R;
```

With these commands the retardation factor is computed for the different sorption alternatives as described above.

The complete code is included in the accompanying software under the name *'pdepetrans.m'*

Exercise 6.3: Compare concentration profiles for the linear isotherm with $R = 2.3$ and the Freundlich isotherm with $\theta = 0.2$, $\rho_b = 1200$ kg/m^3, $K_d = 4 \cdot 10^{-4}$m^3/kg; use the following parameter assignments for further parameters: initial concentration $c_0 = 0.1$ mg/l, $c_{in} = 1$ mg/l, $v = 1$ m/s, $D = 1$ m^2/s and no degradation.

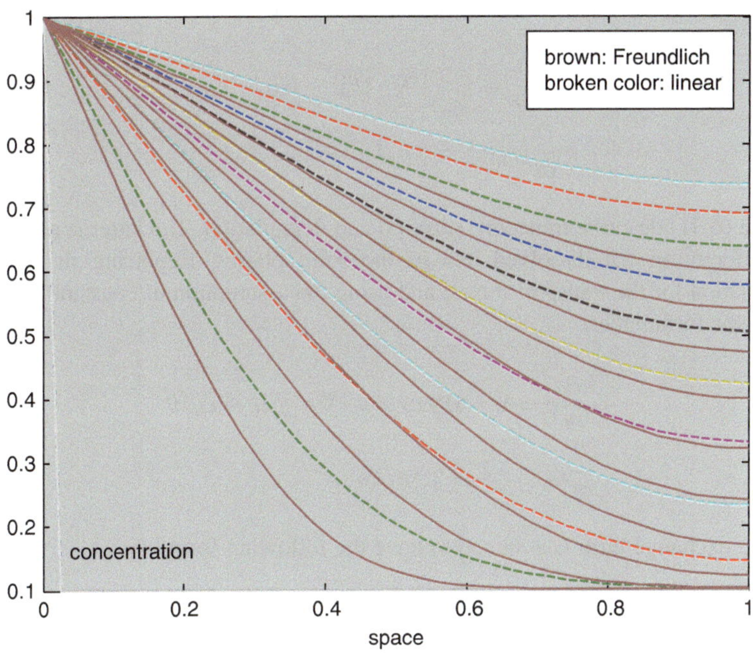

Fig. 6.5 Result of exercise 6.3; both situations are represented by ten concentration curves, visualizing the proceeding front. Dotted lines without markers represent the case for the Freundlich isotherm; lines with markers represent the case for linear sorption with the same marginal retardation factor

In Fig. 6.5 the concentration profiles for linear and Freundlich isotherms are compared. There is the same marginal retardation factor for both cases, i.e. for concentration $c = 1$ the retardation for the Freundlich isotherm is identical to $R = 2.3$ of the linear isotherm. The figure illustrates the higher retardation for the Freundlich isotherm for low concentrations, with the effect that for the same time instant the concentration values for the Freundlich case are always below those of the linear sorption case.

6.5 Slow Sorption

In the derivation of differential equations, as presented above, it was assumed that the interphase exchange processes are fast compared to the other relevant processes. The equilibrium between solid phase and fluid phase concentrations is reached at all times. Such an assumption is valid in many field situations where transport time scales are long, for example, in aquifers or aquatic sediments It is surely not valid in other cases.

Concerning slow sorption one may keep the original set of two differential equations (6.13). In the following we will show how to treat such a system in case of no transport processes for the solid phase ($\mathbf{j}_s = 0$):

$$\begin{aligned} \frac{\partial}{\partial t}(\theta c) &= -\nabla \cdot (\theta \mathbf{j}) - \theta \lambda c - e_{fs} \\ \frac{\partial}{\partial t}(\rho_b c_s) &= -\rho_b \lambda_s c_s + e_{fs} \end{aligned} \tag{6.31}$$

The (6.31) describe transport, sorption and degradation. The latter is allowed to be different in the dissolved and in the solid phases. Replacing the detailed formulation for the transport fluxes, and using the assumption of constant θ and ρ_b, yields the formulation:

$$\begin{aligned} \frac{\partial}{\partial t} c &= \nabla \cdot (\mathbf{D} \nabla c) - \mathbf{v} \cdot \nabla c - \lambda c - e_{fs}/\theta \\ \frac{\partial}{\partial t} c_s &= -\lambda_s c_s + e_{fs}/\rho_b \end{aligned} \tag{6.32}$$

The exchange term is assumed to have the following form:

$$e_{fs} = \kappa_f c - \kappa_s c_s \tag{6.33}$$

The following describes an extension of the already developed M-file to account for slow sorption:

```
function slowsorp
%----------------------------------------------------------------
T = 16;                     % maximum time [s]
L = 8;                      % length [m]
D = 0.1;                    % diffusivity [m*m/s]
v = 0.5;                    % real fluid velocity [m/s]
theta = 0.2                 % porosity [-]
rhob = 1200;                % porous medium bulk density [kg/m*m*m]
kappaf = 0.01;              % transition rate fluid to solid [1/s]
kappas = 0.00;              % transition rate solid to fluid [1/s]
lambdaf = 0;                % decay rate in fluid [1/s]
lambdas = 0;                % decay rate in solid [1/s]
c0f = 0.1;                % initial concentration in fluid [kg/m*m*m]
c0s = 0.01;                 % initial concentration in solid [-]
cin = 1;                    % inflow concentration [kg/m*m*m]

M = 10;                     % number of timesteps
N = 40;                     % number of nodes
%------------------------- output parameters

gplot = 2;                  % =1: breakthrough curves; =2: profiles
```

The output parameters can be adopted from one of the other transport M-files, '*simpletrans.m*', '*analtrans.m*' or '*pdepetrans.m*'. The discretization parameters are also taken analogously as demonstrated above. The execution part consists mainly of the call of the `pdepe`-module. The function names and parameters are of course new for the slow sorption application:

```
t = linspace (T/M,T,M);     % time discretization
x = linspace (0,L,N);       % space discretization
%-------------------execution-------------------------------------
options = odeset;
c = pdepe(0,@slowsorpde,@slowsorpic,@slowsorpbc,x,t,options,...
D,v,theta,rhob,kappaf,kappas,lambdaf,lambdas,[c0f;c0s],cin);

%--------------------- graphical output -------------------------
switch gplot
    case 1
        plot ([0 t],c(:,:,1))          % breakthrough curves
        xlabel ('time'); ylabel ('concentration');
    case 2
        plot (x,c(:,:,1)','--')        % profiles
        xlabel ('space'); ylabel ('concentration');
end
```

The solution is again written in matrix form. As there are two unknown concentrations at each position and for each output time level, there are three indices: one speaks of a tensor of rank 3. Three indices are necessary to mark a single element of `c`. The third index must be 1 or 2, depending on whether to denote the concentration of the dissolved or the adsorbed species. In all output commands, the third index needs to be specified in order to determine which concentration has to be plotted. Where in former codes the variable c was sufficient, now `c(:, :, 1)` has to be inserted. With this command only the fluid phase concentration is plotted. In order to obtain the solid phase concentration use `c(:, :, 2)`.

The idea of two function variables in one vector has to be adopted to understand the functions of the M-file. Where a single value was sufficient in former programs, now two values need to be given; the first for the single phase differential equation, the second for the solid phase differential equation. The functions read as follows:

```
function [c,f,s] = slowsorpde(x,t,u,DuDx,...
D,v,theta,rhob,kappaf,kappas,lambdaf,lambdas,c0,cin)
c = [1;1];
f = [D;0].*DuDx;
s = -[v;0].*DuDx - [lambdaf;lambdas].*u -...
([kappaf -kappas]*u)*[1/theta;-1/rhob];
% --------------------------------------------------------------
function u0 = slowsorpic(x,...
D,v,theta,rhob,kappaf,kappas,lambdaf,lambdas,c0,cin)
u0 = c0;
% --------------------------------------------------------------
function [pl,ql,pr,qr] = slowsorpbc(xl,ul,xr,ur,t,...
D,v,theta,rhob,kappaf,kappas,lambdaf,lambdas,c0,cin)
pl = [ul(1)-cin;0];
ql = [0;1];
pr = [0;0];
qr = [1;1];
```

The coefficient for the time derivative term is 1 in both differential equations. Thus, both components in the column-vector `c` are equal to 1. The flux term `f` contains the negative of Fickian diffusion in the first component and zero in the second (as there is no diffusion in the solid phase). Note that the variable `DuDx` contains the spatial derivatives of the concentrations and is a two-component column vector within the function.

In a similar manner the variable `s` contains the contributions from advection, decay and sorption. The first component of the first term `-v*DuDx(1)` contains the advection term already known from the other M-files. The second component in the first term is zero, as there is no advection in the solid phase.[6] The second term of the `s = ...` assignment includes decay terms, which are allowed to be phase-dependent. If they are phase dependent, the variables `lambdaf` and `lambdas` have to be chosen differently. The last term denotes the interphase exchange. In the coefficient term (in round brackets) the amount of exchange is computed. The exchange needs to be included with a positive sign in the first differential equation and with a negative in the second differential equation. In order to achieve that, one has to multiply with the `[1/theta;-1/rhob]` vector.

For both phases initial conditions are specified in the `slowsorpic` function. Note that `c0` is a two component vector, which contains the specified initial concentrations for both phases. For the boundary conditions the concentration of

[6] In most applications on porous media, the solid phase or porous matrix is assumed to be fixed in the chosen spatial coordinate system. However, in some cases this may not be true. In sediments the solids move with respect to a fixed level in space. If the interface between the sediments and the overlying region of free flow is taken as a reference, one may obtain a situation with no flux of solids. However, this trick works doesn't work if there are temporal changes in sedimentation, or even in steady state, when compaction has to be taken into account.

inflowing fluid is the only parameter required. All other boundary conditions, for solid and fluid phases, are all of no-flow Neumann type.

The complete code is included in the accompanying software under the name '*slowsorp.m*'

Exercise 6.4: Compare concentration profiles for a varying solid to fluid transfer factor κ_f. Which known situation is described by $\kappa_s = 0$? The fluid to solid transfer factor is given by $\kappa_f = 0.02$ 1/d. Use the following values for the other parameters:

$$T = 16 \text{ d}, L = 8 \text{ m}, D = 0.1\text{m}^2/\text{d}, v = 0.5 \text{ m/d}, \lambda = \lambda_s = 0$$
$$\theta = 0.2, \rho_b = 1200 \text{ kg/m}^3, c_{in} = 1 \text{ mg/l}, c_0 = c_{0s} = 0$$

Figure 6.6 shows that for a varying transfer rate the solutions lie between two marginal states. One is given if the transfer factor κ_f is very low. For $\kappa_s = 0$ the solution is identical to the situation with linear decay, as all mass which disappears on the solid surfaces has no way back into the fluid. As the figure shows, this marginal situation already is approached for values $\kappa_s \leqslant 1$. The other marginal state is characterized by an immediate back-reaction. The transfer from solid to fluid is not a limiting factor any more. For the given parameter values this is obviously true for $\kappa_s > 10^5$ 1/d.

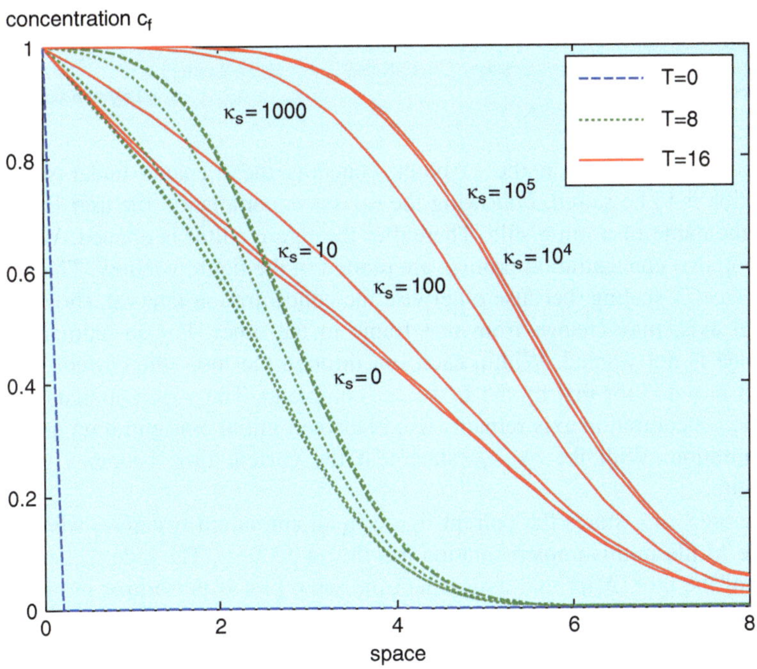

Fig. 6.6 Results of exercise 6.4; on the graphs for $T = 16$ d the corresponding value of the transfer factor κ_f is depicted

6.6 MATLAB® Animations

One can use MATLAB® to produce animations. Using the `getframe` command, single shots from a MATLAB® figure window can be gathered as one animation. After finishing the production, there are commands to store and play the animations.

We demonstrate the procedure, how an animation is produced, for the transport models that were described in Chaps. 4–6. The change of the concentration profile with time is to be illustrated by a sequence of profiles. It is assumed that the results of the simulation are stored on the matrix `c`, which contains concentration profiles in the rows; the different rows represent different time instants. The entire sequence is shown below and is included in the M-files *'simpletrans.m'*, *'analtrans.m'* and *'pdepetrans.m'* (with minor differences concerning `dtout`). The switch variable `ganim` in the input specification is introduced to initiate or not initiate the animation production.

```
if (ganim)
    [FileName,PathName] = uiputfile ('*.mpg');
    figure;
    if (ganim > 1) hold on; end
    for j = 1:size(c,1)
        axis manual;
        plot (x,c(j,:),'r','LineWidth',2);
        YLim = [min(c0,cin) max(c0,cin)];
        legend (['t=' num2str(dtout*(j-1))]);
        Anim(j) = getframe;
        plot (x,c(j,:),'b','LineWidth',2);
    end
    mpgwrite (Anim,colormap,[PathName '/' FileName]);
    movie (Anim,3);                                    % play animation
end
```

The first command in the `if`-block concerns the filename under which the animation is to be stored. Following the `uiputfile` command,[7] the user is asked to input the name of a 'mpg'-file. Thereafter the figure editor is opened. Within the `for`-loop two concentration profiles are plotted in the figure window. The axes are set to `manual` scaling, because otherwise the concentration interval, shown on the vertical axis, may change from one frame to the other. For an animation such a change is not wanted. Within each run through the loop, the current profile is plotted in red color first by the first `plot` command. The `YLim` command ensures that the concentration axis remains fixed between initial concentration and inflow concentration. With the `legend` statement the current time becomes visible in the figure.

The `getframe` stores the current figure in an animation structure, which in the sample M-file in this implementation has the name `Anim`. The index `j` denotes the index of the plot. After that assignment the same plot is performed in blue color,

[7] With the `uiputfile` command a file name is specified using a file-select box.

overwriting the red curve, before proceeding in the same manner with the next concentration profile.

Note that there is another option connected with the `ganim` parameter. If the user chooses `ganim` >1, the final blue colored profiles are not deleted. Thus the history of the profile development remains visible in the following single plots of the animation.

After the end of the loop, the entire animation is stored under the given file name. Here we demonstrate the `mpgwrite` command, which does not belong to core MATLAB®. However, everyone is free to use the command; the corresponding M-file can be downloaded from Mathworks web-site; see: http://www.mathworks.com/matlabcentral/fileexchange. Don't forget to specify the directory path where the corresponding files have to be saved (using the `addpath` command or the 'Set Path...' sub-entry of the 'File' menu.)

The `movie` command starts the movie. The second formal parameter in that call corresponds to the repetition time. If at that place a negative value is specified, each animation is shown forward then backward. It is possible to influence the speed of the animation by specification of a third formal parameter, which represents the number of frames per second.

An example animation with `ganim = 2` is included in the accompanying software under the name *'animation.mpg'*.

The final frame is shown in Fig. 6.7. As an exercise, the user may extend the animation letting the profiles run through a cycle of colors.

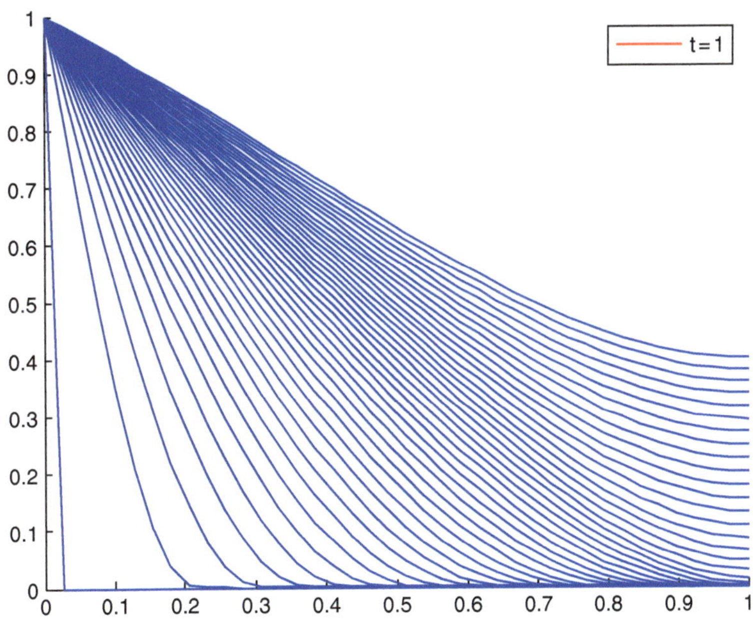

Fig. 6.7 Final frame of example animation

References

Appelo CAJ, Hendriks JA, van Veldhuizen M (1993) Flushing factors and sharp front solution for solute transport with multicomponent ion exchange. J Hydrol 146:89–113

Engesgaard P, Christensen Th H (1988) A review of chemical solute transport models. Nord Hydrol 19:183–216

Fetter CW (1994) Applied hydrogeology. Prentice Hall, Upper Saddle River, p 691

Geibert W (2001) Actinium-227 as a Tracer for Advection and Mixing in the Deep-Sea, Berichte zur Polar- und Meeresforschung 385, Alfred-Wegener-Inst. für Polar- und Meeresforschung, Bremerhaven, p 112

Hölttä P, Siitari-Kauppi M, Haknanen M, Huitti T, Hautojärvi A, Lindberg A (1997) Radionuclide transport and retardation in rock fracture and crushed rock column experiments. J Contam Hydrol 26:135–145

Johnson DW, Susfalk RB, Swank WT (1998) Simulated effects of atmospheric deposition and species change on nutrient cycling in loblolly pine and mixed deciduous forests. In: Mickler R, Fox S (eds) The productivity & sustainability of southern forest ecosystems in a changing environment. Springer, New York, pp 503–524

Karickhoff SW (1984) Organic pollutant sorption in aquatic systems. J Hydraul Eng-ASCE 110:707–735

Karickhoff SW, Brown DS, Scott TA (1979) Sorption of hydrophobic pollutants on natural sediments. Water Res 13:241–248

Kinzelbach W (1987) Numerische Methoden zur Modellierung des Transports von Schadstoffen im Grundwasser. Oldenbourg, München, p 317, in German

Luo S, Ku T-L, Roback R, Murrell M, McLing TL (2000) In-situ radionuclide transport and preferential groundwater flows at INEEL (Idaho): decay-series disequilibrium studies. Geochim Cosmochim Acta 64:867–881

Ogata A, Banks RB (1961) A solution of the differential equation of longitudinal dispersion in porous media, US Geological Survey, Professional Paper No. 411-A

Postma D, Appelo CAJ (2000) Reduction of Mn-oxides by ferrous iron in a flow system: column experiment and reactive transport modeling. Geochim Cosmochim Acta 64(7):1237–1247

Vulava VM, Kretzschmar R, Rusch U, Grolimund D, Westall JC, Borkovec M (2000) Cation competition in a natural subsurface material: modeling of sorption equilibria. Environ Sci Technol 34:2149–2155

Chapter 7
Transport and Kinetics

7.1 Introduction

Very often biogeochemical reactions in which several species of the environment participate play an important role for the fate and the distribution of species of environmental relevance. Due to reactions a potentially hazardous component can be gradually degraded and may reach concentration levels above limits set by environmental regulations. Another scenario is also important: potentially harmful chemical species may emerge as products of a reaction along a flow path within a compartment.

In this and the following chapter the focus is on modeling the simultaneous action of transport and reactions. It will be shown that for mathematical modeling it is relevant whether reactions are slow or fast in comparison to the considered transport processes. In this chapter we stay with slow reactions, while fast reactions are the topic of the next chapter.

The characteristic time for a slow reaction is at least in the same scale as advection and dispersion/diffusion. In applications at different length and velocity scales in different environmental compartments, the classification of slow and fast reactions may differ significantly. A reaction, which has to be classified as slow in a flowing river, can be fast in aquatic sediments or in aquifers.

The rate of reactions is quite different. According to Cox (1994), the lower limit for the characteristic time lies between 10^{-12} and 10^{-13} s. H-bond formation in metal complexes can be as fast as 10^{-10} s, macromolecular complex formation exceeds 10^{-7} s, and hydrolysis 10^{-3} s. All these processes are surely fast in all environmental systems, unless they are inhibited by specific biogeochemical conditions leading to much higher reaction times.

CO_2 hydration is in the order of 10^3 s, Fe(II) oxidation by O_2 in the order of 10^4 s (Morel and Hering 1993). SO_2 transformation and deposition as H_2SO_4 or SO_4^{2-} in the atmosphere has a time characteristic of 10 h and 33 h respectively (Deaton and

E. Holzbecher, *Environmental Modeling*,
DOI 10.1007/978-3-642-22042-5_7, © Springer-Verlag Berlin Heidelberg 2012

Winebrake 1999), i.e. in the order of 10^5–10^6 s. Photolysis of a pesticide (carbaryl) and Mn(II) oxydation have a similar speed. Hydrolysis of an insecticide (disulfoton) and heterogeneous Mn(II) oxidation can be observed in a time scale of 10^7 s. More than a year can be estimated for the hydrolysis of methyl iodide in freshwater and for homogeneous Mn(II) oxidation (Morel and Hering 1993). Tributylin, an ingredient of anti-fouling paints, is degraded in three steps, of which each has a characteristic time between 1.5 and 3 years (Sarradin et al. 1995).

Morel and Hering (1993) give amino acid racemization as an example for an extremely slow process with a rate coefficient of 10^{14} s. This can surely be classified as a geological time-scale (it can be compared to petroleum formation).

For the modeler, the time scale of the process always has to be related to the typical *time scale of interest* of the problem for which the model is designed. Processes which are much faster than the time scale of interest need not to be resolved in the model – neither processes which are much slower than the time scale of interest. Only processes, for which the characteristic time is similar to the problem time scale need to be treated as kinetic processes. Therefore any kinetics classification has no general validity. It is rather problem and site specific.

Kinetics is a branch of chemistry that deals with reactions, for which the reaction rate has to be given as a function of environmental state variables. The determination of such kinetic rates is the main task of kinetics. It is often a formidable task, because the number of state variables and the range of the validity of experimentally determined reaction rates is not clear a priori.

The implementation of kinetics turns out to be unproblematic within the mathematical framework used in this book. Kinetics deals with reaction rates. The transport differential equations are stated in terms of rates. Thus, in the differential equation a rate r just has to be added, and the resulting transport equation for the concentration c of a single species becomes:

$$\theta\frac{\partial c}{\partial t} = -\nabla\cdot\theta\mathbf{j}(c) + q \quad \text{with} \quad \mathbf{j}(c) = -\mathbf{D}\nabla c + \mathbf{v}c \tag{7.1}$$

Within this formalism one can conceive decay and degradation as special cases of kinetics with $q = -\theta\lambda c^n$ for linear decay and $q = -\theta\lambda c^n$ for general decay of order n. If the factor θ appears in all terms, it can be omitted; this will be assumed for the remainder of this chapter.

In general, several reactants are involved in reactions, and it is often not sufficient to consider a single species in a model. Let's assume that a simple reaction has two reactants a and b and one reaction product c:

$$a + b \to c \tag{7.2}$$

The corresponding system of differential equations is obtained by adding the reaction rate in all three differential equations, of which each represents the mass balance for one of the species. Let's denote the concentrations by the symbols c_a, c_b and c_c:

$$\frac{\partial c_a}{\partial t} = -\nabla \cdot \mathbf{j}(c_a) - r$$
$$\frac{\partial c_b}{\partial t} = -\nabla \cdot \mathbf{j}(c_b) - r$$
$$\frac{\partial c_c}{\partial t} = -\nabla \cdot \mathbf{j}(c_c) + r \tag{7.3}$$

Note that the sign of the reaction rate is negative for reactants (as mass is lost) and positive for the product (as mass is gained).

For a general formulation one may introduce a reaction matrix **S**, which is given in the example by:

$$\mathbf{S} = \begin{pmatrix} -1 & -1 & 1 \end{pmatrix} \tag{7.4}$$

indicating that one molecule of species *a* and one molecule of species *b* yield one molecule of species *c*. The product $\mathbf{S}^T\mathbf{q}$ is a column vector, which gathers all reaction terms, and the system of differential (7.3) can be written briefly as:

$$\frac{\partial \mathbf{c}}{\partial t} = -\nabla \cdot \mathbf{j}(\mathbf{c}) + \mathbf{S}^T r \text{ with } \mathbf{c} = \begin{pmatrix} c_a \\ c_b \\ c_c \end{pmatrix} \tag{7.5}$$

The notation can be used for general systems of a multitude of species that are connected by several reactions. The reaction rates are gathered in a column vector **r**. One can write the entire system by the vector equation:

$$\frac{\partial \mathbf{c}}{\partial t} = -\nabla \cdot \mathbf{j}(\mathbf{c}) + \mathbf{S}^T \mathbf{r} \tag{7.6}$$

This approach will be extended in the following chapter. If transport does not have to be considered, (7.6) reduce to:

$$\frac{\partial \mathbf{c}}{\partial t} = \mathbf{S}^T \mathbf{r} \tag{7.7}$$

which is a system of ordinary differential equations. The solution of such dynamical systems is outlined in Chap. 9.

7.2 Law of Mass Action for Kinetic Reactions

As mentioned above, chemical reactions often are described by a so-called *kinetic* formulation. The term kinetic is used in order to distinguish the description from the equilibrium or thermodynamic formulation. In the latter case an equilibrium condition is used to characterise the relationship between the participating species. The thermodynamic formulation is equivalent to the formulation of isotherms, which was introduced in Chap. 6.

In the kinetics approach the rate term itself is expressed in terms of concentrations of the chemical species. Different formulations of such kinetic laws can be found in concerned publications that are valid for certain reactions under certain conditions and a certain parameter range. The most common formulation is the kinetic version of the *Law of Mass Action.* The equilibrium version of that law is presented in the following chapter.

For the reaction example (7.2) the law of mass action formulation is:

$$q = -\kappa c_a c_b \tag{7.8}$$

with a reaction-characteristic parameter κ. The parameter κ is a characteristic for the 'speed' of the reaction. For fast reactions, κ has a high value. For slow reactions it is a small number.

The increase of the rate with the concentration, as expressed by (7.8), is an expected behavior. For a reversible reaction $a + b \rightleftarrows 2c$, the rate law is given by:

$$q = -\kappa_{\rightarrow} c_a c_b + \kappa_{\leftarrow} c_c^2 \tag{7.9}$$

There appear different reaction parameters for reaction and back-reaction: $\kappa_{\rightarrow}$ and $\kappa_{\leftarrow}$. Moreover, the stoichiometric number for the involvement of a species in the reaction appears in form of an exponent of the concentration; two for species c. The rate increases proportionally to the respective power of the concentration if the stoichiometric number exceeds one. Equations (7.8) and (7.9) are special cases of the general form

$$q = -\kappa_{\rightarrow} \prod_{\text{reactants } i} c_i^{\alpha_i} + \kappa_{\leftarrow} \prod_{\text{products } j} c_j^{\alpha_j} \tag{7.10}$$

where the stoichiometric numbers are denoted as α_i and α_j. Equation (7.10) is not the most general formulation of the law of mass action. In fact, it turns out to be valid only for the small and moderate concentration range. If the fluid is highly mineralized, i.e. when concentrations of dissolved species are high, inhibition due to competition between species has to be taken into account. Then activities replace concentrations in the above expressions. Further details are given in Chap. 9.

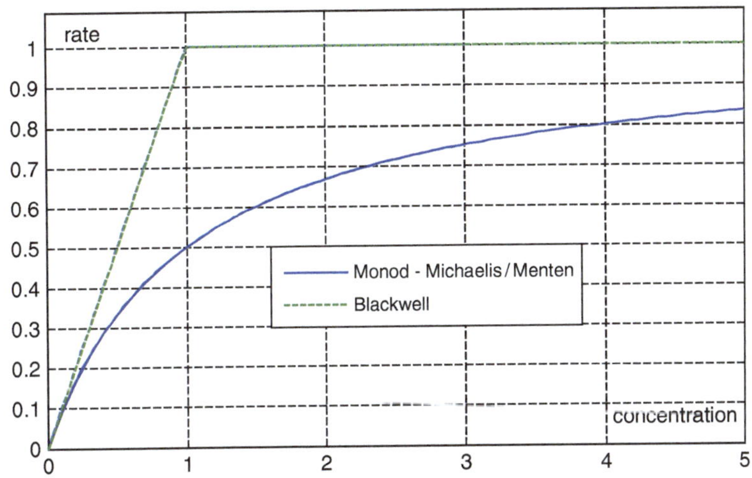

Fig. 7.1 Monod – Michaelis/Menten and Blackwell kinetics (normalized)

7.3 Monod, Michaelis–Menten and Blackwell Kinetics

When biological species are involved, another description for rate is often used. In analogy to an approach proposed by Michaelis[1] and Menten[2] already three decades before (Michaelis and Menten 1913), Monod[3] suggested the following term to describe the growth of bacteria cultures in the 1940s of the twentieth century (Monod 1949) (Fig. 7.1):

$$q = r\frac{c}{c_{\frac{1}{2}} + c} \tag{7.11}$$

Sidebar 7.1: Free Radical Reactions in the Atmosphere
Understanding the behavior of free radicals in the atmosphere is of paramount importance for the understanding of lifetime and hence of spatial scales of pollutant transport. Free radicals participate in photochemical reactions, which are initiated by light. Most free radical species have short life spans. However, they can promote the conversion of ozone to oxygen and thus take part in the catalytic cycle of ozone destruction. The most important radical acting in the lower atmosphere is the hydroxyl radical OH. A system of free radical reactions involving OH is given by

(continued)

[1] Leonor Michaelis (1875–1949), German-American biochemist.

[2] Maud Leonora Menten 1879–1960), Canadian physician.

[3] Jacques Lucien Monod, 1910–1976, French biochemist.

$$H + NO_2 \rightarrow OH + NO$$
$$2OH \rightarrow H_2O + O$$
$$O + OH \rightarrow O_2 + H$$

with reaction constants κ_1, κ_2 and κ_3 (Bradley et al. 1973). Neglecting transport, the entire set of reaction equations is as follows:

$$\frac{\partial c_H}{\partial t} = -\kappa_1 c_H c_{NO_2} + \kappa_3 c_O c_{OH}$$
$$\frac{\partial c_{H_2O}}{\partial t} = \kappa_2 c_{OH}^2$$
$$\frac{\partial c_{NO}}{\partial t} = \kappa_1 c_H c_{NO_2}$$
$$\frac{\partial c_{NO_2}}{\partial t} = -\kappa_1 c_H c_{NO_2}$$
$$\frac{\partial c_O}{\partial t} = \kappa_2 c_{OH}^2 - \kappa_3 c_O c_{OH}$$
$$\frac{\partial c_{O_2}}{\partial t} = \kappa_3 c_O c_{OH}$$
$$\frac{\partial c_{OH}}{\partial t} = \kappa_1 c_H c_{NO_2} - \kappa_2 c_{OH}^2 - \kappa_3 c_O c_{OH}$$

with $\kappa_1 = 2.9$, $\kappa_2 = 0.155$ and $\kappa_3 = 1.1$.

For high concentrations a maximum reaction rate r is approached, while for low concentrations q is proportional to c (with proportionality constant $r/c_{\frac{1}{2}}$). If the concentration c of the species coincides with the half-concentration parameter $c_{\frac{1}{2}}$, half of the maximum rate is reached.

In some computer codes, the transition between linear and constant rate appears at a specified characteristic value of the concentration. Such a distinction between low and high concentration situations is used as a simpler alternative to Monod kinetics. It is referred to as *Blackwell* kinetics and applied by van Cappellen and Wang (1995) among others.

7.4 Bacteria Populations

In models one may also consider bacteria populations explicitly. If there is a high abundance of bacteria, degradation processes are favoured. Vice versa, the abundance of fuel favours bacteria population growth.

For a description that takes more details into account concerning the bacterial degradation, a formulation can be used in which the bacteria population X appears as another variable. In addition to the differential equation for the chemical species, another differential equation for the biological species has to be added. As extension of equation **(3.20)** it is possible to write:

$$R\frac{\partial c}{\partial t} = \nabla \cdot (\mathbf{D}\nabla c) - \mathbf{v} \cdot \nabla c - \alpha X \frac{c}{c+\beta}$$
$$\frac{\partial X}{\partial t} = \gamma X \frac{c}{c+\beta} - \delta X^n \tag{7.12}$$

with retardation factor R, dispersion tensor $\mathbf{D}$, velocity $\mathbf{v}$ and parameters α, β, γ and δ. The degradation rate for the substrate is linearly dependent on X. For high concentrations c, a maximum rate of αX is reached. Half of that maximum is given for the substrate concentration $C=\beta$. Such behavior is described by the Monod term, the last term in the first equation of system (7.12). The same functional dependency is used to describe the growth of the bacteria population where the coefficient γ includes the relation between bacteria population and substrate concentration. The last term in the second equation of system (7.12) accounts for the decline of the bacteria population or natural death of the bacteria, for which two additional parameters, δ and n, are introduced. Bacteria are assumed not to migrate with flow; that's why the dispersion and advection terms are missing in the second equation.

Approaches as in (7.12) are common in biogeochemical modeling. The use of a linear term for the decline of X is a common approach used in biogeochemical modeling (Lensing 1995; Tebes-Stevens et al. 1998). For $n = 2$, the approach coincides with the so-called logistic equation, which is most popular in the biological and ecological sciences; see Chap. 9). Marsili-Libelli (1993) refers to the given approach with a first order growth term in X and a degradation term with a free exponent as *Richards dynamics*. Even more general approaches with free exponents in growth as well as in decay terms are examined by Savageau (1980).

An alternative formulation of similar complexity is obtained when the second equation of the system (7.12) is replaced by

$$\frac{\partial X}{\partial t} = \gamma X \frac{c}{c+\beta} - \delta_1 X - \delta_2 X^2 \tag{7.13}$$

In (7.13) a linear and a quadratic decay term are included. In a discussion of various different mortality terms in ecological models, Fulton et al. (2003) favour such an approach stating that the linear term represents 'basal' mortality, while the quadratic term is due to predators which are not explicitly represented in the model.

A further approach for modeling the bacteria population was suggested by Schäfer et al. (1998), following Kindred and Celia (1989), using an inhibition term for the bacteria population:

$$\frac{\partial X}{\partial t} = \gamma X \frac{c}{c+\beta} \frac{\varepsilon}{X+\varepsilon} - \delta X \tag{7.14}$$

where ε denotes one additional parameter. The inhibition factor $\varepsilon/(X+\varepsilon)$ has also to be included in the decay term of the substrate equation.

If the one-dimensional formulation of the differential (7.12) is sufficient, one may write:

$$\begin{aligned} R\frac{\partial c}{\partial t} &= \frac{\partial}{\partial x}\left(\alpha_L v \frac{\partial c}{\partial x}\right) - v\frac{\partial c}{\partial x} - \alpha X \frac{c}{c+\beta} \\ \frac{\partial X}{\partial t} &= \gamma X \frac{c}{c+\beta} - \delta X^n \end{aligned} \tag{7.15}$$

Concerning the formulation (7.15) it is assumed that dispersion dominates over diffusion (see Chap. 3): molecular diffusivity is omitted. If the velocity and also the dispersivity are constants, as for example in column experiments, the coefficients α_L and v can be taken out of the brackets of the first term of the right hand side.

The simulation of the transient change of concentration and/or population of biological species, described by a set of 1D equations, can be performed by using the MATLAB® `pdepe` solver that was already described in chap. 4. In the sequel, as another MATLAB® application, we determine the degradation characteristics by evaluating the steady-state solution.

7.5 Steady States

In order to determine the degradation rate it may be sufficient to examine the steady state. From the 1D formulation for the unsteady situation a set of ordinary differential equations emerges. Two equations result for the system described by the (7.15):

$$\begin{aligned} \frac{\partial}{\partial x}\left(\alpha_L v \frac{\partial}{\partial x} c\right) - v\frac{\partial}{\partial x} c - \alpha X \frac{c}{c+\beta} &= 0 \\ \gamma X \frac{c}{c+\beta} - \delta X^n &= 0 \end{aligned} \tag{7.16}$$

If bacteria populations remain above zero, a steady-state value for X, in dependence of c can be extracted from the second equation:

$$X = \left(\frac{\gamma}{\delta}\frac{c}{c+\beta}\right)^{\frac{1}{n-1}} \tag{7.17}$$

The formula (7.17) is introduced in the first equation of system (7.16):

$$\alpha_L v \frac{\partial^2 c}{\partial x} - v\frac{\partial c}{\partial x} - \alpha\left(\frac{\gamma}{\delta}\frac{c}{c+\beta}\right)^{\frac{1}{n-1}}\frac{c}{c+\beta} = 0 \tag{7.18}$$

which is a single differential equation of second order. As an equivalent system of two differential equations one obtains:

$$\begin{aligned}\frac{\partial c}{\partial x} &= c' \\ \frac{\partial c'}{\partial x} &= \frac{c'}{\alpha_L} + \frac{\alpha}{\alpha_L v}\left(\frac{\gamma}{\delta}\frac{c}{c+\beta}\right)^{\frac{1}{n-1}}\frac{c}{c+\beta}\end{aligned} \tag{7.19}$$

In cases in which dispersive fluxes are small, one obtains a single first order differential equation:

$$\frac{\partial c}{\partial x} = -\frac{\alpha}{v}\left(\frac{\gamma}{\delta}\frac{c}{c+\beta}\right)^{\frac{1}{n-1}}\frac{c}{c+\beta} \tag{7.20}$$

or

$$\frac{\partial c}{\partial x} = -\eta\left(\frac{c}{c+\beta}\right)^{n/(n-1)} \tag{7.21}$$

with $\eta = \dfrac{\alpha\gamma^{1/(n-1)}}{v\delta^{1/(n-1)}}$. Formula (7.21) can be re-written as:

$$\int\left(1+\frac{\beta}{c}\right)^{n/(n-1)} dc = -\eta(x-x_0) \tag{7.22}$$

For $n = 2$ (7.22) can be integrated analytically. One obtains:

$$x - x_0 = -\frac{1}{\eta}\left(c + 2\beta\log(c) - \frac{\beta^2}{c} - c_{in} - 2\beta\log(c_{in}) + \frac{\beta^2}{c_{in}}\right) \tag{7.23}$$

where c_{in} is the inflow concentration at $x = x_0$. Expression (7.23) is an implicit formula for c as function of x for given parameters and boundary condition. Provided a value for x is known, the corresponding c can be determined by a zero-finding algorithm. See Sidebar 7.2 for an application using the MATLAB® `fzero` command.

With the same assumptions the approaches (7.13) and (7.14) also lead to ordinary differential equations:

$$\frac{\partial c}{\partial x} = -\frac{\alpha\gamma}{v\delta_2}\left(\frac{c}{c+\beta}\right)^2 + \frac{\alpha\delta_1}{v\delta_2}\frac{c}{c+\beta} \tag{7.24}$$

and

$$\frac{\partial c}{\partial x} = -\eta_1\left(\eta_2\frac{c}{c+\beta} - 1\right) \tag{7.25}$$

respectively.

Sidebar 7.2: Gadolinium-DTPA Steady Transport and Degradation
Diethylene-triamine-pentaacetic acid (DTPA) is used in the pulp and paper industry, where its application increased dramatically with the introduction of H_2O_2 as a substitute for the bleaching agent chlorine. Chelating agents, particularly DTPA, are added to bind heavy metals and thus prevent the decomposition of H_2O_2 during the bleaching process (van Dam et al. 1999). After processing, DTPA remains a component of the effluent reaching sewage treatment plants and downstream surface water bodies such as rivers and lakes, as well as connected aquifers. Finally the substance enters the water supply systems.

Gadolinium (Gd) is a rare earth element (REE), which rarely occurs in natural environments. Therefore Gd has become an indicator for human impact in metropolitan areas; Gd is widely used in medical applications. Since 1988 most contrast agents for magnetic resonance imaging in medicine contain gadolinium. Gd is complexed with DTPA to form Gd-DTPA, an aqueous soluble stable complex. Gd-DTPA does not accumulate in the human body but is eliminated without significant chemical change via the kidneys within a day.

Gd-DTPA reaches sewage treatment plants after having passed the sewage systems. The sewage treatment processes are not sufficient to degrade the Gd-DTPA complex. In densely populated areas increased concentrations of Gd-DTPA can be found in surface water bodies that are partially recharged by effluents of sewage plants. Even aquifers contain Gd-DTPA introduced by infiltration of surface water. Such a path is of particular concern, where drinking water is pumped from well galleries in the vicinity of surface water bodies, where bank filtration is used for public water supply.

(continued)

In the sequel, we deal with a column experiment that was set up in order to identify the DTPA degradation processes in porous media. The reported column experiment was performed in a series of columns with an entire flowpath length of 30 m. The set-up and the evaluation of the experiment was already reported by (Holzbecher et al. 2005). Inflow velocity and dispersion coefficient were obtained from the evaluation of a tracer experiment. Mean interstitial velocity and longitudinal velocity for the entire duration of the experiment are $v = 0.86$ m/d and $\alpha_L = 0.06$ m. Advection dominates within the system.

Concentrations were measured at several locations along the flowpath. Figure 7.2 depicts the measured *breakthrough curves* for Gd concentration at four selected positions, 0.22 m, 10 m, 20 m and 30 m from the inlet to the first column. After an initial time period with initial zero concentrations, c increases fast when the front approaches. Clearly, the length of the initial time depends on the position of the observation point. Following the sharp increase, finally a level is reached which is constant over time. The concentration level of that steady period depends on the position along the flowpath. Minor fluctuations are due to difficulties maintaining a constant concentration at the inlet. In Fig. 7.2 on the first breakthrough curve a rectangle indicates the time period with steady concentration.

The constant concentration levels in dependence of travel distance are depicted as dots in Fig. 7.3. The figure also shows two curves obtained by two different modeling approaches. If the approach (7.12) is assumed to be valid, according to the presented derivation, one has to solve (7.21).

For $n = 2$, one has the alternative to solve the implicit formula (7.23), which is implemented by using the MATLAB® `fzero` command. The complete M-file sequence is:

```
xdata = [0.22 0.42 0.84 1.66 3.33 5 10 15 20 25 30];
ydata = [54.7 54.3 54.2 54.0 52.8 52.4 51.0 49.9 48.7 47.6 46.2];
cin = 55;
aeta = 4304;
beta = 5937;
for i=1:size(xdata,2)
    yfunc(i) = fzero (@GD,cin,odeset,cin,beta,aeta,xdata(i));
end
plot (xdata,ydata,'o',xdata,yfunc);
xlabel ('distance [m]'); ylabel ('Gd concentration [ng/ml]');

function y = GD(c,cin,beta,aeta,x);
y = x+(c+2*beta*log(c)-beta*beta/c-cin-...
2*beta*log(cin)+beta*beta/cin)/aeta;
```

(continued)

The complete code is included in the accompanying software under the name *'GdDTPA.m'*. Figure 7.3 depicts the result of the computation labelled by 'n = 2'. Estimated parameter values for β and η were adopted from Holzbecher et al. (2005).

Additionally, Fig. 7.3 depicts the result of another model run in which the parameter n was included in the estimation. For the solution, the steady state equations (7.21) were modeled using a MATLAB® solver for ordinary differential equations (see Chap. 9). The optimum fit was obtained for $n = 1.029$, β=16.2 and $\eta = 5425$. How parameter estimations can be performed in core MATLAB® is shown in Chap. 10.

The second run obviously represents much better the curvature in the observed data than the first run. A more detailed model may even improve the fit, i.e. the correspondence between measured and modelled data. Also the approaches (7.13) and (7.14) deliver better results than the run with a fixed $n = 2$ (Holzbecher et al. 2005). Which of the approaches is more realistic can only be judged by including non-mathematical findings of the applied sciences.

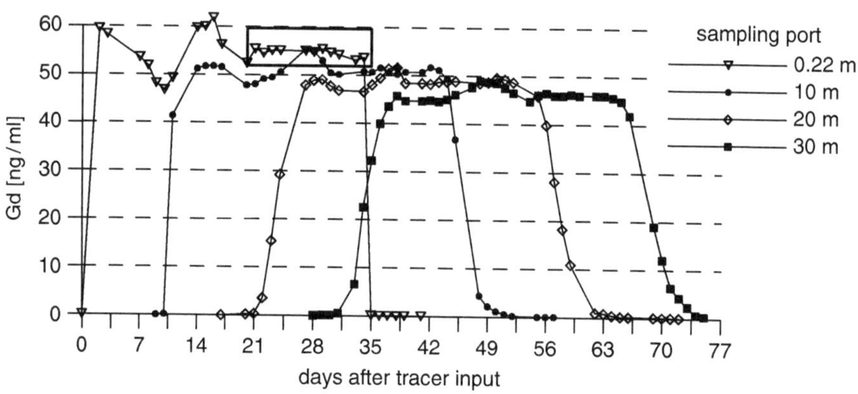

Fig. 7.2 Observed breakthrough curves (selected) in Gd-DTPA column experiment

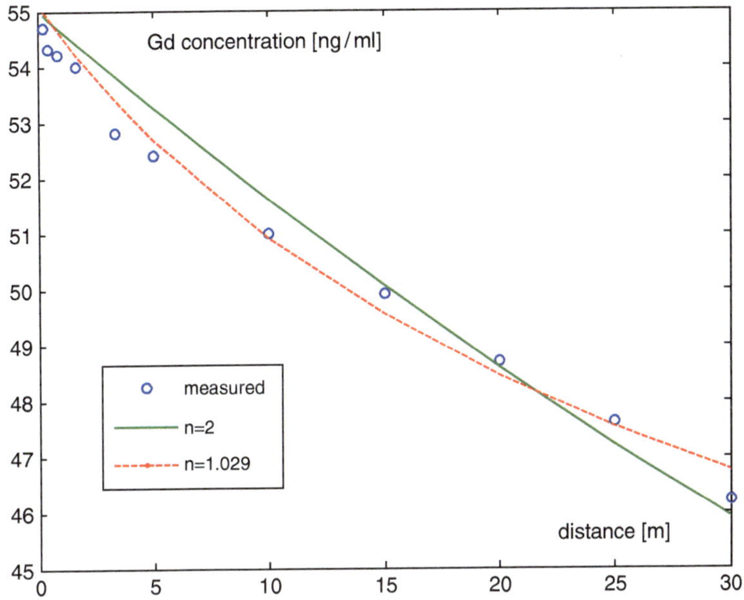

Fig. 7.3 Results for steady state modeling of the Gd-DTPA column experiment

References

Bradley JN, Hack W, Hoyermann K, Wagner HG (1973) Kinetics of the reaction of hydroxyl radicals with ethylene and with C_3 hydrocarbons. J Chem Soc Faraday Trans 1 69:1889–1893

Cox BG (1994) Modern liquid phase kinetics. Oxford Science Publ, Oxford, p 92

Deaton ML, Winebrake JI (1999) Dynamic modeling of environmental systems. Springer, New York, p 194

Fulton E, Smith A, Johnson C (2003) Mortality and predation in ecosystem models: is it important how these are expressed? Ecol Mod 169:157–178

Holzbecher E, Knappe A, Pekdeger A (2005) Identification of degradation characteristics – exemplified on Gd-DTPA in a large experimental column. Environmental Modeling Assessment 10(1):1–8

Kindred JS, Celia M (1989) Contaminant transport and degradation two: conceptual model and test simulation, Water Res Res 25:1149–1159

Lensing HJ (1995) Numerische Modellierung mikrobieller Abbauprozesse im Grundwasser, vol 51. Inst. für Hydrologie und Wasserwirtschaft Universität Karlsruhe, Karlsruhe (in German)

Marsili-Libelli S (1993) Ecological modeling. In: Zanetti P (ed) Environmental modeling, vol 1. Comp Mech. Publ, Southampton, pp 273–337

Michaelis L, Menten ML (1913) Die Kinetik der Invertinwirkung. Biochemische Zs 49:333–369

Monod J (1949) The growth of bacterial cultures. Ann Rev Microbiol 3:371–394

Morel F, Hering J (1993) Principles and applications of aquatic chemistry. Wiley, New York, p 588

Sarradin P-M, Lapaquellerie Y, Satrc A, Latouche C, Astruc M (1995) Long term behavior and degradation kinetics of tributylin in a marina sediment. Sci Total Environ 170:59–70

Savageau MA (1980) Growth equations: a general equation and a survey of special cases. Math Biosci 8:267–278

Schäfer D, Schäfer W, Kinzelbach W (1998) Simulation of reactive processes related to biodegradation in aquifers one. Structure of the three- dimensional reactive transport model. J Cont Transp 31:167–186

Tebes-Stevens C, Valocchi AJ, van Briesen JM, Rittmann BE (1998) Multicomponent transport with coupled geochemical and microbiological reactions: model description and example simulations. J Hydrol 209:8–26

van Cappellen P, Wang Y (1995) Metal cycling in surface sediments: modeling the interplay of transport and reaction. In: Allen HE (ed) Metal contaminated aquatic sediments. Ann Arbor Press, Chelsea, pp 21–64

van Dam RA, PorterNA AJT, Holdway DA (1999) Stability of DTPA and iron(III)-DTPA under laboratory ecotoxicological conditions. Water Res 33:1320–1324

Chapter 8
Transport and Equilibrium Reactions

8.1 Introductory Example

The situation that environmentally relevant species take part in chemical reactions, while being transported through a compartment of the environment, was already treated in Chap. 7. In this chapter the same situation is taken up again with the difference concerning the time scale of the reactions. Here we deal with reactions which are fast in comparison to transport processes.

The situation that chemical reactions are fast compared to other environmental processes is met quite often but on very different time scales. The scale difference is related to the fact that the transport time scale deviates significantly in different compartments. There are systems which are almost in a no-flow state. In deep underground reservoirs transport is measured in geological time scales, and most chemical transformations can be assumed to be fast in comparison. In near-surface aquifers velocities are often in the range of several meters per year or higher, and some chemical processes may not reach their equilibrium. Sedimentation rates in the deep ocean or in lakes are in the order of several mm per year; this sets the scale for fast and slow reactions in those systems. Therefore, each environmental compartment has its own time-scale and is related to chemical processes in a different manner.

Reactions with a characteristic time, which is in the same scale as transport or even slower, can be included in the mathematical description as shown in the previous chapter. The reaction rate has to be formulated in dependence of concentrations and maybe some other state variables, like temperature, and added as another term in the differential equation. As the reaction contributes to the mass balance, which is expressed by the differential equation, the mathematical formulation is straight forward. More difficult is the determination of rate expressions that are relevant in practice from the chemical point of view. But this has to be left to chemists and is outside of the scope of this book.

For fast reactions, the rate law is not relevant. Instead, the equilibrium characteristic is brought into play. A widely accepted formulation is given by the law of

E. Holzbecher, *Environmental Modeling*,
DOI 10.1007/978-3-642-22042-5_8,

mass action, which is presented in sub-chapter 8.2. Here, the mathematical framework is demonstrated in a special introductory example.

Let's take a reaction in which n_a molecules of species a react with n_b molecules of species b to produce n_c parts of species c:

$$n_a a + n_b b \rightleftarrows n_c c \tag{8.1}$$

In connection with transport the development of the system of three species can be described by the set of three differential equations:

$$\begin{aligned}
\frac{\partial c_a}{\partial t} &= \nabla \bullet \mathbf{j}(c_a) - n_a r \\
\frac{\partial c_b}{\partial t} &= \nabla \bullet \mathbf{j}(c_b) - n_b r \\
\frac{\partial c_c}{\partial t} &= \nabla \bullet \mathbf{j}(c_c) + n_c r
\end{aligned} \tag{8.2}$$

This is a slight extension of the system (7.3), where all stoichiometric numbers n_a, n_b and n_c were set to 1. As in the previous chapters, the vector $\mathbf{j}$ denotes the flux due to transport processes, and the symbol in brackets denotes the species which is transported. The last term in all three differential equations represents the reaction. The notation is analogous to the description introduced in Chap. 7. r denotes the reaction rate. The problem is that the reaction rate r is not known in equilibrium reactions. The rate r has to be eliminated from the mathematical description, which is achieved by an appropriate gathering of the equations (sums of first and third, as well as second and third equations):

$$\begin{aligned}
\frac{\partial c_a}{\partial t} + \frac{n_a}{n_c}\frac{\partial c_a}{\partial t} &= \nabla \bullet \mathbf{j}(c_a) + \frac{n_a}{n_c}\nabla \bullet \mathbf{j}(c_c) \\
\frac{\partial c_b}{\partial t} + \frac{n_b}{n_c}\frac{\partial c_b}{\partial t} &= \nabla \bullet \mathbf{j}(c_b) + \frac{n_b}{n_c}\nabla \bullet \mathbf{j}(c_c)
\end{aligned} \tag{8.3}$$

In favour of simplicity, we assume that the flux term is linear, i.e. that dispersivities, diffusivities and fluid fluxes are independent of the concentrations. Then fluxes on the left side can be gathered in a single term:

$$\begin{aligned}
\frac{\partial}{\partial t}\left(c_a + \frac{n_a}{n_c}c_c\right) &= \nabla \bullet \mathbf{j}\left(c_a + \frac{n_a}{n_c}c_c\right) \\
\frac{\partial}{\partial t}\left(c_b + \frac{n_b}{n_c}c_c\right) &= \nabla \bullet \mathbf{j}\left(c_b + \frac{n_b}{n_c}c_c\right)
\end{aligned} \tag{8.4}$$

In order to obtain a suitable formulation for three unknown variables c_a, c_b and c_c, these two equations are complemented by the mathematical formulation of the equilibrium state, which is a mathematical function including c_a, c_b and c_c.

According to the law of mass action (see sub-chapter 8.2), the equilibrium condition for the given reaction is given by the equation:

$$\frac{c_c{}^{n_c}}{c_a{}^{n_a} c_b{}^{n_b}} = K \tag{8.5}$$

with a specific reaction-dependent equilibrium constant K.

The general solution procedure for the entire system is as such: solve the differential (8.4) in order to obtain the variables $A := c_a + \frac{n_a}{n_c} c_c$ and $B := c_b + \frac{n_b}{n_c} c_c$ as functions of time and space! In a second step, determine for each location and time instant the three unknown values c_a, c_b and c_c from the three known values A, B and K!

In order to perform this task, we utilize the resulting explicit formulae for c_a and c_b as functions of $c := c_c$ as well as A and B: $c_a = A - (n_a/n_c)c_c$, $c_b = B - (n_b/n_c)c_c$. Thus we may re-write (8.5) as:

$$\frac{c^{n_c}}{\left(A - \frac{n_a}{n_c} c\right)^{n_a} \left(B - \frac{n_b}{n_c} c\right)^{n_b}} = K \tag{8.6}$$

or:

$$f(c) := c^{n_c} \left(A - \frac{n_a}{n_c} c\right)^{-n_a} \left(B - \frac{n_b}{n_c} c\right)^{-n_b} - K = 0 \tag{8.7}$$

The further task is to find the zero of a non-linear equation. The standard procedure for such a computation is Newton's method (see Sidebar 8.1).

Sidebar 8.1: Newton[1]'s Method

Newton's method is a mathematical standard method for the determination of the zeros of a function. Let's introduce the method for functions of one variable $f(x)$ first before extending it to several dimensions.

Newton's method is an iterative method; i.e. starting from an initial guess, a new approximation for the zero is computed in each iteration step. The formula for the next approximation of the zero is:

$$x \leftarrow x - \frac{f(x)}{f'(x)}$$

(continued)

[1] Isaac Newton (1642–1727), English scientist and philosopher.

where $f'(x)$ denotes the derivative. It can be shown easily that the new x is the position where the tangent on the function graph at the old x position meets the x-axis.

The iteration stops, if the change of approximation within the last iteration is lower than a specified tolerance (1), or if the maximum number of iterations is reached (2). The following command sequence demonstrates the method by calculating the zero of the cosinus, i.e. $\pi/2$.

```
toll = 1.e-7;          % tolerance
nmax = 20;             % max. no. of iterations
x = 2.5;               % initial guess
err = toll+1; nit = 0;
while (nit < nmax & err > toll),
    nit = nit+1;
    F = f(x);
    DF = fderiv(x);
    dx = -F/DF;
    err = abs(dx);
    x = x+dx
end
display (['Zero obtained after ' num2str(nit) ' iterations:']);
x

function F = f(x)
F = cos(x);

function DF = fderiv(x)
DF = -sin(x);
```

The complete code is included in the accompanying software under the name *'newtondemo.m'*.

In the first three lines tolerance, maximum number of iterations and initial guess are specified. What follows is the initialization of the error variable `err` and the iteration counter `nit`. In the `while`-loop the new x-position is calculated from the old one. First, the iteration counter is increased. Then, after function and derivative are evaluated, the change of x (the second term in the iteration formula) is computed by the command `dx = -F/DF`. The absolute value of `dx` becomes the error variable `err`. Finally, the new `x` is calculated.

The result of the iteration for the cosinus as demonstration function is $x = 1.5708$, which is already reached after three iterations. There are further iterations, because the required tolerance with 10^{-7} is higher than the digits presented in the MATLAB® command window.

Note that the Newton method requires an explicit formula for the derivative of the function.

Note that the derivative $\partial f/\partial c$ is to be computed in order to apply Newton's method. For function (8.7) the derivative is available, although a bit lengthy to write down. In the following sub-chapter we present a general procedure, which can be applied to general systems of multi-species interacting in several reactions.

8.2 The Law of Mass Action for Equilibrium Reactions

The most common unit for concentrations of components in the fluid phase is [component mass / fluid volume]; in SI units [kg/m^3]. Another way describing concentrations is by using mass fractions [component mass / fluid mass], which in physical units are dimensionless. In chemistry it is common to use molar or molal concentrations, where the number of moles per fluid volume or fluid mass is measured.

From the mathematical point of view a reaction is regarded as the transition from one set of species to another. In the following, the total number of species is denoted by N_s, and the number of reactions by N_r. In a reaction a subset of species has the role of reactants and another set gathers the reaction products. Formally one may write:

$$\sum_{i=1}^{N_s} n_i^{(j)} \bullet component_i \rightarrow \sum_{i=1}^{N_s} m_i^{(j)} \bullet component_i \; j = 1, \ldots N_r \tag{8.8}$$

where the coefficients $n_i^{(j)}$ and $m_i^{(j)}$ are stoichiometric integer numbers. Reversible reactions are characterized by the observation that both the reaction and the back reaction occur. If the reaction is fast in comparison to the other relevant processes (as transport for example), the equilibrium between the reactions in both directions is reached at every time instance. This can be assumed to hold for many geochemical processes, because transport and compaction in the pedosphere are usually slow. According to the *Law of Mass Action* the equilibrium is characterized by a constant K_j:

$$K_j := \frac{\prod_{i=1}^{N_s} a_i^{m_i^{(j)}}}{\prod_{i=1}^{N_s} a_i^{n_i^{(j)}}} \tag{8.9}$$

Equation (8.9) is valid for fast reversible reactions. The K_j's are independent from the geochemical surrounding but depend on temperature and pressure.

a_i denotes the activity of the ith species, which is the product of concentration and a species-dependent activity coefficient γ_i:

$$a_i = \gamma_i \bullet c_i \tag{8.10}$$

For more details concerning activities see Sidebar 8.3 below. For low concentrations, the activity coefficients are close to one and the activities in the law of mass action can be replaced by concentrations. Taking the logarithm[2] of (8.9) one obtains:

$$\sum_{i=1}^{N_s} \left(m_i^{(j)} - n_i^{(j)}\right) \log(a_i) = \log(K_j) \; j = 1, \ldots N_r \tag{8.11}$$

It is convenient to write (8.11) in matrix form. In order to do that, the convention is used that stoichiometric coefficients related to reactants obtain a positive exponent and those related to reaction products obtain a negative exponent:

$$n_{ji} = m_i^{(j)} - n_i^{(j)} \tag{8.12}$$

With the help of matrix $\mathbf{S} = (n_{ji})$ results:

$$\mathbf{S} \bullet \mathbf{log}(\mathbf{a}) = \mathbf{log}(\mathbf{k}) \qquad \text{with } \mathbf{a} = \begin{pmatrix} a_1 \\ \ldots \\ a_{N_s} \end{pmatrix} \text{ and } \mathbf{k} = \begin{pmatrix} K_1 \\ \ldots \\ K_{N_r} \end{pmatrix} \tag{8.13}$$

S has N_r rows and N_s columns. Each line in (8.13) represents one reaction. With respect to the calculation of N_s unknown activities, (8.13) provides N_r conditions.

For low ionic strength, the activity coefficients are approximately one, and it is allowed to replace activities by concentrations:

$$\mathbf{S} \bullet \mathbf{log}(\mathbf{c}) = \mathbf{log}(\mathbf{k}) \tag{8.14}$$

Usually it is required that there is no reaction that can be obtained as combination of the other reactions. The mathematical expression for that requirement is that the rank[3] of matrix **S** is maximal. The maximal rank of the $N_r \times N_s$ reaction matrices is N_r. With MATLAB® it is easy to check if the condition is fulfilled, by using the `rank` command.

[2] In this chapter, we deal with logarithms to the basis 10; in the text we do not introduce a different notation to distinguish from the logarithm to the basis e, which is used in other chapters. However, in MATLAB® the `log10` command has to be taken!

[3] The rank of a matrix is the maximum number of linear independent column or row vectors in the matrix; see textbooks on Linear Algebra.

The following reaction example between hydrogen *H*, nitrotriacetic acid *NTA* (a chelating agent in detergents influencing the metal ion activity in aqueous systems), and cobalt (II) *Co* demonstrates the case:

$$H + NTA \rightleftarrows HNTA$$
$$CoNTA \rightleftarrows Co + NTA$$
$$CoNTA + H \rightleftarrows Co + NTA$$

There are five species involved in three reactions and the corresponding reaction matrix is:

```
S = [-1 -1 1 0 0; 0 1 0 -1 1; -1 0 1 -1 1]
```

In MATLAB® it is easy to check if the above given requirement is fulfilled. The *rank of the matrix* is the maximum number of independent reactions. In the example case MATLAB® shows:

```
rank(S)
ans =
     2
```

In fact, only two of the given reactions are independent. The rank is lower than 3 and thus not maximal. The chemical system is treated in more details by Fang et al. (2003). In order to obtain a maximum rank for the given example, one of the three reactions has to be omitted. The resulting 2 × 5 matrix has maximum rank. Without mentioning we assume a matrix of maximum rank in all following theoretical derivations.

8.3 Speciation Calculations

The equation for a system of species in equilibrium is given by:

$$\mathbf{S}^T \mathbf{r}_{eq} = 0 \tag{8.15}$$

The problem with the system (8.15) is that it can not be computed directly, because the exchange rates of the equilibrium reactions $\mathbf{r}_{eq}$ are neither known, nor are they given by an explicit expression. In order to reach an appropriate formulation, the equations have to be added up in a way that eliminates the unknown reaction rates. There are N_s equations in the system (8.15), and there are N_r equilibrium conditions for the reactions. Usually $N_s > N_r$ holds, i.e. the system is overdetermined.

In order to complete the system (8.15), N_s-N_r additional conditions in form of specified values for *total concentrations* can be given (Steefel and

Sidebar 8.2: Single Reaction Example
For the reaction between species A, B and C

$$2A + B \Leftrightarrow C$$

the reaction matrix **S** is given by

$$\mathbf{S} = (2 \quad 1 \quad -1)$$

and for **U** holds:

$$\mathbf{U} = \begin{pmatrix} 1 & 0 & 2 \\ 0 & 1 & 1 \end{pmatrix}$$

Total concentration u_1 and u_2 are thus given by:

$$u_1 = c_A + 2c_C$$
$$u_2 = c_B + c_C$$

which can be explained as follows: the total concentration u_1 of A is given by the concentration c_A and two times the concentration c_C, because according to the reaction two A-species are within the reaction product C. For the total concentration u_2 of B the two concentrations c_B and c_C have to be summed up, because there is one unit of B to be found in C. This may give an impression why the term 'total concentration' is used.

In this example case the matrix **U** is unique if it is required that **U** is combined by the unit matrix and one additional column fitting to the formulation (8.20).

MacQuarrie 1996). Total concentrations are invariants that are obtained by linear combinations of the species concentrations. Mathematically this can be expressed by (left) multiplication of the (8.15) with a matrix **U,** a matrix with N_s-N_r rows and N_s columns. The vector

$$\mathbf{u} = \mathbf{U} \bullet \mathbf{c} \tag{8.16}$$

contains values of total concentrations. **u** is a column vector with N_s-N_r components. The conditions (8.15) imply:

$$\mathbf{U} \bullet \mathbf{S}^T \bullet \mathbf{r}_{eq} = 0 \tag{8.17}$$

In order to fulfil the conditions (8.17), the matrix **U** is chosen to obey the equation

$$\mathbf{U}\bullet\mathbf{S}^T = 0 \tag{8.18}$$

There are several ways to find such a matrix **U** which is not unique. Saaltink et al. (1998) mention Gram-Schmidt orthogonalization and singular decomposition as alternative methods to construct a matrix **U** with the property (8.18) and provide an example for the latter procedure. In addition, Saaltink et al. (1998) suggest another procedure which can easily be implemented using MATLAB®, as it is formulated in matrix form. The algorithm is based on the partition of the matrix **S** in two sub-matrices:

$$\mathbf{S} = (\mathbf{S}_1|\mathbf{S}_2) \tag{8.19}$$

where $\mathbf{S}_2$ is a regular square matrix with N_r rows and columns. $\mathbf{S}_1$ is a matrix with N_s-N_r rows and N_r columns. **U** is then given by:

$$\mathbf{U} = \left[\mathbf{I}_{N_s-N_r}\middle|\mathbf{S}^{*^\mathbf{T}}\right] \tag{8.20}$$

with

$$\mathbf{S}^* = -\,\mathbf{S}_2{}^{-1}\bullet\mathbf{S}_1 \tag{8.21}$$

and unit matrix $\mathbf{I}_{N_s-N_r}$ with N_s-N_r rows or columns.

In order to perform the matrix operation, $\mathbf{S}_2$ must be invertible. Sometimes some permutations of the species' system are required to achieve that $\mathbf{S}_2$ is regular. It is possible in either case if the matrix **S** has maximum rank (see above). The number of entries in the right sub-matrix of (8.20) is $N_r \times (N_s$-$N_r)$, for which there are $(N_s$-$N_r)$ conditions only.

For a given vector **u** the (8.16) is a system of N_s-N_r equations for the unknown components of the vector **c**. In addition, there are N_r equilibrium conditions. For given total concentrations, gathered in vector **u**, the system

$$\begin{aligned}\mathbf{u} &= \mathbf{U}\bullet\mathbf{c}\\ \mathbf{S}\bullet\log(\mathbf{c}) &= \log(\mathbf{K})\end{aligned} \tag{8.22}$$

has to be solved for **c**. It is a nonlinear system of N_s equations for N_s unknown values. The so-called speciation problem (8.22) can be solved by the Newton-Raphson method, an extension of the Newton method described above. There are N_s-N_r linear balance equations and N_r non-linear equilibrium equations. In MATLAB®, the generalization of the Newton method for vector functions of several independent variables can be implemented for that purpose:

```
toll = 1.e-10;                      % tolerance
nmax = 50;                          % max. no. of iterations
Se = [-1 -1 1];                     % reaction matrix
logc = [1.e-10; 1.e-10; 0];         % initial guess (log)
logK = [-0.93];                     % equilibrium constants (log)
logu = [-0.301; 0];                 % total concentrations (log)

ln10 = 2.3026;
n=size(Se,1); m=size(Se,2);
S1 = Se(:,1:m-n);
S2 = Se(:,m-n+1:m);
S1star=-S2\S1;
U=[eye(m-n),S1star'];

c=exp(ln10*logc);
u(1:m-n,1) = exp(ln10*logu);
err = toll+1; nit = 0;
while (nit < nmax & err > toll),
    nit = nit+1;
    F = [U*c-u; Se*log10(c)-logK];
    DF = [U; Se*diag((1/ln10)./c)];
    dc = -DF\F;
    cn = max(c+dc,0.005*abs(c));
    err = max(abs(cn-c));
    c = cn;
    logc = log10(c);
end
display (['Species concentrations obtained after ' num2str(nit) '
iterations:']);
c
```

The complete code is included in the accompanying software under the name *'Speciation.m'*.

In the input part of the M-file the tolerance, the maximum number of iterations, and an initial guess are specified in a similar way as in the Newton M-file (see Sidebar 8.1). In addition, the reaction matrix, the equilibrium constants for the reactions, and the total concentrations are specified. All concentrations and equilibrium constants are entered using their logarithmic values.

The execution part of the M-file is divided into two parts. In the first part, the initialization is done for the iterations to be performed in the second part. In the first part, the sub matrices $\mathbf{S}_1$ and $\mathbf{S}_2$ of the reaction matrix are determined to be used in the computation of $\mathbf{S}^*$ and $\mathbf{U}$. Total and species concentrations are converted from logarithm to their real values. Error variable and iteration counter are already introduced in the former example, see Sidebar 8.1.

Within the iteration the zero of the function

$$\mathbf{F}(\mathbf{c}) = \begin{pmatrix} \mathbf{U} \cdot \mathbf{c} - \mathbf{u} \\ \mathbf{S} \cdot \log(\mathbf{c}) - \log(\mathbf{K}) \end{pmatrix} \tag{8.23}$$

is computed. It is easy to see that the zero is identical with the solution of system (8.22) if activity corrections are neglected. The derivative of the function **F**, the so called Jacobi matrix, is given by:

$$\mathbf{DF}(\mathbf{c}) = \begin{pmatrix} \mathbf{U} \\ \mathbf{S}/(\mathbf{c}\log(10)) \end{pmatrix} \tag{8.24}$$

In analogy to the Newton algorithm, presented in Sidebar 8.1, the generalization is given by the formula:

$$\mathbf{c} \leftarrow \mathbf{c} - \mathbf{DF}(\mathbf{c})^{-1} \cdot \mathbf{F}(\mathbf{c}) \tag{8.25}$$

In the command sequence, first the second term is evaluated and stored in the `dc` variable. The following command in the listing ensures that the concentrations remain positive.

8.4 Sorption and the Law of Mass Action

One can formally write sorption as an equation of a reaction between sorption sites and free species on one side and sorbed species on the other side. In analogy to the Law of Mass Action, one may note the differential equation for the temporal change as:

$$\begin{aligned} \frac{\partial c}{\partial t} &= \kappa_{\leftarrow} c_s - \kappa_{\rightarrow} c \cdot s \\ \frac{\partial s}{\partial t} &= \kappa_{\leftarrow} c_s - \kappa_{\rightarrow} c \cdot s \\ \frac{\partial c_s}{\partial t} &= \kappa_{\rightarrow} c \cdot s - \kappa_{\leftarrow} c_s \end{aligned} \tag{8.26}$$

where c denotes the concentration in the fluid, s the number of free sorption sites and c_s the number of occupied sorption sites. The terms for transport have been omitted in order to focus on sorption. The equilibrium condition is then given by:

$$\frac{c_s}{c \cdot s} = \frac{\kappa_{\rightarrow}}{\kappa_{\leftarrow}} =: K \tag{8.27}$$

or:

$$\frac{c_s}{c} = K \bullet s \tag{8.28}$$

s here denotes the number of free sites. If no competition between species is taken into account, s can also be expressed as $c_{s,\max} - c_s$, the number of available free sites is diminished by the number of occupied sites. Altogether results:

Sidebar 8.3: Activities and Activity Coefficients

a_i denotes the activity of the ith species in a multi-species system. It is the product of concentration c_i and activity coefficient γ_i that is species-dependent, i.e. for the ith chemical species holds:

$$a_i = \gamma_i \cdot c_i$$

The activity depends on ionic strength μ in the solution, defined by

$$\mu = \frac{1}{2} \sum c_i z_i^2$$

where the sum has to be extended over all charged components. z_i denotes the electrical charge of species i. The logarithm of the activity coefficient can be computed explicitly by a formula like:

$$\log \gamma_i = -A \cdot {z_i}^2 \sqrt{\mu}$$

This is the simplest formulation that can be derived theoretically. For very low values of μ ($<10^{-2.3}$; Sigg and Stumm 1989), the formula is valid with a value of $A = 0.51$ (in water of 25°C, see: Krauskopf and Bird 1995). An extended formula was proposed by Davies:

$$\log \gamma_i = -A \cdot {z_i}^2 \left(\frac{\sqrt{\mu}}{1 + \sqrt{\mu}} - 0.3\mu \right)$$

The Davies equation is usually assumed to be valid for ionic strengths up to $\mu = 0.5$. The coefficient A depends on temperature only. Values for A are given in Table 8.1. The activity coefficient thus depends on temperature, on ionic strength and the electric charge of the species.

Another extended formulation for the relation between activity coefficients and ionic strength is found with reference to Debye[4]-Hückel[5]:

$$\log \gamma_i = -\frac{A \cdot z_i^2 \sqrt{\mu}}{1 + B a_i^0 \sqrt{\mu}} + b_i \mu$$

with coefficients B, ${a_i}^0$ and b_i (Debye and Hückel 1923). The coefficient B depends on temperature only. The ion-size parameter ${a_i}^0$ as well as b_i are species-dependent. The latter formulation is used in the speciation code

(continued)

[4] Peter Debye (1884–1966), Dutch chemist and physicist.

[5] Erich Armand Arthur Joseph Hückel (1896–1980), German chemist and physicist.

PHREEQC (Parkhurst 1995). Another extension uses a different formulation of the last term:

$$\log \gamma_i = -\frac{A \cdot z_i^2 \sqrt{\mu}}{1 + Ba_i^0 \sqrt{\mu}} + B' \mu$$

This is implemented in the CHESS code (van der Lee 1998). The temperature-dependency of the parameters is shown in Table 8.1. If activities are considered, the derivation above leads to equation

$$\mathbf{u} = \mathbf{U} \cdot \mathbf{c}$$
$$\mathbf{S} \cdot [\log(\gamma) + \log(\mathbf{c})] = \log(\mathbf{K})$$

instead of (8.22), where γ denotes the vector of activity coefficients.

Table 8.1 Coefficients in extended equations for activity equations (Berner 1971; van der Lee 1998)

Temperature [°C]	A	B	B'
0	0.4883	0.3253	0.0374
5	0.4921		
10	0.4960		
15	0.5000		
20	0.5042		
25	0.5085	0.3288	0.0410
30	0.5130		
35	0.5175		
40	0.5221		
50	0.5319		
60	0.5425	0.3346	0.0440
100	0.9595	0.3421	0.0460

$$\frac{c_s}{c} = K \cdot \left(c_{s,\max} - c_s\right) \tag{8.29}$$

or, resolved for c_s:

$$c_s = \frac{K c_{s,\max} c}{1 + K \cdot c} \tag{8.30}$$

With parameters $\alpha_{L1} = c_{s,\max}$ and $\alpha_{L2} = K^{-1}$ this is the already mentioned Langmuir isotherm (6.7). A generalization of this formula results for species which participate with more than one unit on the reaction, as for example gaseous O_2. Starting from the reaction terms:

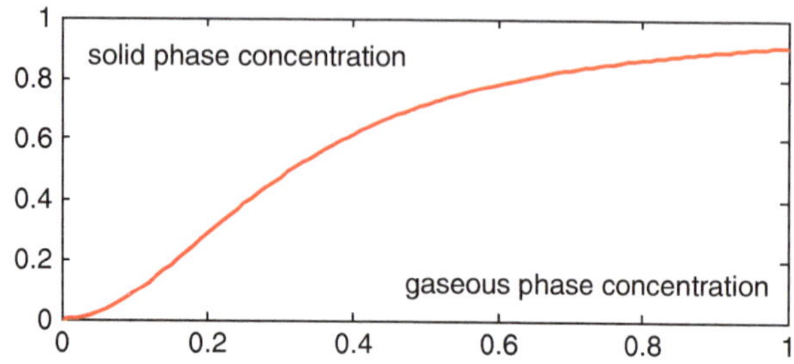

Fig. 8.1 Sorption isotherm for species with stoichiometric coefficient = 2 (see text)

$$\begin{aligned}\frac{\partial c}{\partial t} &= \kappa_{\leftarrow} c_s - \kappa_{\rightarrow} c^2 \cdot s \\ \frac{\partial s}{\partial t} &= \kappa_{\leftarrow} c_s - \kappa_{\rightarrow} c^2 \cdot s \\ \frac{\partial c_s}{\partial t} &= \kappa_{\rightarrow} c^2 \cdot s - \kappa_{\leftarrow} c_s\end{aligned} \tag{8.31}$$

the same arguments as above lead to the isotherm:

$$c_s = \frac{K c_{s,\max} c^2}{1 + K \cdot c^2} \tag{8.32}$$

Figure 8.1 depicts an example of such a sorption isotherm. Formula (8.32) is a special case of the so called Langmuir-Freundlich isotherm (Klug et al. 1998) with exponent two.

8.5 Transport and Speciation

The differential equation for multi-species reactive transport has been written in vector notation (see equation (7.6)):

$$\frac{\partial \mathbf{c}}{\partial t} = -\nabla \bullet \mathbf{j}(\mathbf{c}) + \mathbf{S}^T \mathbf{r} \tag{8.33}$$

The system (8.33) contains one differential equation for each species. It is a system with N_s equations. In case of equilibrium reactions it is not possible to treat the system (8.33) directly, because the exchange rates of the equilibrium reactions $\mathbf{r}$ are neither known nor given by an explicit expression. In order to reach a feasible formulation, the equations have to be summed up in a way that eliminates the last term on the right side of the equation. The procedure was already described in Chap. 8.3.

It is appropriate to multiply the equation from the left by a matrix $\mathbf{U}$ with the property $\mathbf{U} \cdot \mathbf{S}^T = 0$. The matrix $\mathbf{U}$ has N_s-N_r rows and N_s columns and is not unique. The result is:

$$\frac{\partial}{\partial t}\mathbf{Uc} = -\mathbf{U} \cdot \nabla \bullet \mathbf{j}(\mathbf{c}) + \mathbf{U} \cdot \mathbf{S}^T \mathbf{r} \tag{8.34}$$

The last term on the right side can be omitted because of $\mathbf{U} \cdot \mathbf{S}^T = 0$. If the transport terms are linear, multiplication with matrix $\mathbf{U}$ and differentiation can be exchanged. From (8.34) one directly obtains a differential equation for the total concentrations $\mathbf{u} := \mathbf{U} \cdot \mathbf{c}$:

$$\frac{\partial}{\partial t}\mathbf{u} = -\nabla \bullet \mathbf{j}(\mathbf{u}) \tag{8.35}$$

Equation (8.35) represents a system of N_s-N_r transport equations for the N_s-N_r components of $\mathbf{u}$. The problem is thus significantly simpler than the original system with N_s coupled equations. The gain is paid by the need for speciation calculations. When the vector $\mathbf{u}$ is calculated, the N_s species have to be computed in a second step with the help of (8.22).

The formulation of the entire problem, given by (8.35) and (8.22), offers various advantages. The given formulation has the following properties:

1. Transport and reaction modeling are not coupled, because transport can be solved independently from the speciation; solving the transport problem, the knowledge of the $\mathbf{c}$-vector is not required
2. The transport problem consists of N_s-N_r linear differential equations
3. The differential equations of the transport problem are independent from each other
4. The reaction problem consists of a nonlinear system of N_s equations which have to solved for each node

For the solution of such a system it is justified to apply a *sequential non-iterative approach* (SNIA, see: Steefel and MacQuarrie 1996): transport is solved first in order to obtain $\mathbf{u}$, from which $\mathbf{c}$ is determined in a second step. In the first step N_s-N_r independent linear transport equations have to be solved.

In the case of reactive transport with equilibrium equations, the SNIA approach causes no additional errors in contrast to other approaches. If the type of boundary conditions is the same for all total concentrations, it is even sufficient to solve the transport equation only once. When the transport step is completed, the speciation has to be calculated for each block (or node). These computations are independent from each other, i.e. the speciation in one block does not depend on the speciation in any other block of the model region.

The de-coupled solution procedure is only possible, because within the formulation itself (8.35) and (8.22) transport and equilibrium geochemistry are not coupled. The sequential (de-coupled) treatment of transport and speciation is not

Table 8.2 Carbon chemistry equilibrium constants

Reactions and equilibrium constants (log) for T = 25°C	
$H^+ + OH^- \leftrightarrow H_2O$	−14
$H^+ + CO_3{}^{2-} \leftrightarrow HCO_3{}^-$	−10.329
$Ca^{2+} + HCO_3{}^- \leftrightarrow CaHCO_3{}^+$	−1.106
$2H^+ + CO_3{}^{2-} \leftrightarrow H_2CO_3$	−16.7
$Ca^{2+} + CO_3{}^{2-} \leftrightarrow CaCO_3$	−8.48

always possible. When kinetic reactions have to be considered additionally, the equations remain coupled and the numerical solution is not as easy (Holzbecher 2005).

The situation is also more complex when species in different phases are involved. In the mathematical formulation species in different phases can not be expressed by the same transport operator, as it is assumed in (8.33).

For demonstration purposes we selected an example from carbon chemistry. A formulation is chosen in which seven species are involved in five reactions. These are ordered in the species vector as follows:

$$\left(H^+, HCO_3{}^-, Ca^{2+}, OH^-, H_2CO_3, CO_3{}^{2-}, CaHCO_3{}^+\right)$$

The major reactions for the carbon species and the equilibrium constants are gathered in Table 8.2. Note that two additional species appear in the reactions: water, H_2O, and calcite, $CaCO_3$. Those two species are not included in the model, because it is assumed that both are available from an infinite pool. In that case their concentrations can be set to unity. Water is the medium within which other species are dissolved. Concerning calcite, it is assumed that the flow passes through a calcareous or limy formation, a fracture, a pipe or a karst system.

The following reaction matrix corresponds with the reaction system:

$$\mathbf{S} = \begin{pmatrix} 1 & 0 & 0 & 1 & 0 & 0 & 0 \\ 1 & -1 & 0 & 0 & 0 & 1 & 0 \\ 0 & 1 & 1 & 0 & 0 & 0 & -1 \\ 2 & 0 & 0 & 0 & -1 & 1 & 0 \\ 0 & 0 & 1 & 0 & 0 & 1 & 0 \end{pmatrix} \tag{8.36}$$

The matrices $\mathbf{S}_1$ and $\mathbf{S}_2$ are given by:

$$\mathbf{S}_1 = \begin{pmatrix} 1 & 0 \\ 1 & -1 \\ 0 & 1 \\ 2 & 0 \\ 0 & 0 \end{pmatrix} \quad \mathbf{S}_2 = \begin{pmatrix} 0 & 1 & 0 & 0 & 0 \\ 0 & 0 & 0 & 1 & 0 \\ 1 & 0 & 0 & 0 & -1 \\ 0 & 0 & -1 & 1 & 0 \\ 1 & 0 & 0 & 1 & 0 \end{pmatrix} \tag{8.37}$$

and for the transformation matrix **U** results:

$$\mathbf{U} = \begin{pmatrix} 1 & 0 & 1 & -1 & 1 & -1 & 1 \\ 0 & 1 & -1 & 0 & 1 & 1 & 0 \end{pmatrix} \tag{8.38}$$

Thus there are two totals which are a combination of the species concentrations:

$$\begin{pmatrix} Tot1 \\ Tot2 \end{pmatrix} = \begin{pmatrix} H^+ + Ca^{2+} - OH^- + H_2CO_3 - {CO_3}^{2-} + {CaHCO_3}^+ \\ {HCO_3}^- - Ca^{2+} + H_2CO_3 + {CO_3}^{2-} \end{pmatrix}$$

For each of the totals a differential equation is solved. The entire system is reduced to the solution of the transport equation for two totals and the specification calculations. The connection with transport was tested for a situation of fracture within calcareous rock, entered by water which was under-saturated with respect to calcite. Such a system was studied by Saaltink et al. (2001). Some parameter values were adopted from that study.

A list of parameters is given in Table 8.3. A 1D porous fracture of 100 m length is modeled with water entering at a Darcy velocity of 2 m/a. Initial and input concentrations for both species have to be specified as well.

The MATLAB® model for the problem is implemented as a combination of the `pdepe` solver and the speciation calculations. `pdepe` is used for the usual transport equation as described in Chap. 4 in detail. The speciation has to be performed for each geochemical species set at each node and each time instance of interest.

The simulation using MATLAB® shows the propagation of the front for all species. Figure 8.2 depicts pH as an example. The pH rises, because carbon species, entering the aqueous system by calcite dissolution, bind more of the available H^+-species. With pH = 9.7 the inflowing front thus shows higher values than the initial system with pH = 8.75.

The given procedure can also be applied to systems in which both kinetic and equilibrium reactions are expected. The mathematical formalism has to be slightly extended, as it was demonstrated by Holzbecher (2005). Without going into details, we demonstrate the procedure for a situation derived from the example above.

Table 8.3 Parameter for calcite dissolution equilibrium test-case

Variable	Value	Unit
Length	100	m
Maximum simulated time	5	a
Porosity	0.1	–
Darcy velocity	2	m/a
Diffusivity	200	m^2/a
Initial concentration Tot1	−4.019	log mol/l
Initial concentration Tot2	−3.018	log mol/l
Input concentration Tot1	−4.365	log mol/l
Input concentration Tot2	−5.421	log mol/l

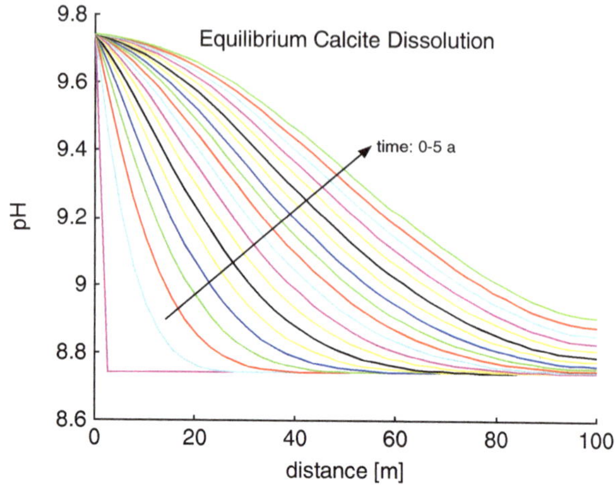

Fig. 8.2 Calcite dissolution example, equilibrium case

Saaltink et al. (2001) already consider the case in which the calcite dissolution reaction (the last in Table 8.2) is slow compared to all other processes.

In that case, the reaction matrix for equilibrium reactions contains four rows only. for $\mathbf{S}_1$ and $\mathbf{S}_2$ results:

$$\mathbf{S}_1 = \begin{pmatrix} 1 & 0 & 0 \\ 1 & -1 & 0 \\ 0 & 1 & 1 \\ 2 & 0 & 0 \end{pmatrix} \quad \mathbf{S}_2 = \begin{pmatrix} 1 & 0 & 0 & 0 \\ 0 & 0 & 1 & 0 \\ 0 & 0 & 0 & -1 \\ 0 & -1 & 1 & 0 \end{pmatrix} \tag{8.39}$$

Using (8.20) and (8.21) and MATLAB® it is easy to calculate $\mathbf{U}$:

$$\mathbf{U} = \begin{pmatrix} 1 & 0 & 0 & -1 & 1 & -1 & 0 \\ 0 & 1 & 0 & 0 & 1 & 1 & 1 \\ 0 & 0 & 1 & 0 & 0 & 0 & 1 \end{pmatrix} \tag{8.40}$$

Now the total concentrations are different combinations of the species:

$$\begin{pmatrix} TotH \\ TotC \\ TotCa \end{pmatrix} = \begin{pmatrix} H^+ - {CO_3}^{2-} + H_2CO_3 - OH^- \\ {HCO_3}^- + H_2CO_3 + {CO_3}^{2-} + {CaHCO_3}^+ \\ Ca^{2+} + {CaHCO_3}^+ \end{pmatrix}$$

There are three initial and inflow concentrations required for the totals *TotH*, *TotC* and *TotCa*. The values proposed by Saaltink et al. (2001) are provided in Table 8.4. The kinetic transfer coefficient α varies over 4 orders of magnitude. It is used in the kinetic rate law given by:

$$r_{kin} = r = \alpha(1 - \sigma) \tag{8.41}$$

where σ denotes the calcite saturation index.

Table 8.4 Parameter for calcite dissolution simulation (CAL – see: Saaltink et al. 2001)

Variable	Value	Unit
Initial concentration *TotH*	−7.978	log mol/l
Initial concentration *TotC*	−3.018	log mol/l
Initial concentration *TotCa*	−3.019	log mol/l
Inflow concentration *TotH*	−5.496	log mol/l
Inflow concentration *TotC*	−5.421	log mol/l
Inflow concentration *TotCa*	−4.398	log mol/l
Kinetic transfer coefficient(s)	$9.939\ 10^{-4(1)7}$	mol/(l a)

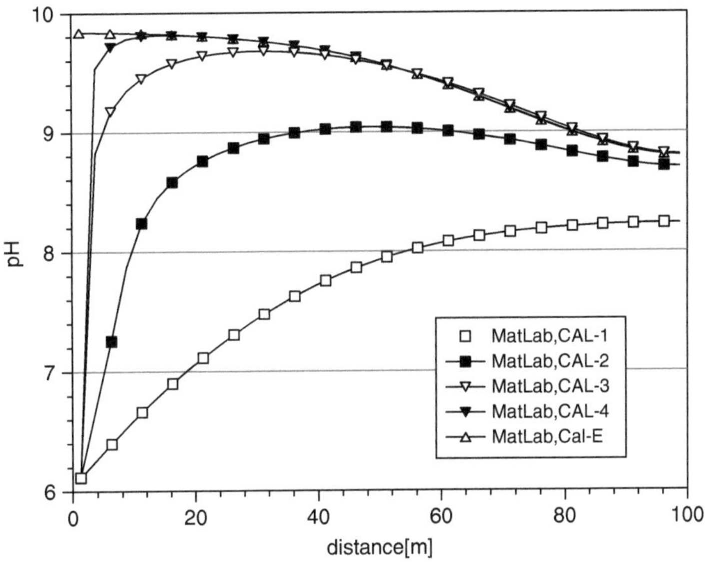

Fig. 8.3 Calcite dissolution example; results for different calcite dissolution kinetics

Figure 8.3 illustrates the results for pH in dependence of the kinetic transfer coefficient. For the cases CAL-1–CAL-4, the kinetics rate is increased by a factor of 10, taking the values given in Table 8.4. In the CAL-1 case, the process of calcite dissolution is too slow to have any effect on pH. There is a front of low pH penetrating the system.

For enhanced kinetics the already mentioned rise of pH becomes more and more pronounced. Due to increased calcite disolution, H^+ ions are increasingly consumed by the dissolved carbon species. As a result, the pH is increased where the inflowing water dominates. The equilibrium situation is approached gradually with a front of high pH entering the fracture.

The MATLAB® simulation is again based on a combination of the `pdepe` solver and speciation calculations based on the Newton method. Holzbecher (2006) extended the presented approach for the simulation of the horizontal and vertical concentration distribution within a fracture. The 2D flow field is computed following the Hagen-Poiseuille analytical solution (see Chap. 11). The 2D advection-dispersion equation with an anisotropic dispersion tensor is solved using COMSOL Multiphysics[6]. The reaction term (8.41) is computed at each node and each time level. In order to evaluate formula (8.41), the concentrations of the species need to be known. They are obtained within a MATLAB® speciation M-file, which is called by COMSOL Multiphysics. The M-file looks as follows:

```
function [q] = source(t,x,y,u,u2,u3)
U = [1 0 0 -1 1 -1 0; 0 1 0 0 1 1 1; 0 0 1 0 0 0 1];
Se = [1 0 0 1 0 0 0; 1 -1 0 0 0 1 0; 0 1 1 0 0 0 -1; 2 0 0 0 -1 1 0];
logK = [-14; -10.329; -1.106; -16.7];
Skin = [0 0 1 0 0 1 0];
c = [3.1913e-6; 3.7928e-6; 3.9992e-5; 3.1329e-9; 9.9985e-11; 9.9985e-
  11; 9.9985e-11];
pkin = 9.939e-4;
toll = 1e-10; nmax = 100;

for i=1:max(size(u))
    err = toll+1; nit = 0;
    while (nit < nmax & err > toll * max(abs(c))),
        nit = nit+1;
        F = [U*c-[u(i); u2(i); u3(i)]; Se*log10(c)-logK];
        DF = [U; Se*diag((1/2.3026)./c)];
        dc = -DF\F;
        cn = max(c+dc,0.005*abs(c));
        err = max(abs(cn-c));
        c = cn;
        logc = log10(c);
    end
    sp = exp(2.3026*(Skin*logc+8.48));
    q(i) = pkin*(ones(size(sp))-sp);
    if isnan(q(i)) q(i) = 0; end
end
```

Using the software combination of this book, it is thus possible to treat reactive transport in several space dimensions.

References

Berner RA (1971) Principles of chemical sedimentology. McGraw-Hill, New York, p 240

Debye P, Hückel E (1923) Zur Theorie der Elektrolyte. Physikalische Zeitschrift 23:185–206, in German

Fang Y, Yeh G-T, Burgos WD (2003) A general paradigm to model reaction-based biogeochemical processes in batch systems. Water Res Res 39(4):1083. doi:10.1029/2002WR001694

Holzbecher E (2005) Reactive transport – concepts and numerical approaches. In: Ingham DB, Pop I (eds) Transport phenomena in porous media, vol III. Elsevier, Oxford, pp 305–340

Holzbecher E (2006) On the coupling of transport and chemical speciation calculations with FEMLAB. Chem Eng Technol 29(5):5–7, Supplement 'Trends in Chemical Engineering'

Klug M, Sanches M, Laranjeira M, Fávere V, Rodrigues C (1998) Análise das isotermas de adsorção de Cu(II), Cd(II), Ni(II) e Zn(II) pela N-(3,4-dihidroxibenzil) quitosana empregando o método da regressão não linear. Química Nova 21(4):410–413, in Potuguese

[6] See: http://www.comsol.com/.

Krauskopf KB, Bird DK (1995) Introduction to geochemistry. McGraw Hill, New York, p 647
Parkhurst DL (1995) PHREEQC: a computer program for speciation, reaction-path, advective transport, and inverse geochemical calculations, Lakewood. U.S. Geological Survey, Water Resources Investigation Report 95–4227, p 143
Saaltink MW, Ayora C, Carrera J (1998) A mathematical formulation for reactive transport that eliminates mineral concentrations. Water Res Res 34(7):1649–1656
Saaltink MW, Carrera J, Ayora C (2001) On the behavior of approaches to simulate reactive transport. J Contaminant Hydrol 48:213–235
Sigg L, Stumm W (1989) Aquatische Chemie. vdf-Verlag, Zurich, p 498, in German
Steefel CI, MacQuarrie KTB (1996) Approaches to modeling of reactive transport in porous media. In: Lichtner PC, Steefel CI, Oelkers EH (eds) Reactive transport in porous media, reviews in mineralogy, vol 34, Mineral. Soc.of Am., pp 83–129
van der Lee J (1998) Thermodynamic and mathematical concepts of CHESS, Ècole des Mines de Paris. Technical Report LHM/RD/98/39, p 99

Chapter 9
Ordinary Differential Equations: Dynamical Systems

Ordinary differential equations (ode) are differential equations for functions which depend on one independent variable only. These 'odes' are simpler than partial differential equations which contain more than one independent variable. In almost all models or simulations independent variables are either time and/or space.

In environmental modeling, two situations can be distinguished in which odes appear. The first situation deals with systems in which spatial differences can be neglected and the temporal development is questioned. In chemistry, the continuously stirred reactor is an often used concept for which an approach is allowed with time t as the only independent variable.

One such situation a was already described In Chap. 5.1, with degradation or decay as the only relevant process for the transient change of some species concentration. Below (Chap. 10.1) an example is presented dealing with the determination of a reaction kinetic, using data from batch experiments. In the field situation such ideal systems are seldom appropriate, but sometimes the assumption of no space-dependence may be approximately fulfilled. The long-term accumulation of a substance in lakes can be modeled with the idea of an ideally mixed reservoir, for example.

In a second different constellation time is neglected and a steady state is sought for a system which can be described by a single space variable. Such models are common for aquatic sediments, where parameters and variables show characteristic changes normal to the water-sediment interface, usually in vertical direction. Also in streams the one-dimensional approach can be applied under certain circumstances: the space coordinate is taken along the steady streamline following the river downstream. Surface water infiltrating an aquifer has a pronounced direction along the flow path. Here the 1D formulation is justified, because the transverse gradients almost vanish.

Aside of analytical solutions, two numerical solvers of MATLAB® are introduced in this chapter, one (`ode15s`) designed for the solution of initial value problems, the other (`bvp4c`) for the solution of boundary value problems (bvp). In initial value problems boundary conditions are formulated for one value of the independent variable only (typically: $t = 0$), whereas in boundary value problems

E. Holzbecher, *Environmental Modeling*,
DOI 10.1007/978-3-642-22042-5_9, © Springer-Verlag Berlin Heidelberg 2012

there are conditions required at both ends of the interval of the independent variable. A special MATLAB® related textbook on modeling with odes is published by Shampine et al. (2003).

9.1 Streeter-Phelps Model for River Purification

A relatively simple model concerning decreased concentration of oxygen downstream from a polluting discharge and the recovery to background level was proposed by Streeter and Phelps already in 1925. Although the application is based on several assumptions, such type of modeling is used as part of regulations for sewage facilities. Bacteria gradually degrade the organic matter contained in the discharge along the course of the river downstream. The most relevant bacteria prefer aerobic conditions, i.e. they also rely on oxygen (DO = dissolved oxygen). The change of these two components (organic matter and DO) along the flow path is simulated in the following.

In the model, the concentration of the organic pollutant is measured as biodegradable oxygen demand (BOD) as a proxy. Two parameters are connected to the BOD behavior: the inflow rate f_{BOD} (kg/m^3/s) and the degradation rate k_1 (1/s). Degradation concerns both BOD and DO, which is expressed in the system of two ordinary differential equations

$$\begin{aligned} \frac{\partial c_{BOD}}{\partial t} &= f_{BOD} - k_1 c_{BOD} \\ \frac{\partial c_{DO}}{\partial t} &= k_2\left(c_{DO,sat} - c_{DO}\right) - k_1 c_{BOD} \end{aligned} \tag{9.1}$$

The oxygen concentration c_{DO} is additionally determined by the reaeration process. Reaeration brings oxygen back into the water, for which various processes may be relevant. One of the most important is the contact with atmospheric air. If the contact time is long enough, the equilibrium between partial pressure of oxygen in air$p_{O_2}$and c_{DO} is established. According to Henry's Law[1] such equilibria are characterized by a component specific ratio of dissolved concentration and partial pressure. As in the earth atmosphere $p_{O_2}$is fixed with approximately 0.21 atm, c_{DO} can reach a value of 12.9 mg/l,[2] which is the saturation limit for oxygen in water.

However, the saturation limit in a natural river may be somewhat smaller as a result of other processes that influence the oxygen balance. Aquatic plants produce oxygen, aquatic fauna consumes oxygen. Moreover, oxygen is needed by bacteria which are busy degrading natural organic matter at the bottom of the water

[1] William Henry, 1775–1836, English chemist.

[2] Henry's Law constant for the equilibrium between gaseous and aqueous phase oxygen at 5°C is 61.2 mg/l/atm.

body and in the sediment compartment. Thus, the oxygen saturation limit $c_{DO,sat}$ depends on local circumstances and is not an universal (not even a terrestrial) constant.

However, the reaeration process will make oxygen concentration converge to the limit. Therefore in the most simple approach the difference from the limit $c_{DO,sat} - c_{DO}$ determines the rate of DO-change aside from the kinetic parameter k_2.

The system (9.1) of two differential equations is quite simple to solve. The first equation contains only one dependent variable (c_{BOD}) and has an exponential function as analytical solution. With that solution one can regard the second equation as a differential equation for c_{DO}. Together, both (9.1) form a linear system. A method for the solution of general linear systems is presented below (Chap. 18). The following lines are a special implementation for the Streeter-Phelps system. The presented numerical approach can also be extended to non-linear systems of equations, which may appear if there are complex parameter dependencies.

```
T = 25;                    % maximum time [T]
k1 = 0.3;                  % deoxygenation rate coefficient [1/T]
k2 = 0.4;                  % reaeration rate coefficient [1/T]
DOsat = 11;                % DO saturation [M/L^3]
BODin = 7.33;              % initial BOD [M/L^3]
DOin = 8.5;                % initial DO concentrations [M/L^3]
fBOD = 1;                  % natural BOD inflow [M/(L^3*T)]
N = 60;                    % discretization of time

%---------------------- execution --------------------------------

% BOD = y(1), DO = y(2)
options = odeset('AbsTol',1e-20);
[t,y] = ode15s(@SP,[0 T],[BODin; DOin],options,k1,k2,DOsat,fBOD);

%---------------------- graphical output -------------------------

plot (t,y);

legend ('BOD','DO');
xlabel ('traveltime'); ylabel ('concentration');
grid;

%---------------------- function ---------------------------------

function dydt = SP(t,y,k1,k2,DOsat,fBOD)
dydt = zeros(2,1);
dydt(1) = fBOD-k1*y(1);
dydt(2) = k2*(DOsat-y(2))-k1*y(1);
```

The complete code is included in the accompanying software under the name *'StreeterPhelps.m'*

Input values for the example are adopted from Deaton and Winebrake (2000). Deoxygenation and reaeration rate coefficients are given in 1/d. BOD inflow

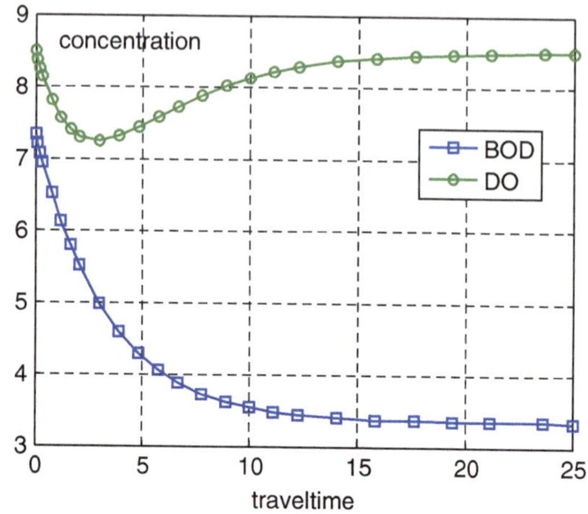

Fig. 9.1 Model results for river purification

is in mg/(l·d). All concentrations are in mg/l. The steady state is given by $c_{BOD} = f_{BOD}/k_1 = 3.33$ mg/l and $c_{DO} = c_{DO,sat} - f_{BOD}/k_2 = 8.5$ mg/l.

Figure 9.1 shows the result for the increased input values with $c_{BOD} = 7.5$ mg/l. BOD concentration decreases within a time period of 15 days. The content of oxygen decreases within the first 3 days if degradation exceeds reaeration. After that follows a second time period in which reaeration is dominant leading to a gradual recovery of the DO level back to the natural state.

The Streeter-Phelps model is based on several conditions. Diffusion and dispersion processes are neglected. There is no distinction of concentrations within the river cross-section. The system of differential (9.1) is based on a Lagrangian[3] description, which is a formulation for the concentration along the flow path. In the Lagrangian description the advection terms disappear, whereas they remain in the alternative Eulerian description. The Eulerian approach, which was introduced in Chaps. 3 and 4, is based on the conception of a fixed space and delivers the following set of equations

$$
\begin{aligned}
\frac{\partial c_{BOD}}{\partial t} &= \nabla \bullet \mathbf{j}(c_{BOD}) + f_{BOD} - k_1 c_{BOD} \\
\frac{\partial c_{DO}}{\partial t} &= \nabla \bullet \mathbf{j}(c_{DO}) + k_2\left(c_{DO,sat} - c_{DO}\right) - k_1 c_{BOD}
\end{aligned} \tag{9.2}
$$

Also with respect to biogeochemistry the Streeter-Phelps model has to be extended to capture a more detailed behavior. Vanrolleghem et al. (2000) present and discuss several generalizations of the simple Streeter-Phelps approach. In order to consider photosynthesis-respiration, the model contains four more variables: ammonium, nitrate, phosphorus and algae. The mathematical description takes

[3] Joseph-Louis Lagrange, 1763–1813, French mathematician.

into account that phosphorus and nitrate are produced by the degradation of organic matter. Algae growth on the other hand consumes nutrients, nitrate-NO_3 and phosphorus-HPO_4, and produce oxygen.

Even more complex models may consider anoxic and anaerobic degradation as well as growth and respiration of different bacteria populations (Vanrolleghem et al. 2000).

9.2 Details of Michaelis–Menten or Monod Kinetics

The Michaelis–Menten kinetics for the description of rate limited chemical and biochemical degradation processes was introduced in Chap. 7. It can be written as

$$\frac{\partial s}{\partial t} = -r := -\frac{k_1 s}{k_2 + s} \tag{9.3}$$

with maximum rate k_1 and half-degradation concentration k_2. s denotes the concentration of a chemical species, here called the substrate. It will be demonstrated that the kinetics (9.3) may result from a sequence of reactions, in which aside from the reactant and the final reaction product an enzyme e and an intermediate product i are involved. The system of chemical reactions can be noted as:

$$s + e \rightleftarrows i \rightarrow p + e$$

The substrate s and enzyme e are connected with an intermediate species i by an equilibrium reaction. Moreover, the intermediate i is broken down into the final reaction product p with the enzyme e as a by-product in an irreversible reaction. According to the law of mass action, the set of differential equations for the reaction system is:

$$\begin{aligned} \frac{\partial s}{\partial t} &= -k_1 s e + k_3 i \qquad & \frac{\partial e}{\partial t} &= -k_1 s e + k_3 i + k_2 i \\ \frac{\partial i}{\partial t} &= k_1 s e - k_3 i - k_2 i \qquad & \frac{\partial p}{\partial t} &= k_2 i \end{aligned} \tag{9.4}$$

with rate coefficients k_1, k_2 and k_3. Transport is neglected in this system. It can be shown (Fife 1979; Morel and Hering 1993) that for long times the behavior of the four-species system can simpler be described by the following Michaelis–Menton kinetics:

$$\frac{\partial s}{\partial t} = -\frac{k_1 k_2 (i_0 + e_0)}{k_2 + k_3 + k_1 s} s \tag{9.5}$$

Both alternative systems are implemented in MATLAB® using the `ode15s` solver. After the input section the M-file looks like this:

```
%----------------------execution----------------------------------
% substrate= y(1),enzyme2= y(2),intermediate= y(3), product= y(4)

options = odeset('AbsTol',1e-20);
[t,y] = ode15s(@detail,[0 T],[s0; e0; i0; p0],options,k);
[tt,z] = ode15s(@lumped,[0 T],s0,options,e0,i0,k);

%--------------------- graphical output --------------------------
plot(t,y(:,1:4)); hold on;
plot(tt,z(:,1),'--',tt,s0-z(:,1)+p0,'--');
legend ('substrate','enzyme','intermediate','product',...
'lumped substrate','lumped product');
grid;

%--------------------- functions -------------------------------
function dydt = detail(t,y,k)
r1 = k(1)*y(1)*y(2);
r2 = k(3)*y(3);
r3 = k(2)*y(3);
dydt = zeros(4,1);
dydt(1) = -r1 + r2;
dydt(2) = -r1 + r2 + r3;
dydt(3) = r1 - r2 - r3;
dydt(4) = r3;

function dzdt = lumped(t,z,e0,i0,k)
dzdt(1) = -k(1)*k(2)*(e0+i0)*z/(k(2)+k(3)+k(1)*z);
```

The system of four species is solved in the `y`-vector, and the system is specified in the `detail`-function. The substrate development according to the Michaelis–Menton kinetics is given in variable `z` and specified in the `lumped`-function.

The complete code is included in the accompanying software under the name *'MichaelisMenten.m'*

For the input parameters $s_0 = 1, e_0 = 0.2, i_0 = 0.1$ and $p_0 = 0.3$ and the rate coefficients $k_1 = 1, k_2 = 0.15, k_3 = 0.4$ the result is depicted in Fig. 9.2. The strong initial change is obviously not captured by the Michaelis–Menten approach. But for long time periods the common Michaelis–Menten model is a good approximation, although several initial conditions and rate constants are lumped into two parameters. The difference between the detailed and the lumped model is visualized in Fig. 9.3. For long times the relative error for the substrate is almost 100%. That must not be crucial, because the high relative error coincides with low concentrations, as the comparison between the two figures shows. However, if the initial concentrations are several orders of magnitude above a limit value, low concentrations may still be harmfull and the error of 100% is not tolerable.

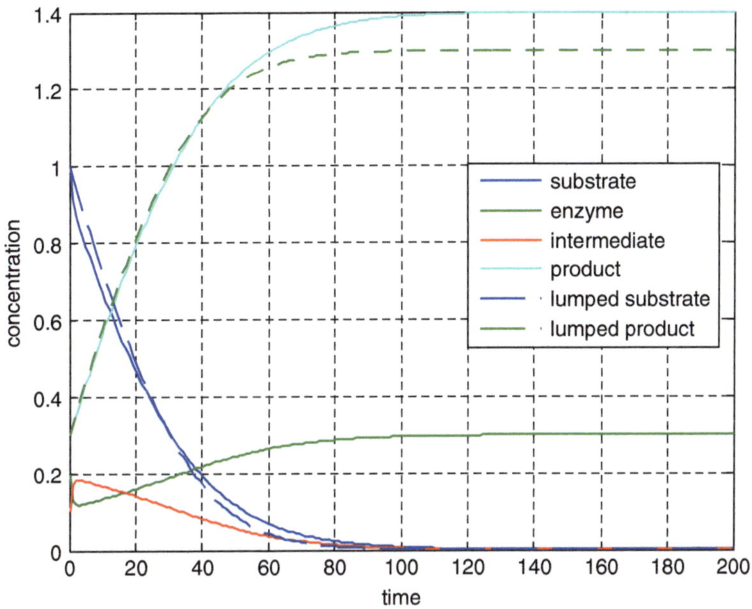

Fig. 9.2 Comparison of a detailed four-species model and the approximation using Michaelis-Menten kinetics

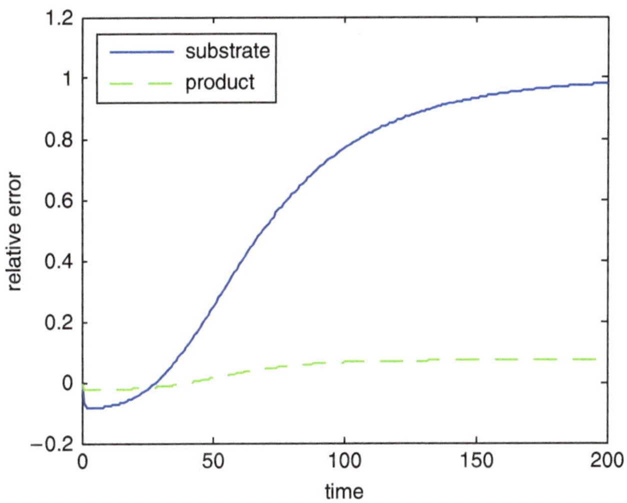

Fig. 9.3 Relative error of a lumped Michaelis-Menten model, in comparison with a detailed four-species model

9.3 1D Steady State Analytical Solution

Steady-state or stationary means time-independent. In real systems steady state is achieved when the relevant timescale for the problem is too long compared to the internal time scale.

The transport equation for the steady state is simply obtained by omitting the storage term on the left side, which includes the time-derivative $\partial c/\partial t$. The right hand side of the 1D transport equation has thus been set to zero:

$$\frac{\partial}{\partial x} D \frac{\partial c}{\partial x} - v \frac{\partial c}{\partial x} + q = 0 \tag{9.6}$$

Equation (9.6) is an ordinary differential equation, as only space derivatives in a single direction (x) appear. For constant parameter values, the solutions of such an ordinary differential equation can be written explicitly. In the first step to obtain a solution we neglect sources and sinks, i.e. the last term in (9.6) is omitted: $q = 0$. If there is a Dirichlet condition at one side and a Neumann condition at the other side of a finite system with length L

$$c(0) = c_{in} \qquad \partial c/\partial x(L) = 0 \tag{9.7}$$

the solution is obviously given by the constant function $c = c_{in}$, because all derivatives vanish. For constant parameters D and v and Dirichlet boundary conditions on both sides

$$c(0) = c_{in} \qquad c(L) = c_1 \tag{9.8}$$

the solution is:

$$c(x) = c_{in} + (c_1 - c_{in}) \frac{1 - \exp(v \cdot x/D)}{1 - \exp(v \cdot x/D)} \tag{9.9}$$

or in dimensionless form:

$$\frac{c(\xi) - c_{in}}{c_1 - c_{in}} = \frac{1 - \exp(Pe \cdot \xi)}{1 - \exp(Pe)} \tag{9.10}$$

with dimensionless space variable ξ, dimensionless Péclet number $Pe = v \cdot L/D$ for dimensionless concentration on the left side of (9.10). Note that we here allow the Péclet number to have a sign depending on flow direction. Results for selected values of the Péclet number are shown in Fig. 9.4. In all cases a steeper gradient can be observed near the outflow boundary; i.e. for negative Pe on the left, and for positive Pe on the right boundary. The deviation in the slope between inflow and outflow side increases with the absolute value of Pe. The linear profile, which

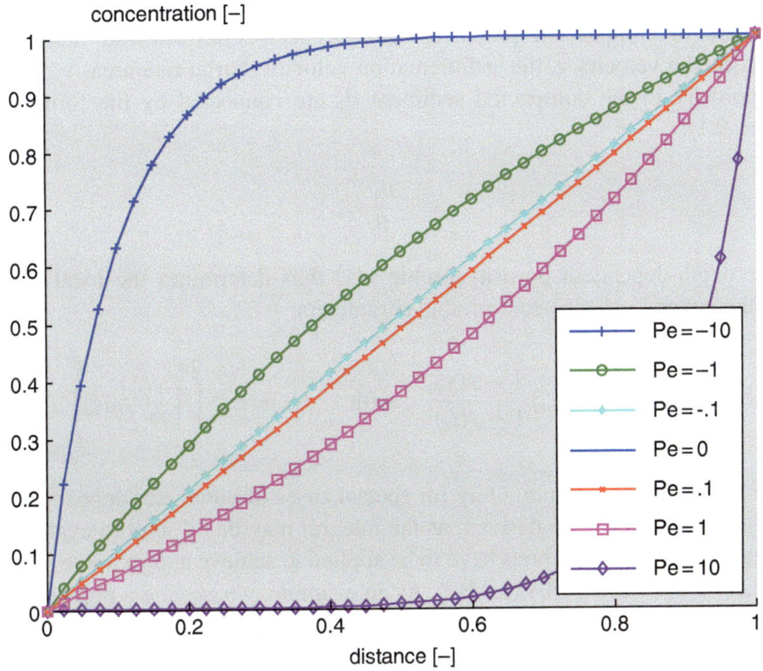

Fig. 9.4 The steady-state solution of the advection-diffusion equation in dimensionless form in dependence of the dimensionless Péclet number

results for the pure diffusion situation, is not plotted as it corresponds with the main diagonal of the coordinate system. Clearly, the figure is symmetric with respect to the diagonal.

The figure is obtained by the following command sequence and some post-processing in the figure editor on the graphs' appearance (introduction of markers and colors):

```
Pe = [-10,-1,-.1,0,.1,1,10];
x = [0:0.025:1];
figure; hold on;
for i = 1:size(Pe,2)
  plot (x,(1-exp(Pe(i).*x))./(1-exp(Pe(i).*ones(1,size(x,2)))));
end
legend ('Pe=-10','Pe=-1','Pe=-.1','Pe=0','Pe=.1','Pe=1','Pe=10');
```

The complete code is included in the accompanying software under the name *'sttransanal.m'*

For aquatic sediments the fluid velocity in the pores decreases with depth. Near the bottom the velocity is smaller, because the fluid obeys an additional moment against the sedimentation direction. The reason is the shrinking size of the pore space due to compaction processes. Holzbecher (2002) showed that in such a situation the velocity v, the sedimentation velocity (burial rate/area) v_{bur} and the final porosity in the compacted sediment θ_1 are connected by the formula (see Sidebar 9.1):

$$v = \frac{\theta_1}{\theta} v_{bur} \tag{9.11}$$

The depth-dependent porosity profile $v(x)$ thus determines the local velocity. Then the general solution for variable parameters

$$c(x) = c_{in} + (c_1 - c_{in}) \frac{1 - J(x)}{1 - J(L)} \quad \text{with} \quad J(x) = \exp\left(\int_0^x \frac{v(\xi)}{D(\xi)} d\xi \right) \tag{9.12}$$

has to be taken into account. Only for special cases of parameter dependencies an analytical solution can be derived, as the integral may be difficult to evaluate. For that reason numerical methods have to be applied to achieve approximate solutions. This can be done using MATLAB®. As an example, we assume a porosity profile which changes exponentially between a high value θ_0 at the sediment-water interface and the final value θ_1:

$$\theta(x) = \theta_1 + (\theta_0 - \theta_1) \exp(-\alpha x) \tag{9.13}$$

Solutions with MATLAB® for the variable parameter situation are given below.

Sulphate Reduction

Dealing with sulphate reduction in aquatic sediments, Berner (1964) introduces a simple model on organic matter. Organic carbon is utilized by bacteria in the biochemical degradation process. It is assumed that '*the rate of metabolism of the bacteria is directly proportional to the concentration of utilizable organic matter*'. Taking into account the sedimentation velocity w, which is derived from the burial rate, he solves the differential equation for the steady state of organic matter in the solid phase:

$$-w \frac{\partial}{\partial z} c_{org,s} - \lambda c_{org,s} = 0 \tag{9.14}$$

with degradation constant of organic matter λ in the physical unit [time^{-1}]. Equation (9.14) is a differential equation for the concentration of organic matter in the solid phase $c_{org,s}$. Such an equation was already treated in chap. 4 (equation (4.6)). The equation reflects that under steady state conditions there is a balance between degradation of organic matter and sedimentation. It can be derived from the general

Sidebar 9.1: Velocity Profiles in Steady Compacted Aquatic Sediments
Sedimentary deposits of all kinds near the surface often show a characteristic decrease of the pore space in vertical direction. The process that causes the decrease is compaction. Due to rearrangement of sediment particles and to compression of particles, i.e. the porosity, the volumetric share of pore space, decreases with distance from the interface.

Here we take the Lagrangian approach, where the origin of the coordinate system is fixed at the sediment interface and is thus moving in space. In an Eulerian system, which is fixed in space, no steady state can be expected for compacting sediments. A constant burial velocity v_{bur} and a constant porosity φ_0 at the interface are assumed, a necessary prerequisite for a steady-state in the Lagrangian system.

The mass conservation equation for the fluid filling the pore space is given by the continuity equation

$$\frac{\partial}{\partial x}(\theta \rho v) = 0$$

(compare Chaps 3 and 4), which in case of constant fluid density ρ reduces to

$$\frac{\partial}{\partial x}(\theta v) = 0$$

The solution obviously is

$$\theta v = C$$

with integration constant C. The constant C can be evaluated in the deep sediments:

$$C = \theta_1 v_1$$

with v_1 denoting the fluid velocity in the deep sediments, where porosity does not change any more. v_1 can be expressed differently by taking into account that the burial velocity at all locations is given by

$$v_{bur} = (1 - \theta(x))v_{sed}(x) + \theta(x)v(x)$$

where v_{bur} denotes the burial velocity and v_{sed} the velocity of the sediment (solid) phase. The latter changes with depth. In the deep sediments, out of reach for compaction processes, holds

$$v(\infty) = v_{sed}(\infty) = v_1$$

(continued)

It follows:

$$v_1 = v_{bur};$$

and thus:

$$C = \theta_1 v_{bur}$$

The final result is

$$v = \frac{\theta_1}{\theta} v_{bur}$$

(see also: Holzbecher 2002).

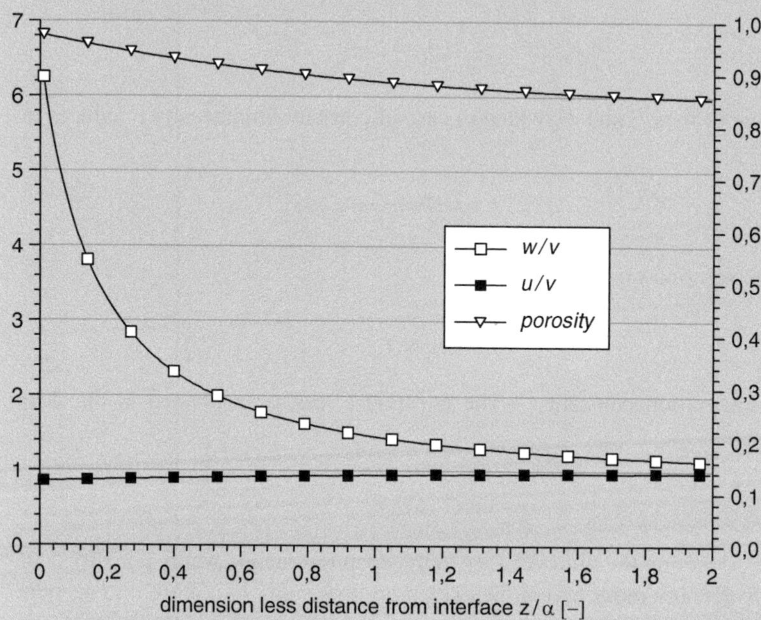

Figure pore velocity u, porous media velocity w, in relation to interface velocity v; in case of compaction; velocity ratios w/v, u/v are measured on the left axis, porosity on the right axis (modified from: Holzbecher 2002)

transport equation (**3.19**) by restriction to the one-dimensional steady state situation, when diffusion due to bioturbation, additional source- and sink-terms and inter-phase exchanges are assumed to be irrelevant. Furthermore, both remaining relevant terms are divided by bulk density ρ_b, which they have in common. For the spatial coordinate we use z here in order to distinguish the vertical direction.

Concerning sulphate, Berner (1964) takes into account the processes of diffusion, fluid flow and degradation. The resulting differential equation for the 1-dimensional steady state is:

$$D\frac{\partial^2}{\partial z^2}c_S - u\frac{\partial}{\partial z}c_S - \lambda_S c_{org,s} = 0 \tag{9.15}$$

where c_S denotes sulphate concentration. D is the relevant diffusion coefficient, u denotes fluid flow and λ_S the kinetic degradation coefficient. Equation (9.15) is a differential equation for sulphate concentration c_S. If there is no compaction and no connection to the ambient groundwater regime, one can set $v = u = w$. The general solution of the given system for both components is:

$$\begin{aligned} c_{org,s} &= c_{org,s0}\exp\left(-\frac{\lambda}{v}z\right) \\ c_S &= C_0 + C_1\exp\left(\frac{v}{D}z\right) - \frac{v^2 c_{org,s0}}{v^2 + D\lambda}\exp\left(-\frac{\lambda}{v}z\right) \end{aligned} \tag{9.16}$$

with integration constants C_0 and C_1. The particular solution given by Berner (1964) results for $C_1 = 0$:

$$\begin{aligned} c_{org,s} &= c_{org,s0}\exp\left(-\frac{\lambda}{v}z\right) \\ c_S &= (c_{S0} - c_{S\infty})\exp\left(-\frac{\lambda}{v}z\right) + c_{S\infty} \end{aligned} \tag{9.17}$$

where the subscript '0' denotes the concentration at the sediment-water interface, while the subscript '∞' represents the concentration in the deep sediment. Berner (1964) presented the solution (9.17) to describe sulphate reduction in maritime sediments and later used it to describe a part of the nitrogen cycle (Berner 1971). If the nitrogen cycle is concerned, c_S in the equations given above has to be replaced by the concentration of total ammonia.

As soon as the sulphate disappears, the solution of (9.16) produces negative concentrations. Boudreau and Westrich (1984) suggest the use of the Monod or Michaelis–Menton kinetics (see Chaps. 7 and 9 above) to describe sulphate reduction and organic carbon content. In the here used notation one obtains:

$$\begin{aligned} -w\frac{\partial}{\partial z}c_{org,s} - \frac{k c_{org,s} c_S}{K_S + c_S} &= 0 \\ \frac{\partial}{\partial z}D\frac{\partial}{\partial z}c_S - u\frac{\partial}{\partial z}c_S - \frac{f k c_{org,s} c_S}{K_S + c_S} &= 0 \end{aligned} \tag{9.18}$$

where k denotes the reaction parameter, which is a characteristic for the 'speed' of the reaction. K_S is a 'half reaction concentration', which means that for given $c_{org,s}$ the reaction rate takes half of its maximum value. f is a parameter relating the mass of organic matter to the sulphate mass. In f the stoichiometric relation in the sulphate reduction process has to be taken into account. The bulk density and porosity also play a role if organic matter in the solid phase reacts with sulphate in the fluid phase.

The system (9.18) of two ordinary differential equations has no analytical solution. Results can only be obtained by numerical methods. The system can be treated using the MATLAB® `bvp4c` function for the solution of boundary value problems (bvp). The function is designed for solving problems which can be formulated in the following notation (see MATLAB® help on 'bvp'):

$$\begin{aligned} y' &= f(x, y, p) \\ 0 &= bc(y(a), y(b), p) \end{aligned}$$

where the first line characterizes the differential equation telling that the first derivative y' of the unknown variable y is a function of y itself, of the independent variables x and of a parameter set p. As always in MATLAB®, y as well as p can be vectors. The second line characterizes the boundary conditions telling that these are specified at the boundaries $x = a$ and $x = b$.

The first differential equation in the system (9.18) is of first order and can easily be brought into the required form:

$$\frac{\partial}{\partial z} c_{org,s} = -\frac{1}{w} \frac{k c_{org,s} c_S}{K_S + c_S} \tag{9.19}$$

The second equation has a second derivative and can be the brought into the necessary form by a simple trick. Another variable is introduced: $c_S{}'$, which is the first derivative of the unknown function c_S. The second-order differential equation can thus be written as a system of two first order systems:

$$\begin{aligned} \frac{\partial}{\partial z} c_S &= c_S{}' \\ \frac{\partial}{\partial z} c_S{}' &= \frac{1}{D}\left(u \frac{\partial}{\partial z} c_S{}' + \frac{f k c_{org,s} c_S}{K_S + c_S}\right) \end{aligned} \tag{9.20}$$

Note that for the entire formulation of the problem there are three unknown functions: $c_{s,org}$, c_S and $c_S{}'$. The vector y in the MATLAB® notation has three components, which have to be taken into account in the formulation of the system of differential equations and of the boundary conditions. The following m-code excerpt shows how the MATLAB® routine for boundary value problems can be utilized to solve the system (9.19) and (9.20).

```
%--------------------execution-------------------------------

solinit = bvpinit (linspace(0,L,N),[cin; 0]);
sol = bvp4c(@bw,@bwbc,solinit,odeset,D,v,k,KS,f,cin);

%-------------------- graphical output -----------------------

plotyy (sol.x,sol.y(1,:),sol.x,sol.y(2,:));
legend ('Corg','SO_4'); grid;

%--------------------functions-------------------------------
function dydx = bw(x,y,D,v,k,KS,f,cin)

monod = k*y(2)/(KS+y(2));
dydx = zeros (3,1);
dydx(1) =  -monod*y(1)/v;
dydx(2) =  y(3);
dydx(3) =  (v*y(3)+f*monod*y(1))/D;

function res = bwbc (ya,yb,D,v,k,KS,f,cin)
res = [ya(1:2)-cin; yb(3)];
```

The complete code is included in the accompanying software under the name *'boudreau_westrich.m'*.

The initial commands, in which the parameters `L, v, D, k, KS, f, cin` and `N` are specified, are omitted in the listing above. Before the call of the solution routine in the second line, an initial guess of the unknown functions is required. This is done by the `bvpinit` command. The first formal parameter in the command is the vector of x-values, for which the variable-values are to be computed. The second formal parameter consists of three values (`cin` is a two-component column vector). Each of these three is a guess of a constant valued function.

In the `bvp4c` call there is a list of formal parameters. First in the list are the function calls: (1) the function in which the system of differential equations is specified, (2) the function in which the boundary conditions are specified. `solinit` is the initial guess obtained from the previous command. `odeset` yields the standard options for the solution routine. What follows as formal parameters is the set of parameters for the described example.

In the `bw` function the differential equations are specified, following (9.19) and (9.20). `y` denotes the vector of unknown variables. Thus `y(1)` denotes the concentration of organic matter, `y(2)` the sulphate concentration and `y(3)` the derivative of sulphate concentration. The Monod-term for sulphate is calculated under `monod`.

The `bwbc` function specifies the boundary conditions. At the left side, i.e. at the lowest x-value (a) given above, the condition is computed using the `ya` variable. The variable `yb` is responsible for the condition at the highest x-value (b) . The condition is formulated in a way making the residual `res` vanish. Thus the first two variables at boundary $x = a$ take the values given in the `cin` vector, while the third variable has a vanishing value at boundary $x = b$. The physical meaning of the latter condition is that there is no diffusive flux because of a vanishing concentration gradient.

As output, concentrations of organic carbon and sulphate are plotted in one figure with two *y*-axes, one on the left and one on the right side. MATLAB® provides the `plotyy` command for that task.

The problem of negative concentrations, which was recognized in Berner's approach according to (9.14) and (9.15), is overcome in the presented model. In fact, the sink term in the sulphate equation is responsible for that improvement. Some other approaches concerning the sink term of organic matter can be found in the literature. In their model on oxygen penetration depths and fluxes, Cai and Sayles (1996) use a simpler linear degradation term for c_{org} and the corresponding analytical solution of exponential decline. The argument for such a simplified approach is that degradation of organic matter occurs anyway. If the aerobic pathway with oxygen as an elector donor is not sufficient, other anaerobic pathways for decomposition usually take over. The following sub-chapter shows approaches how such a redox sequence can be treated in detail.

9.4 Redox Sequences

Redox conditions play an important role in environmental systems, as they are a determining factor for population growth or decline of bacteria and microbes. Depending on the local redox state, conditions may favour or disfavour the existence of certain microbial cultures which are able to degrade hazardous or otherwise harmful organic substances.

Redox zones are usually observed in aquatic sediments at the bottom of surface water bodies, in aquifers with infiltrating river water, and downstream from contaminated sites or landfills.

There are six major chemical pathways being responsible for the degradation of organic matter in environmental systems. These are oxic respiration, denitrification, manganese oxide and iron (hydr)oxide reduction, sulphate reduction and methanogenesis. The transfer of electrons plays a crucial role in all six redox reactions. In each reaction one specific substance acts as electron donor: oxygen in oxic respiration for example. All half-reactions, including the donors, are listed in Table 9.1. Each half reaction is completed by another half-reaction in which an electron acceptor consumes the electrons. If degradation processes are concerned, organic matter is the electron acceptor. As a matter of fact, microorganisms do the work in most redox reactions. A detailed approach including bacteria populations

Table 9.1 Primary redox (half) reactions, according to Hunter et al. (1998)

1. Oxic respiration	$O_2 + 4H^+ + 4e^- \rightarrow 2H_2O$
2. Denitrification	$NO_3^- + 6H^+ + 5e^- \rightarrow 0.5N_2 + 3H_2O$
3. Manganese oxide reduction	$MnO_2 + 4H^+ + 2e^- \rightarrow Mn^{2+} + 2H_2O$
4. Iron (hydr)oxide reduction	$Fe(OH)_3 + 3H^+ + e^- \rightarrow Fe^{2+} + 3H_2O$
5. Sufate reduction	$SO_4^{2-} + 9H^+ + 8e^- \rightarrow HS^- + 4H_2O$
6. Methanogenesis	$CO_2 + 8H^+ + 8e^- \rightarrow CH_4 + 2H_2O$

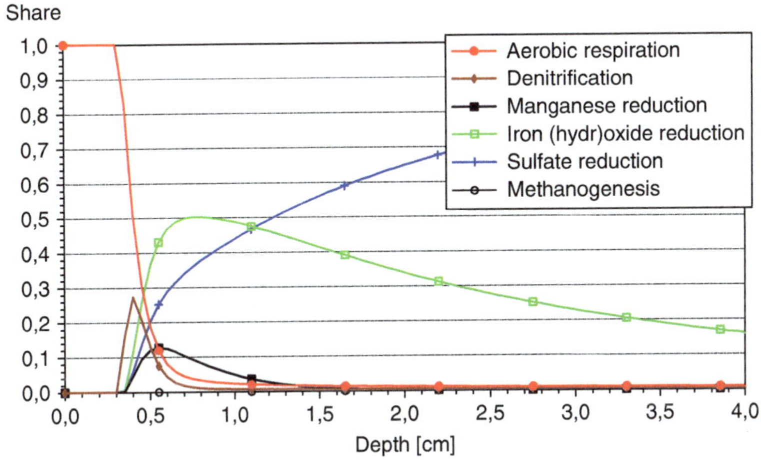

Fig. 9.5 Redox-sequence in aquatic sediments, see van Cappellen and Wang (1995, 1996); recalculated by the author

requires some knowledge of their consumption and reproduction behavior, which is seldom available.

It is an important characteristic of redox reactions that most environments show a distinctive preference for a special redox reaction, depending on the biogeochemical conditions. The numbering in Table 9.1 provides the preference priority. As long as oxygen is abundantly available in a system, all other redox reactions play a minor role. If there is no or only a small amount of oxygen present, nitrate takes the leading role. If there is also no nitrate, manganese and iron oxide reduction become important and so forth.

The example in Fig. 9.5 results from measurement and modeling studies in aquatic sediments, published by van Cappellen and Wang (1995, 1996). The relevant redox parameters change within the first centimeters below the sediment water interface ('depth' in the figure). The figure depicts the share of specific redox processes on the total reaction dynamics as a function of depth below the water sediment interface. In the upper half centimeter, in the aerobic zone where oxygen is present, aerobic respiration dominates over all other processes. Below follows a zone where the share of reactions consuming oxygen becomes less important, increasingly admitting all other redox interactions to take place. Denitrification is the first competing process but is relevant only in a narrow zone of a few millimeters length and always remaining with a share of less than 30%. Iron and manganese reduction become relevant at the same depth, but, due to the higher abundance of iron (hydr)oxides, the former redox process remains more relevant within the rest of the depth interval. Sulphate reduction is the only process that gains relevance with increasing depth. Obviously, in this set-up sulphur is available at such high concentrations that methane formation is suppressed everywhere.

In other environmental compartments the extensions of redox zones may have a different length scale than in the given example. Holzbecher et al. (2001) present a study on river water infiltrating an aquifer where redox zones are in the range of 10–100 m. Further studies on bank filtration were published by Doussan et al. (1997). A similar range is often met downstream of contaminated sites and landfills (Keating and Bahr 1998; Lu et al. 1999).

The approach, presented above for sulphate reduction as one of the dominant redox processes, can be extended to treat such a sequence of redox reactions. This is achieved by the introduction of so-called inhibition terms. If the abundance of species with concentration c inhibits some other kinetics, the formula for the latter should contain the inhibition factor $\frac{k}{k+c}$ with a suitable value for the inhibition parameter k. Clearly, the factor is small for high values of c, and approaches one for low values.

The idea is demonstrated on a rudimentary redox model containing the first three redox reactions of Table 9.1.

The command listing for the file, included in the accompanying software under '*redoxsteady.m*', is as follows:

```
L = 100;                     % length [m]
v = 1;                       % velocity [m/s]
D = 0.02;                    % diffusivity [m*m/s]
lambda = 0.01;               % organic carbon degradation parameter [1/m]
k1 = [0.1; 1; 0.2];          % 1. Monod parameter [m/s]
k2 = [0.035; 1];             % 2. Monod parameter [kg/m*m*m]
k3 = [3.5e-3; 1];            % inhibition coefficient [kg/m*m*m]
corg = 1;                    % organic carbon concentration at interface
cin = [4; 3; 0.001];         % interface concentrations [kg/m*m*m]
N = 100;                     % number of nodes

%----------------------execution-----------------------------------

x = linspace(0,L,N);
solinit = bvpinit (x,[cin; zeros(3,1)]);
sol = bvp4c(@redox,@bcs,solinit,odeset,D,v,lambda,k1,k2,k3,corg,cin);

%---------------------- graphical output --------------------------
plot (x,corg*exp(-lambda*x),sol.x,sol.y(1:3,:));
legend ('C_{org}','O_2','NO_2','Mn'); grid;

%----------------------functions-----------------------------------
function dydx = redox(x,y,D,v,lambda,k1,k2,k3,corg,cin)

c0 = corg*exp(-lambda*x);
monod = k1.*y(1:2)./(k2+y(1:2));
monod(3) = k1(3);
inhib = k3./(k3+y(1:2));
dydx = zeros (6,1);
dydx(1) =  y(4);
dydx(4) =  (v*y(4)+c0*monod(1))/D;
dydx(2) =  y(5);
dydx(5) =  (v*y(5)+c0*monod(2)*inhib(1))/D;
dydx(3) =  y(6);
dydx(6) =  (v*y(6)-c0*monod(3)*inhib(1)*inhib(2))/D;

function res = bcs (ya,yb,D,v,lambda,k1,k2,k3,corg,cin)
res = [ya(1:3)-cin; yb(4:6)-zeros(3,1)];
```

As the main part of the M-file and the function calls coincide with the previous example, the explanation of the code is restricted to the function section. Note that in this example there are six unknown functions, which are the concentrations of the three redox species that are included in the model with their derivatives. All functions are included in the vector `y`. `y(1)` is the oxygen concentration, `y(4)` its derivative, `y(2)` is the nitrate concentration, `y(5)` its derivative, `y(3)` is the manganese concentration and `y(5)` its derivative.

In the variable `c0` the analytical solution for organic matter is implemented. The `lambda` value used here coincides with the ratio λ/w of formula (9.14). In the `monod` vector three Monod terms are calculated and in the `inhib` vector the two inhibition factors for the two redox processes with higher preference.

The derivative specifications for the first three components of `y` are the defining formulae for the derivatives of the concentrations. The other three specifications represent the differential equations in the following form:

$$\begin{aligned}
\frac{\partial}{\partial z}{c_{O_2}}' &= \frac{1}{D}\left(u\frac{\partial}{\partial z}{c_{O_2}}' + c_{org,s}\frac{k_{O_2}c_{O_2}}{K_{O_2}+c_{O_2}}\right)\\
\frac{\partial}{\partial z}{c_{NO_2}}' &= \frac{1}{D}\left(u\frac{\partial}{\partial z}{c_{NO_2}}' + c_{org,s}\frac{\bar{k}_{O_2}}{\bar{k}_{O_2}+c_{O_2}}\frac{k_{NO_2}c_{NO_2}}{K_{NO_2}+c_{NO_2}}\right)\\
\frac{\partial}{\partial z}{c_{Mn}}' &= \frac{1}{D}\left(u\frac{\partial}{\partial z}{c_{Mn}}' - c_{org,s}\frac{\bar{k}_{O_2}}{\bar{k}_{O_2}+c_{O_2}}\frac{\bar{k}_{NO_2}}{\bar{k}_{NO_2}+c_{NO_2}}\frac{k_{Mn}c_{MnO_2}}{K_{Mn}+c_{MnO_2}}\right)
\end{aligned} \qquad (9.21)$$

The degradation of oxygen is given by the product of organic matter concentration with the Monod term: `c0*monod(1)`. The coefficients k_{O_2}, k_{NO_2} and k_{Mn} are stored in the `k1` vector, the parameters K_{O_2}, K_{NO_2} and K_{Mn} in the `k2` vector and the inhibition parameters $\bar{k}_{O_2}$and $\bar{k}_{NO_2}$ in the `k3` vector. The degradation term for nitrate in addition to the Monod term includes the inhibition term for oxygen `inhib(1)`. The corresponding term for manganese has another additional term: `inhib(2)`, the inhibition term for nitrate. Note that the sign of the reaction term in the equation for manganese is opposite to that of the other two substances, as free manganese ions are produced in the redox reaction, while oxygen and nitrate are consumed. That is the reason why the Monod-term for the manganese redox reaction is different. Equations (9.21) also contain the concentration of the reaction product MnO_2. In order to keep the number of parameters small, it is assumed that the manganese pool in the porous matrix remains at a constant high level for this model. For $c_{MnO_2} >> K_{Mn}$ the Monod term can be approximated by k_{Mn} (Fig. 9.6).

Biochemical redox reactions, which are strongly coupled with the degradation of organic matter, are taken into account by the formulation:

$$q_i = -\lambda_{org}c_{org}f_i \qquad (9.22)$$

with concentration of organic matter c_{org} and degradation constant λ_{org}. The factor f_i is a measure for the relative share of the ith redox process on the total degradation

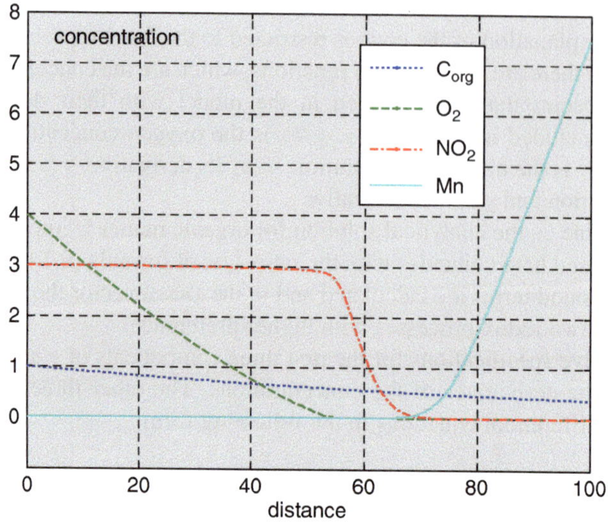

Fig. 9.6 Result of MATLAB® example for the redox sequence example

of organic matter. Following van Capellen and Wang (1995), the factor f_i takes a form that is similar to Monod kinetics:

$$f_i = \left(1 - \sum_{j=1}^{i-1} f_j\right) \frac{c_i}{c_i - c_{\lim,i}} \tag{9.23}$$

where a limit concentration $c_{\lim,i}$ appears, which is specific to components. The term in the brackets ensures that the sum of all contributions does not exceed 1. All six major redox processes are thus implemented in the model: organic respiration, denitrification, manganese-, iron- and sulphate reduction and methanogenesis. More details of the model are given by Holzbecher et al. (2001). The graphical output of the program is depicted in Fig. 9.6.

References

Berner RA (1964) An idealized model of dissolved sulphate distribution in recent sediments. Geochim Cosmochim Acta 28:1197–1503

Berner RA (1971) Principles of chemical sedimentology. McGraw-Hill, New York, p 240

Boudreau BP, Westrich JT (1984) The dependence of bacterial sulphate reduction on sulphate concentration in marine sediments. Geochim Cosmochim Acta 48:2503–2516

Cai W-J, Sayles FL (1996) Oxygen penetration depths and fluxes in marine sediments. Marine Geochemistry 52:123–131

Deaton ML, Winebrake JI (2000) Dynamic modeling of environmental systems. Springer, New York, p 194

Doussan C, Poitevin G, Ledoux E, Detay M (1997) River bank filtration: modeling of the changes in water chemistry with emphasis on nitrogen species. J Contaminant Hydrol 25:120–156
Fife PC (1979) Mathematical aspects of reacting and diffusing systems, lecture notes in biomathematics. Springer, Berlin/Heidelberg/New York, p 185
Holzbecher E (2002) Advective flow in sediments under the influence of compaction. Hydrol Sci J 47(4):641–649
Holzbecher E, Maßmann G, Horner C, Pekdeger A et al (2001) Redox-processes in the oderbruch aquifer. In: Gehrels H (ed) Impact of human activity on groundwater dynamics, vol 269. IAHS-Publ, Wallingford, pp 229–238
Hunter KS, Wang Y, van Cappellen P (1998) Kinetic modeling of microbially-driven redox chemistry of subsurface environments: coupling transport, microbial metabolism and geochemistry. J Hydrol 209:53–80
Keating EH, Bahr J (1998) Reactive transport modeling of redox geochemistry: approaches to chemical disequilibrium and reaction rate estimation at a site in northern Wisconsin. Water Res Res 34(12):3573–3584
Lu G, Clement TP, Zheng C, Wiedemeier TH (1999) Natural attenuation of BTEX compounds: model development and filed-scale application. Groundwater 37(5):707–717
Morel F, Hering J (1993) Principles and applications of aquatic chemistry. Wiley, New York, p 588
Shampine LF, Gladwell I, Thompson S (2003) Solving ODEs with MATLAB. Cambridge University Press, Cambridge, p 263
Streeter HW, Phelps EB (1925) A study of the pollution and natural purification of the Ohio river. US Public Health Service, Washington DC, Bulletin 146
van Cappellen P, Wang Y (1995) Metal cycling in surface sediments: modeling the interplay of transport and reaction. In: Allen HE (ed) Metal contaminated aquatic sediments. Ann Arbor Press, Chelsea, pp 21–64
Van Cappellen P, Wang Y (1996) Cycling of iron and manganese in surface sediments: a general theory for the coupled transport and reaction of carbon, oxygen, nitrogen, sulfur, iron, and manganese. Am J Sci 296:197–243
Vanrolleghem P, Borchardt D, Henze M, Rauch W, Reichert P, Shanahan P, Somlyódy L (2000) River water quality model no. 1 (RWQM1): III. Biochemical submodel selection, 1st World Congress of the International Water Association (IWA), Paris

Chapter 10
Parameter Estimation

10.1 Introduction

In most application fields one frequently finds models that are applied with a different goal than described in the previous chapters. The purpose of modeling was defined as prediction in a general sense. Models show how an environmental system develops, starting from an initial state, restricted by some boundary conditions under the assumption that some parameters are well defined and well known. That usual procedure of simulation was demonstrated above.

Mostly, before being able to start any prediction run, parameters turn out to be the problem: values for the parameters need to be known. There are various ways to obtain parameter values. They can be taken from literature. There are well known constants: the acceleration due to gravity (=9.807 m^2/s), for example, can be treated as a constant in environmental problems. Values can be taken from tables: thermodynamic data are found in steam tables, for example; and reaction constants can be found in concerned data-bases. Some parameter values are reported in text-books, reports and in journal publications. Under certain conditions, parameter values can be obtained from experiments, i.e. from a controlled environment which is similar to field situation. Some parameters can be measured on-site directly, like spatial extensions, time, temperature etc.

After utilizing all these possibilities, there may be still some parameters left. These need to be determined by parameter estimation. Parameter estimation can be performed using the model in question (then we speak of calibration) or for an especially designed experimental set-up. When observations of one or more variables are available, the model can be run with different parameter values in order to check, which parameter value fits best to the observations. Following such a procedure, the original role of parameters and variables is exchanged: now the parameters are unknown, while variables are known. For that reason parameter fitting is also named *inverse modeling*.

Example: Microcystins (MCYST) are a group of toxic substances produced by cyanobacteria ('blue-green-algae'). In case of cyanobacterial blooms microcystin

E. Holzbecher, *Environmental Modeling*,
DOI 10.1007/978-3-642-22042-5_10,

concentrations in surface waters may reach values far above the value proposed as provisional guideline for drinking water by the WHO of 1 μg/l for MCYST-LR. When a well is installed in the vicinity of a river or lake shore, part of the pumped water originates from surface water, which is called bank filtrate. When bank filtrate is utilized for drinking water, it has to be ensured that concentrations in pumped water are below the threshold. For that reason, it is important to understand the sorption and degradation processes during the sub-surface passage of the bank filtrate water. Batch experiments using surface water and characteristic porous materials are a first approach for such an examination.

In a batch experiment microcystin, dissolved in water, was brought into contact with porous sediments. The original concentration was 1 μg/l. In aqueous sediments microcystin is subject to sorption and biodegradation. In order to obtain retardation factors and degradation rates batch experiments can be run. The decrease of the concentration was measured at several time instants after the first contact. The following table and figure shows example concentrations for five instants of time (Grützmacher 2006) (Table 10.1) (Fig. 10.1):

In MATLAB® the values are entered as two line vectors:

```
tfit = [0.25 1 2 4 8];
cfit = [0.7716 0.5791 0.4002 0.1860 0.1019];
```

Table 10.1 Example data-set from microcystin batch experiment

Time (h)	0.25	1	2	4	8
Concentration (μg/l)	0.9405	0.5537	0.3994	0.1509	0.0592

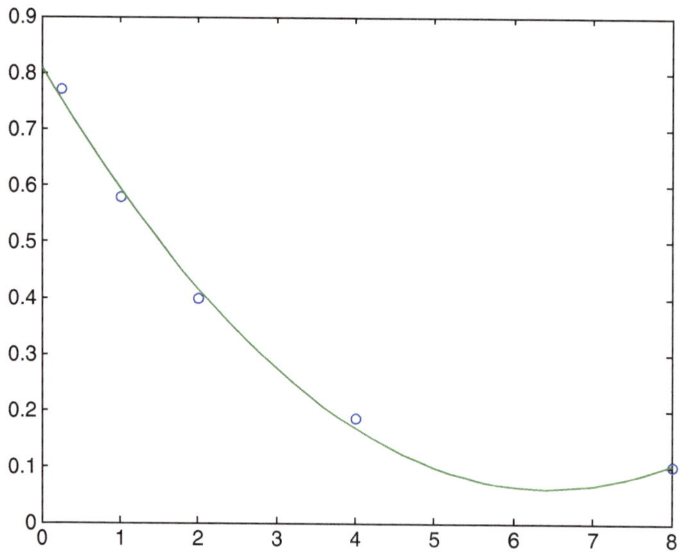

Fig. 10.1 Quadratic curve fitting for example batch experiment

10.2 Polynomial Curve Fitting

In order to outline the basic concepts of inverse modeling, we first focus on the simple situation, in which the dependent variable obeys an analytical formula with respect to a variable time t. Imagine the dependent variable being the concentration of a pollutant. The reader may think of t as time. For a first approach it is even assumed that this formula is a polynomial. The coefficients of the polynomial, which are connected to the parameters, are unknown and have to be determined by the inverse modeling procedure.

The best fit is that polynomial (represented by a set of coefficients that are the parameters), for which the deviation between given values and modelled values is minimal. For such polynomial curve fitting MATLAB® has the `polyfit` command. Use the following command to find the best fit for the data-set given above under the assumption that the formula is quadratic:

```
p = polyfit (tfit,cfit,2)
p =
    0.0181    -0.2326     0.8099
```

The last formal parameter in the brackets on the right side specifies the degree of the polynomial, i.e. the exponent of the highest power term. It needs to be above one and lower or equal to the number of given measurements. MATLAB® returns the coefficients of the polynomial in an array, here `p`. The answer shown above corresponds to the polynomial:

$$c(t) = 0.0181t^2 - 0.2326t + 0.8099 \tag{10.1}$$

Polynomials are evaluated by the `polyval` command. Use the following command in order to compute the values of the polynomial, given by (10.1), and just obtained by parameter fitting:

```
c = polyval(p,[0:.2:8])
c =
  Columns 1 through 7
    0.8099    0.7641    0.7198    0.6769    0.6354    0.5954    0.5568
  Columns 8 through 14
    0.5197    0.4840    0.4498    0.4170    0.3857    0.3557    0.3273
  Columns 15 through 21
    0.3003    0.2747    0.2506    0.2279    0.2067    0.1869    0.1685
  Columns 22 through 28
    0.1516    0.1362    0.1222    0.1096    0.0985    0.0888    0.0806
  Columns 29 through 35
    0.0738    0.0684    0.0645    0.0621    0.0611    0.0615    0.0634
  Columns 36 through 41
    0.0667    0.0715    0.0777    0.0853    0.0944    0.1050
```

The first formal parameter of `polyval` is a line vector and is interpreted as the polynomial whose coefficients are the elements of the vector. The second formal parameter is the line vector of instants at which the polynomial is evaluated. The output is a corresponding vector with values of the polynomial. The `plot` command yields a visual comparison of the original values and the fitted curve:

```
plot(tfit,cfit,'o',[0:.2:8],c,'-');
```

Note that the fitting procedure is based on a quantitative measure for the quality of an approximation. Such different evaluations are based on the *residual vector*, showing the difference between given and modelled values for all measurements:

```
c = polyval(p,tfit);
resc = cfit-c
resc =
        0.0187   -0.0163   -0.0168    0.0175   -0.0031
```

There is no unique measure for the quality of a fit. One can use for example the mean absolute error $\frac{1}{N}\sum\left|c(t_{fit})-c_{fit}\right|$, the mean quadratic error $\frac{1}{N}\sqrt{\sum\left(c(t_{fit})-c_{fit}\right)^2}$, the maximal absolute error $\max\left\{\left|c(t_{fit})-c_{fit}\right|\right\}$ or the maximal quadratic error $\max\left\{\left(c(t_{fit})-c_{fit}\right)^2\right\}$. There are even other measures possible.

It is most common to check the quadratic error. The mean quadratic error for the given approximation is obtained by the command:

```
sqrt(sum (resc.*resc))/5
ans =
    0.0070
```

One may also use the square root of the sum of the squares of the residuals

```
normc = sqrt(sum(resc.*resc))
normc =
    0.0348
```

This is the so called norm of the residuals which can also be obtained by the commands:

```
normc = norm(resc)
```

or

```
normc = norm(resc,2)
```

For more details concerning the `norm` command, which can be used for alternative quality measurements, see the MATLAB® help. For every polynomial other than the best fit, the deviation between measured and calculated values, quantified in the residual norm, will be higher. Lets make one check:

```
c = polyval([0.0015   -0.1078   2.0315],tfit);
normc = norm (cfit-c)
normc =
    2.9623
```

The residual norm for the chosen quadratic polynomial is far above the residual of the best fit .

Alternatively, curve fitting can be performed from the MATLAB® figure editor. Use the plot command first in order to perform the fitting:

```
plot (tfit,cfit,'o')
```

Then click the 'Basic Fitting' submenu in the 'Tools' menu of the figure editor to obtain the box depicted in Fig. 10.2.

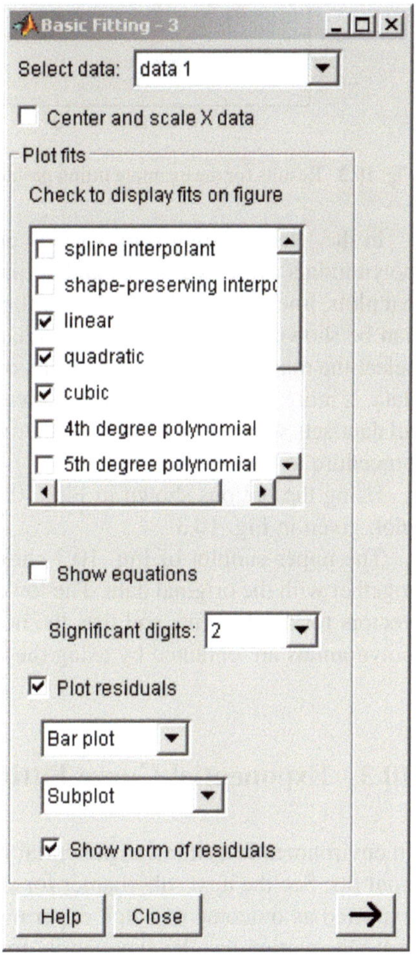

Fig. 10.2 The 'Basic fitting' command box

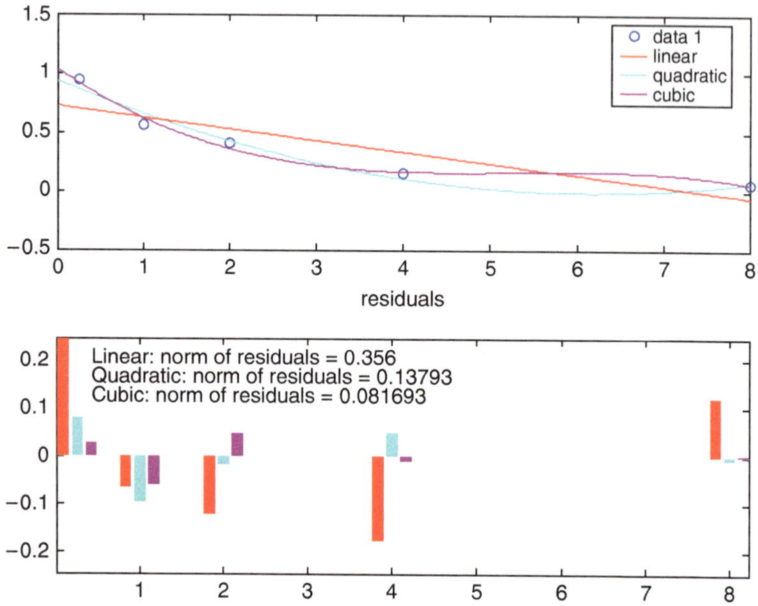

Fig. 10.3 Results for the example fitting problem, using the 'Basic fitting' tool under MATLAB®

In the 'Basic Fitting' box various options are available: polynomial fitting for polynomials of degrees 1–10 and spline interpolation. Residuals can be shown as bar plots, line plots or scatter plots, either in a subplot or a separate figure. Equations can be shown in the curve plot, residual norms in the residual plot. The user may select the number of significant digits of the fitting and may center and scale x-axis data. If more than one data-set is depicted in the figure, fitting can be performed for all data-sets separately. Moreover, there is the option to save the results of the fitting procedure to the workspace.

Using the options shown in Fig. 10.2 for the example data-set, one obtains the plot, given in Fig. 10.3.

The upper subplot of Fig. 10.3 shows the linear, quadratic and cubic best fits together with the original data. The lower subplot depicts histograms of the residual vectors for all three fits and lists the norm of the residuals. The coefficients of the polynomials are obtained by using the → button in the 'Basic Fitting' box.

10.3 Exponential Curve Fitting

In environmental systems exponential fits are often more appropriate than polynomial fits. See the next sub-chapter for an argument, why exponential curves can be expected as outcome of batch experiments. Generally, there may be some reason that the solution has the exponential form:

$$c(t) = c_0 \exp(-\lambda t) \tag{10.2}$$

There are two parameters: the initial concentration c_0 and the decay constant λ. Note that the fitting procedure can be performed for any dependent variable c and the independent variable t; both symbols do not necessarily represent concentration and time in this chapter.

As the `polyfit` command works with polynomials, one may use the idea that the exponential curve is a linear curve in logarithmic representation. Thus the `polyfit` command can be used for the logarithm of the concentration vector:

```
p = polyfit(tfit,log(cfit),1)
```

```
p =
   -0.2619   -0.3387
```

The best fit for the logarithm is thus given by:

$$\log(c) = -0.2619t - 0.3387 \tag{10.3}$$

or:

$$\begin{aligned} c(t) &= \exp(-0.3387)\exp(-0.2619t) \\ &= 0.7127\exp(-0.2619t) \end{aligned} \tag{10.4}$$

Comparison with formula (10.2) shows that the second line of (10.4) is the aimed formulation with $c_0 = 0.8236$ and $\lambda = 0.3487$. In Fig. 10.4 the result is again plotted together with the original data:

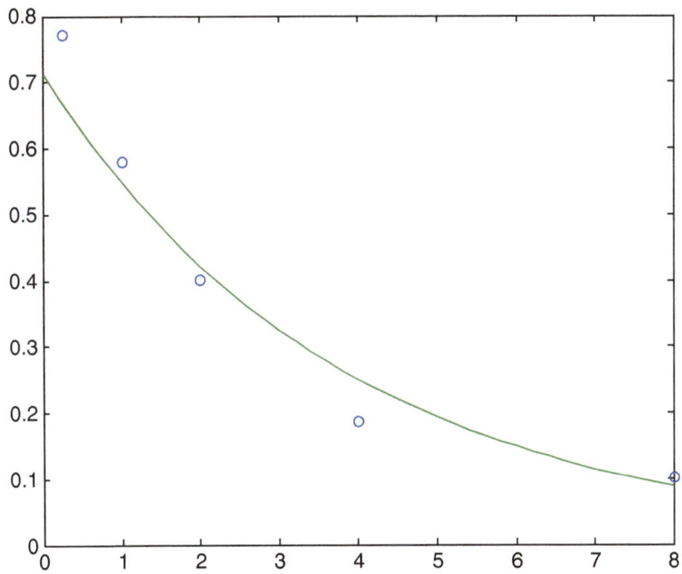

Fig. 10.4 Exponential fit for example batch experiment

```
c = exp(polyval(p,[0:.2:8]));
plot(tfit,cfit,'o',[0:.2:8],c,'-');
```

Lets check again the quadratic difference between observed and modelled data:

```
c = polyval(p,tfit);
normc = norm (cfit-c)

normc =
    3.6427
```

and check:

```
c = exp(-tfit);
normc = norm (cfit-c)

normc =
    0.3915
```

Obviously the chosen function

$$c(t) = \exp(-t) \tag{10.5}$$

is a much better approximation for the example data set, showing that the coefficients obtained before for the exponential fit do not represent the best fit. The explanation for that apparent contradiction is nearby: the linear curve is found under the condition that the sum of squares of the quadratic logarithmic deviations $\sum \left(\log(c_{fit}) - \log(c(t_{fit}))\right)^2$is minimized and not the sum of the quadratic deviations $\sum \left(c_{fit} - c(t_{fit})\right)^2$. In the next sub-chapter, we present a better procedure for the determination of the optimal exponential curve.

10.4 Parameter Estimation with Derivatives

MATLAB® offers the possibility to find the best exponential fit by another procedure, which is demonstrated in the following. The method can be extended for parameter fitting for arbitrary curves, i.e. it is not restricted to polynomial or exponential curve fitting.

Remember that it is the goal to minimize the norm of the residual vector:

$$\|res\| = \sqrt{\sum \left(c(t_{fit}) - c_{fit}\right)^2} \tag{10.6}$$

The square root operation within the norm formula does not change the result of the task and can be omitted. Thus, the task can also be formulated without the square-root: the objective is to minimize the term e understood as a function of the parameter λ:

$$e(\lambda) = \sum \left(c(t_{fit}, \lambda) - c_{fit}\right)^2 \tag{10.7}$$

Note that c also is conceived as a function of the parameter λ. A necessary condition for the minimum value of λ is that the derivative of e with respect to λ becomes zero:

$$\frac{\partial e}{\partial \lambda} = 2\sum \left(c(t_{fit}, \lambda) - c_{fit}\right)\frac{\partial c}{\partial \lambda}(t_{fit}, \lambda) = 0 \tag{10.8}$$

Obviously the leading two of condition (10.8) can be omitted. The last factor can be obtained from (10.2),

$$\frac{\partial c}{\partial \lambda} = -tc_0 \exp(-\lambda t) = -t \cdot c \tag{10.9}$$

leading to the following formulation of the condition:

$$\sum \left(c(t_{fit}, \lambda) - c_{fit}\right)c(t_{fit}, \lambda)t_{fit} = 0 \tag{10.10}$$

Using the vector notation, the last formula can also be written as:

$$\left(\mathbf{c}(\mathbf{t}_{fit}, \lambda) - \mathbf{c}_{fit}\right)\left(\mathbf{c}(\mathbf{t}_{fit}, \lambda) \bullet \mathbf{t}_{fit}\right)^T = 0 \tag{10.11}$$

One can define the right hand side as a function, and the conditions (10.10) or (10.11) are fulfilled for the zero of that function. In order to find the zero of a function, MATLAB® provides the `fzero` command.

`fzero` starts a MATLAB® algorithm for the computation of the zeros of a function $f(x)$, i.e. to find a value x_0 with $f(x_0) = 0$. If `fzero` is called, at least two parameters have to be given by the user. One concerns the function f, the other is a starting value for x_0:

```
fzero(@f,x0);
```

The function `f` can be any function which MATLAB® knows. For example, one obtains the well-known zero of the cosine-function by the command:

```
fzero(@cos,0.11)
ans =
    1.5708
```

As expected, the zero is $\pi/2$. The input of another start value may yield another zero of the cosinus-function, as demonstrated by the following command:

```
fzero(@cos,4.11)

ans =
    4.7124
```

Of course, it is possible to determine the zeros of user specified functions. In the following, the example name `myfun` is used as name for a user-specified function. The function with name `myfun` needs to be specified in an M-file. Here we write the function *f* together with the `fzero` command in the same M-file:

```
function parest0
x0 = fzero(@myfun,1.)

function f = myfun(x);
f = sin(x) + cos(x)*cos(x);
```

The command sequence is stored in a function M-file. It is not possible to use the same sequence in a script-file. When running, the M-file produces the result of $x_0 = -0.6662$ in the MATLAB® command window, which is a zero of the function $\sin(x) + \cos^2(x)$.

Parameter estimation is performed in the same manner. The function `f` for the exponential fit needs to be calculated according to formula (10.11):

```
function par_est
% parameter estimation with derivatives
% for exponential fit for lambda
lambda = fzero(@myfun,0.05)

function f = myfun(lambda);
tfit = [.25 1 2 4 8];
cfit = [0.7716 0.5791 0.4002 0.1860 0.1019];
c = exp(-lambda*tfit);         % equation for c
f = (cfit-c)*(c.*tfit)';       % specify function f to vanish
```

In the fourth line of the M-file, the `fzero`-function is called for the function `myfun` and the starting value of 0.11 for λ. In the first line of the function, the given values are specified as two line vectors: `tfit` for the values of the independent variable and `cfit` for the values of the dependent variable. These two vectors may represent measured concentration values for a certain substance at different time instants. The next line computes the values of the function at the specified time instants for a given value of λ. The last line corresponds to formula (10.10). The term $c(t_{fit})t_{fit}$ is evaluated element-wise and then transformed into a column vector (using the ’-operator!). The usual vector-multiplication of the line vector $c(t_{fit}) - c_{fit}$ with the column vector $\left(c(t_{fit})t_{fit}\right)^T$ yields the summation required according to the formula.

Running the M-file delivers the result: λ=0.3329. The following M-file *‘par_est.m’* is slightly extended to perform further post-processing tasks:

```
function par_est
% parameter estimation with derivatives
% for exponential fit for lambda
global tfit cfit c0

% specify fitting data
tfit = [0.25 1 2 4 8];
cfit = [0.7716 0.5791 0.4002 0.1860 0.1019];
c0 = 0.8; lambda0 = 0.5;

lambda = fzero(@myfun,lambda0);
normc = norm(cfit-c0*exp(-lambda*tfit));
display (['Best fit for lambda = ' num2str(lambda)]);
display (['Norm of residuals =' num2str(normc)]);
tmax = tfit(size(tfit,2));
t = [0:0.01*tmax:tmax];
figure; plot (tfit,cfit,'or',t,c0*exp(-lambda*t),'-');
legend ('given','modelled');
text(0.5*tmax,c0*0.7,['\lambda: ' num2str(lambda)]);
text(0.5*tmax,c0*0.8,['norm of residuals: ' num2str(normc)]);

function f = myfun(lambda);
global tfit cfit c0
c = c0*exp(-lambda*tfit); %solve linear decay eq. for c with c(0)=c0
f = (cfit-c)*(c.*tfit)';  % specify function f to vanish
```

The corresponding M-file *'parest.m'* is included in the accompanying software.

The vectors of fitting data `tfit` and `cfit` as well as the initial concentration `c0` are now specified in the input section of the M-file. Also the initial value for λ (in the M-file called `lambda0`) is specified at the beginning. `normc`, computed directly after the calculation of λ, represents the norm of the residuals. The next two lines initiate output of `lambda` and `normc` in the MATLAB® command window. The last four commands in the main module produce a plot showing given values and best fit curve, the best fit value and the norm of residuals. The following figure is the plot obtained for the microcystins example data introduced at the beginning of this chapter (Fig. 10.5).

The optimum λ is 0.3329 and the norm of residuals is 0.0646. Obviously, the function with the new value of λ is a much better fit than the exponential fit of the previous sub-chapter that was obtained by polynomial curve fitting.

Using the calculated λ one may have the idea to improve the fit further by changing c_0 in formula (10.2). For this purpose, the *'par_est.m'* file has to be changed slightly in order to optimize for c_0 and not for λ. Instead of condition (10.8) one obtains:

$$\frac{\partial e}{\partial c_0} = 2\sum \left(c(t_{fit}) - c_{fit}\right)\frac{\partial c}{\partial c_0}(t_{fit}) = 0 \tag{10.12}$$

The factor 2 can be omitted. Evaluation of the last factor delivers the condition:

$$\sum \left(c(t_{fit}) - c_{fit}\right)\exp(-\lambda t_{fit}) = 0 \tag{10.13}$$

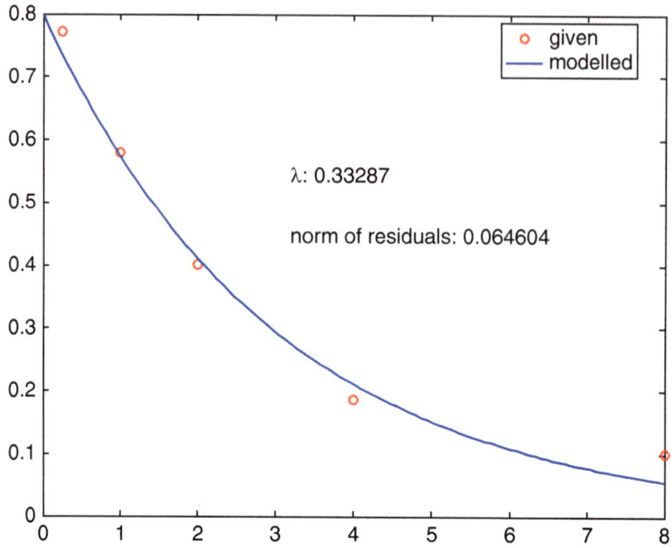

Fig. 10.5 Exponential fit for parameter λ using the demonstration data set

In the following M-file the zero of that function is determined. The sum of (10.13) is evaluated in the last line. In all other parts the M-file resembles the *'par_est.m'* example given above.

```
function par_esta
% parameter estimation with derivatives
% for exponential fit for c0
global tfit cfit lambda

% specify fitting data
tfit = [0.25 1 2 4 8];
cfit = [0.7716 0.5791 0.4002 0.1860 0.1019];
lambda = .3329; c00 = 1.;

c0 = fzero(@myfun,c00);
normc = norm(cfit - c0*exp(-lambda*tfit));
display (['Best fit for c0= ' num2str(c0)]);
display (['Norm of residuals= ' num2str(normc)]);
tmax = tfit(size(tfit,2));
t = [0:0.01*tmax:tmax];
figure; plot (tfit,cfit,'or',t,c0*exp(-lambda*t),'-');
legend ('given','modelled');
text(0.5*tmax,c0*0.7,['c_0: ' num2str(c0)]);
text(0.5*tmax,c0*0.8,['norm of residuals: ' num2str(normc)]);

function f = myfun(c0);
global tfit cfit lambda

c = c0*exp(-lambda*tfit);
    %solve linear decay equation for c with c(0)=c0
cc0 = exp(-lambda*tfit); % equation for dc/dc0
f = (c-cfit)*cc0';       % specify function f to vanish
```

The corresponding M-file *'paresta.m'* is included in the accompanying software.

The result is $c_0 = 0.816$ with an improved residual norm of 0.061. The reader may check the result as an exercise.

It is also possible to combine the last two M-files to estimate both c_0 and λ. Then, two function modules have to be used within the M-file. In one function (`myfun`) λ is optimized. Within that function, the second one (`myfun2`) is called, in which c_0 is optimized.

```
function par_estb
% parameter estimation with derivatives
% for exponential fit with c0 amd lambda as parameters
global tfit cfit c0
c0 = 0.816; lambda = 0.333;      % start values

% specify fitting data
tfit = [0.25 1 2 4 8];
cfit = [0.7716 0.5791 0.4002 0.1860 0.1019];

lambda = fzero(@myfun,lambda);
normc = norm(cfit - c0*exp(-lambda*tfit))
display (['Best fit for lambda= ' num2str(lambda)]);
display (['Best fit for c0= ' num2str(c0)]);
display (['Norm of residuals= ' num2str(normc)]);
tmax = tfit(size(tfit,2));
t = [0:0.01*tmax:tmax];
plot (tfit,cfit,'or',t,c0*exp(-lambda*t),'-');
legend ('given','modelled');
text(0.5*tmax,c0*0.8,['\lambda: ' num2str(lambda)]);
text(0.5*tmax,c0*0.7,['c_0: ' num2str(c0)]);
text(0.5*tmax,c0*0.6,['norm of residuals: ' num2str(normc)]);

function f = myfun(lambda);
global tfit cfit c0

options = optimset;
c0 = fzero(@myfun2,c0,options,lambda);
display (['Best fit for c0 = ' num2str(c0)]);

c = c0*exp(-lambda*tfit);    % solve linear decay eq. for c with c0
clambda = -c.*tfit;          % equation for dc/dlambda
f = (c-cfit)*clambda';       % specify function f to vanish

function f = myfun2(c0,lambda);
global tfit cfit

c = c0*exp(-lambda*tfit); %solve linear decay eq. for c with c(0)=c0
cc0 = exp(-lambda*tfit);  % equation for dc/dc0
f = (c-cfit)*cc0';        % specify function f to vanish
```

The corresponding M-file *'parestb.m'* is included in the accompanying software.

The formal parameter `lambda` needs to be added in the call of the second function. In order to do that, options must be transferred, too. The standard options set is obtained by using the `options = optimset` command.

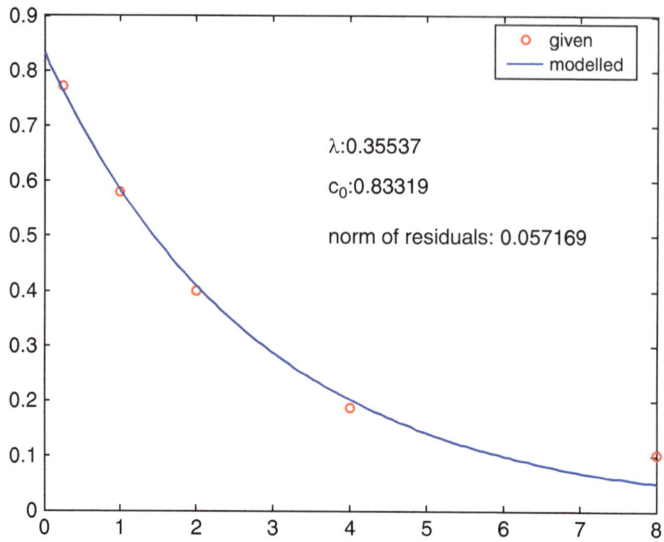

Fig. 10.6 Exponential fit for λ and c_0

The result of the estimation procedure for both parameters is $\lambda = 0.355$ and $c_0 = 0.833$. Figure 10.6 depicts the result with the improved norm of residuals equal to 0.057. For the chosen data set the change of parameters in the last two steps is relatively marginal, but in general that can be very different. It can be crucial to apply a procedure that really delivers optimal approximation.

Example 1

For many substances in many environmental compartments it can be assumed that sorption processes are fast in comparison to degradation processes. If this is true, the first drop of the concentration from the original concentration `cref` in solution (in the example 1 μg/l) to a value of `c0` can be attributed to sorption: part of the total mass available attaches to the surfaces of the porous material. The slow decline observed thereafter is due to degradation processes. Concerning degradation we assume a linear degradation characteristic (compare Chap. 5). The temporal development of the system can be described by the differential equation:

$$\frac{\partial c}{\partial t} = -\lambda c \tag{10.14}$$

Note that the retardation factor disappears, as R is a coefficient in both relevant terms in the equation. The solution of the differential equation is

$$c(t) = c_0 \exp(-\lambda t) \tag{10.15}$$

for initial concentration c0. The degradation rate λ can directly be obtained from the optimized parameter set of the previous example. The retardation factor R is obtained from a mass balance at time $t = 0$:

$$\theta c_{in} = \theta c_0 + \rho_b c_s \tag{10.16}$$

The initial concentration c_{in} after the fast 'splitting' is divided in a fluid and a solid phase part. We re-write the right hand side and utilize formula (6.15) under the assumption of a linear isotherm:

$$\theta c_{in} = \theta c_0 \left(1 + \frac{\rho_b c_s}{\theta c}\right) = \theta\, R c_0 \tag{10.17}$$

from which we get an initial guess for the retardation, even without the observation of any temporal changes:

$$R = \frac{c_{in}}{c_0} \tag{10.18}$$

According to formula (10.18), microcystin has a retardation of $R = 1.2$ in the above described batch experiment.

Example 2

For nuclide chains the decay rates can be subsequently determined by using the described parameter estimation strategy. We exemplify this here for a system of two species, one the mother and one the daughter nuclide for example. The system of differential equations is given by:

$$\frac{\partial}{\partial t}\begin{pmatrix} c_1 \\ c_2 \end{pmatrix} = \begin{pmatrix} -\lambda_1 & 0 \\ \gamma\lambda_1 & -\lambda_2 \end{pmatrix}\begin{pmatrix} c_1 \\ c_2 \end{pmatrix} \tag{10.19}$$

γ denotes a partition parameter, which is lower than 1, if only a part of specie 1 decays into specie 2. The analytical solution for given initial conditions c_{10} and c_{20} is than given by:

$$\begin{aligned} c_1(t) &= c_{10}\exp(-\lambda_1 t) \\ c_2(t) &= c_{20}\exp(-\lambda_2 t) + c_{10}\frac{\gamma\lambda_1}{\lambda_2 - \lambda_1}(\exp(-\lambda_1 t) - \exp(-\lambda_2 t)) \end{aligned} \tag{10.20}$$

In the first step we determine the decay constant for specie 1, using the condition (10.8) with the derivative (10.9). After that we proceed with specie 2. We can also use condition (10.8). However, the derivative to be used looks a bit more complicated:

$$\frac{\partial c_2}{\partial \lambda_2} = -c_{20} t \exp(-\lambda_2 t) + c_{10} \frac{\gamma\lambda_1}{\lambda_2 - \lambda_1}\left(t\exp(-\lambda_2 t) - \frac{1}{\lambda_2 - \lambda_1}(\exp(-\lambda_1 t) - \exp(-\lambda_2 t))\right) \tag{10.21}$$

For the implementation of this formula use the following function

```
function f = myfun2(lambda2);
global tfit c2fit c0 c02 lambda1 g
t = tfit;
c1 = exp(-lambda1*t);
c2 = exp(-lambda2*t);
dl = lambda2-lambda1;

clambda2 = (g*lambda1*c0/dl-c02)*t.*c2 - (g*lambda1*c0/dl/dl)*(c1-c2);
c2 = c02*c2+g*lambda1*c0*(c1-c2)/dl;
```

More on explicit solutions for decay chains is noted by Yuan and Kernan (2007). Bauer *et al.* (2001) as well as Guerrero *et al.* (2009) focus on the transport of decay chains.

10.5 Transport Parameter Fitting

The described algorithm can also be applied for the estimation of transport parameters. Here this is demonstrated for 1D transport as described by the Ogata-Banks solution:

$$c(x,t) = c_0 + \frac{c_{in}}{2}\left(\text{erfc}\left(\frac{x - vt}{2\sqrt{Dt}}\right) + \exp\left(\frac{v}{D}x\right)\text{erfc}\left(\frac{x + vt}{2\sqrt{Dt}}\right)\right) \tag{10.22}$$

The situation is examined in which the velocity v is the most unknown parameter. Such a situation can be met quite often in the description of environmental systems.

In order to apply the procedure introduced above, the derivative of the solution due to velocity is needed. The derivative of the complementary error function is given by:

$$\frac{\partial}{\partial x}\text{erfc}(x) = -\frac{2}{\sqrt{\pi}}\exp(-x^2) \tag{10.23}$$

By application of the chain rule results:

$$\frac{\partial c}{\partial v}(x,t) = c_{in}\left(\begin{array}{l}\frac{t}{2\sqrt{\pi Dt}}\left[\exp\left(-\left(\frac{x-vt}{2\sqrt{Dt}}\right)^2\right) - \exp\left(\frac{v}{D}x\right)\exp\left(-\left(\frac{x+vt}{2\sqrt{Dt}}\right)^2\right)\right]\\ +\frac{x}{2D}\exp\left(\frac{v}{D}x\right)\operatorname{erfc}\left(\frac{x+vt}{2\sqrt{Dt}}\right)\end{array}\right) \tag{10.24}$$

The derived equations are implemented in the following subroutine. As in the examples mentioned, the measurements are represented in the two vectors `xfit` and `cfit`. They are assumed to be measured at time `T` after the start of migration. `D` denotes the relevant diffusivity, `c0` is the initial and `c1` the inflow concentration.

```
function f = myfun(v);
global xfit cfit T D c0 c1

e=diag(eye(size(xfit,2))); h=1./(2.*sqrt(D*T));
arg1 = h*(xfit-v*T*e'); arg2 = h*(xfit+v*T*e'); arg3 = (v/D)*xfit;

% solve advection diffusion equation for c with c(t=0)=c0 and c(x=0)=c1
c = c0 + 0.5*c1*(erfc(arg1)+(exp(arg3).*erfc(arg2)));

% compute derivative of solution due to v
cv = c1*((T*h/sqrt(pi))*(exp(-arg1.*arg1)-exp(arg3).*...
exp(-arg2.*arg2))+0.5*(xfit/D).*exp(arg3).*erfc(arg2));

% specify function f to vanish
f = 2*(c-cfit)*cv';
```

The algorithm is demonstrated for an example of chloride concentrations measured in a sediment core of the Marmara Sea (Pekdeger 2006). It is assumed that a concentration profile, as it is observed today, results from two interacting processes. Sedimentation compounds the sediment layer. For the sake of simplicity a constant sedimentation rate is assumed which corresponds to a sedimentation velocity v. The second process is diffusion.

When there are no horizontal differences in variables or parameters, a 1D description can be used. The sediment water interface reduces to a single position. One may choose the origin to be located at the sediment water interface for the complete simulation. Note that this point moves in a coordinate system that is fixed to the earth, but that does not change the validity of the differential equations. The change of the concentration profile can be described by the usual transport equation with effective diffusivity D and positive sedimentation velocity v.

The main program is given in the following:

```
function par_estc
% transport parameter estimation with derivatives
global xfit cfit T D c0 c1

% Example values for Marmara Sea Sediment Core
T = 6.3e11;  % [s] 20.000 years
D = 1.0e-5;  % [cm*cm/s]
c0 = 0;      %
c1 = 619;    % [mmol/l]
xmax = 4000; % [cm]

% specify fitting data
xfit = [0 20 40 60 100 120 140 160 215 255 275 300 375 450 525 600 750
   1050 1200 1350 1650 1950 2250 2550 2700 3000 3450 3900];
cfit = [619 597 608 615 619 615 621 571 621 618 619 625 577 612 608 612
   609 590 582 582 556 494 457 489 487 444 381 371];

v = fzero(@myfun,0.2e-8,options);
display (['Best fit for v = ' num2str(v)]);

x = [0:xmax/400:xmax];
h = 1./(2.*sqrt(D*T)); e = diag(eye(size(x,2)));
plot (xfit,cfit,'o',x,c0+0.5*c1*(erfc(h*(x-v*T*e'))+...
   (exp((v/D)*x)).*erfc(h*(x+v*T*e'))),'-');
legend ('given','modelled');
xlabel ('depth [cm]'); ylabel ('chloride concentration [mmol/l]');
text(0.1*xmax,c1*0.65,['sedimentation velocity [cm/a]: '
   num2str(v*3.15e7)]);
e = diag(eye(size(xfit,2)));
normc = norm(cfit-c0+0.5*c1*(erfc(h*(xfit-v*T*e'))+...
   (exp((v/D)*xfit)).*erfc(h*(xfit+v*T*e'))));
text(0.1*xmax,c1*0.6,['norm of residuals: ' num2str(normc)]);
```

The corresponding M-file *'parestc.m'* is included in the accompanying software.

The value for `T` represents 20,000 years. This is approximately the time when a fresh water lake, located where the Sea of Marmara is found today, was flooded from the rising Mediterranean. The value for `D` is a standard first guess for molecular diffusivity, which is valid for many substances dissolved in water. It is assumed that before the flooding the sediment pores were filled with water of low chloride concentration. A more realistic low non-zero value could have been taken, but this has no significant influence on the model results, as the inflow concentration for saline water with 619 mmol/l is quite high. The maximum length for the simulation corresponds to 40 m slightly exceeding the maximum depth of the measurement locations.

What follows is already the `fzero`-command, which here is the optimization routine. Starting value for the velocity is $2 \cdot 10^{-9}$ cm/s. All further commands concern the design of the graphic in which measured data and the modelled curve can be compared, and where the result of the estimation procedure is depicted in addition. The output is shown in Fig. 10.7.

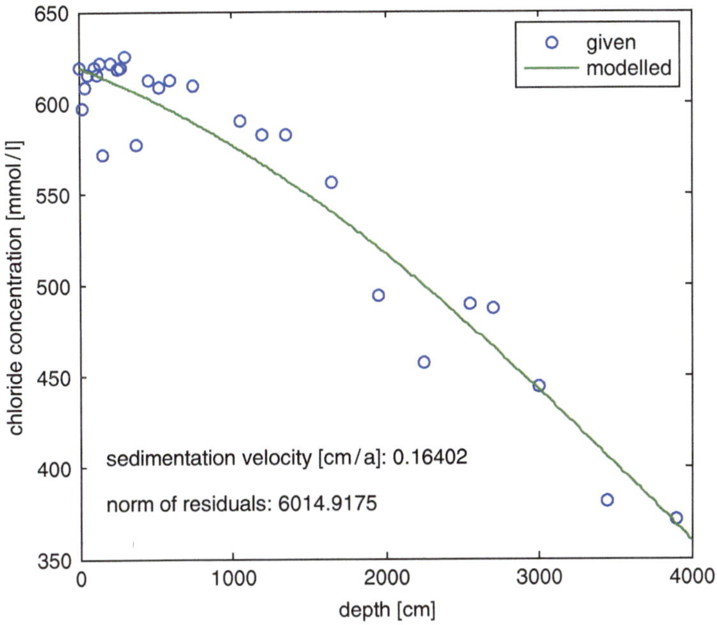

Fig. 10.7 Exponential fit for λ and c_0

10.6 General Procedure

The described zero search procedure for the computation of the optimal fit can also be applied to parameters of differential equations, i.e. if the solution is not given by an explicit formula like in the examples treated above. The value for diffusivity is determined in a new demonstration example, (1) by using analytical formulae, and (2) by using differential equations.

Example: Diffusivity Estimation

An environmental system is considered with two processes governing the distribution of a (bio) chemical species. To keep it simple, we examine a one-dimensional set-up. There are two reservoirs separated by a barrier. The barrier does not allow fluid flow, but chemical species may penetrate by means of diffusion. The driving force for diffusion is the fact that the two reservoirs have a different level of concentration c. Additionally, there is a constant source or sink for the examined species: it is produced or consumed passing through the barrier at a constant rate Q. The differential equation for the steady state of such a system is:

$$\frac{\partial}{\partial x}\left(D\frac{\partial c}{\partial x}\right) + Q = 0 \tag{10.25}$$

The diffusivity D is to be estimated based on a measurement within the barrier. The condition for the best estimate is that the following function e is minimized:

$$e = \sum \left(c(x_{fit}) - c_{fit}\right)^2 \tag{10.26}$$

Like the previous tasks the problem is solved by considering the derivative $\partial e/\partial D$, which is required to be zero for the best fit diffusivity:

$$\frac{\partial e}{\partial D} = 2\sum \left(c(x_{fit}) - c_{fit}\right)\frac{\partial c}{\partial D}(x_{fit}) = 0 \tag{10.27}$$

The derivative $\partial c/\partial D$, which appears in (10.27), fulfills a differential equation that is obtained by differentiation of (10.25) with respect to D:

$$\frac{\partial}{\partial x}\left(D\frac{\partial}{\partial x}\frac{\partial c}{\partial D} + \frac{\partial c}{\partial x}\right) = 0 \tag{10.28}$$

or, taking again (10.25) into account:

$$\frac{\partial}{\partial x}\left(D\frac{\partial}{\partial x}\frac{\partial c}{\partial D}\right) = \frac{Q}{D} \tag{10.29}$$

In order to test the approach we can work with analytical solutions of (10.25) and (10.29). The general solution for c is given by:

$$c(x) = -\frac{Q}{2D}x^2 + C_1x + C_0 \tag{10.30}$$

With boundary conditions

$$c(0) = 1 \text{ and } \frac{\partial c}{\partial x}(1) = 0 \tag{10.31}$$

one obtains: $C_0 = 1$ and $C_1 = Q/D$. Analogously, the general solution of (10.29) is given by:

$$\frac{\partial c}{\partial D}(x) = \frac{Q}{2D^2}x^2 + D_1x + D_0 \tag{10.32}$$

As the Dirichlet condition for c at the left boundary is independent of D one obtains the boundary conditions:

$$\frac{\partial c}{\partial D}(0) = 0 \quad \text{and} \quad \frac{\partial}{\partial x}\frac{\partial c}{\partial D}(1) = 0 \tag{10.33}$$

The second condition follows from the chain rule $\partial c/\partial D = (\partial c/\partial x)(\partial x/\partial D)$ and the second boundary condition given in (10.31). It follows: $D_0 = 0$ and $D_1 = -Q/D^2$.

The following M-file demonstrates the procedure for an example data set (see Rom 2005):

```
function par_est2
% parameter estimation with derivatives
% Idea from FEMLAB - there R instead of Q
% see COMSOL News, Nr. 1, 2005, page 15
global xfit cfit Q

% specify fitting data
xfit = [0.05:0.1:0.95];

  cfit = [0.9256859756097451 0.7884908536585051 0.6665396341462926
0.559832317073104 0.4683689024389414 0.39214939024380824
0.33117378048770196 0.28544207317062964 0.25495426829258294
0.23971036585356142];
Q = -2;
D = fzero(@myfun,2);
display (['Best fit for D = ' num2str(D)]);
x = [0:0.01:1];
plot (xfit,cfit,'o',x,-(Q/D/2)*x.*x + (Q/D)*x + 1,'-');
legend ('given','modelled');
xlabel ('x'); ylabel ('c');

function f = myfun(D);
global xfit cfit Q

% solve diffusion equation for c with c(0)=1 and dc/dx(1)=0
c = -(Q/D/2)*xfit.*xfit + (Q/D)*xfit + 1;

% solve Poisson equation for dc/dD (cD) with boundary conditions
cD = (Q/D/D/2)*xfit.*xfit - (Q/D/D)*xfit;

% specify function de to vanish
f = 2*(c-cfit)*cD';
```

The corresponding M-file *'parest2.m'* is included in the accompanying software.

The main program follows the same line given by the *'par_est'* M-files presented in the previous subchapters. The `fzero` function is called with an initial guess $D = 2$. The function `myfun` consists of three commands. The first evaluates c as formulated in (10.30); the second evaluates the derivative $\partial c/\partial D$ following (10.32); the third specifies function **f** according to formula (10.27).

For $Q = -2$ the procedure delivers the correct result of $D = 1.312$! The best fit is visualized in Fig. 10.8:

Hitherto, the analytical solution of the problem has been utilized again. The first two commands in the function `myfun` can be replaced by calls of differential equation solvers. In the first command, the (10.25) needs to be solved with regard to the boundary conditions (10.31); in the second command, it is the differential (10.29) with regard to conditions (10.33). The function **f** then becomes:

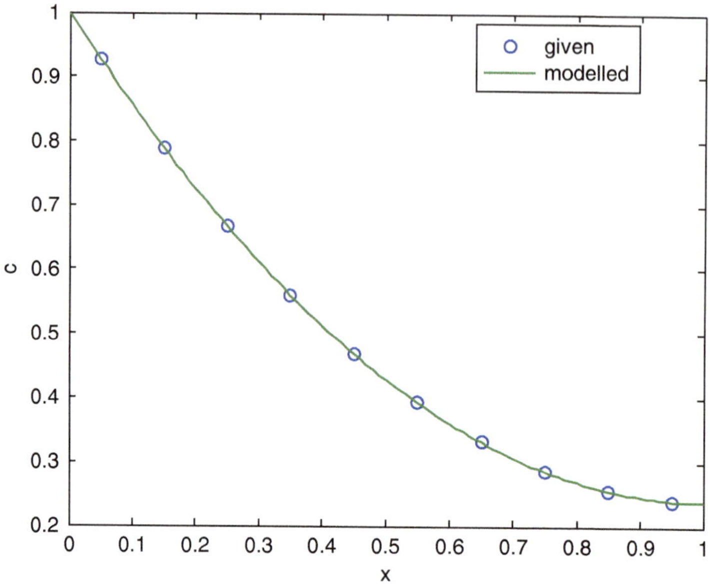

Fig. 10.8 Optimal fit for diffusivity estimation example problem

```
function par_est2a
....
function f = myfun(D);
global xfit cfit Q
options = bvpset;

% solve diffusion equation for c with c(0)=1 and dc/dx(1)=0
solinit = bvpinit([0 xfit 1],@guess);
c = bvp4c (@mat4ode,@mat4bc,solinit,options,Q/D,1);

% solve Poisson equation for dc/dD (cD) with boundary conditions
solinit = bvpinit([0 xfit 1],@guess1);
cD = bvp4c (@mat4ode,@mat4bc,solinit,options,Q/D/D,0);

% specify function f to vanish
f = 2*(c.y(1,2:size(c.y,2)-1)-cfit)*cD.y(1,2:size(c.y,2)-1)';

function dydx = mat4ode(x,y,Q,c0)
dydx = [y(2); -Q];

% -------------------------------------------------------------
function res = mat4bc(y0,y1,Q,c0)
res = [y0(1)-c0; y1(2)];

% -------------------------------------------------------------
function v = guess(x)
v = [x*(x-2)+1; 2*(x-1)];

% -------------------------------------------------------------
function v = guess1(x)
v = [x*(x-2); 2*(x-1)];
```

The corresponding M-file *'parest2a.m'* is included in the accompanying software.

The main body of the M-file is omitted. In the **f**-function module we obtain two solutions by calling the `bvp4c` command for the solution of boundary value problems. Both differential equations are of Poisson type $\frac{\partial^2 c}{\partial x^2} = const$; therefore, it is possible to use the same module (`mat4ode`) with different formal parameters. It is an alternative to use two different functions. The second order differential equation is re-written as two first order differential equations with $c_1 = c$:

$$\frac{\partial c_1}{\partial x} = c_2 \quad \frac{\partial c_2}{\partial x} = const \tag{10.34}$$

In the `mat4ode` sub-module the vector **y** has two elements representing c_1 and c_2. The `mat4bc` module specifies the boundary conditions (10.31) and (10.33), which in terms of c_1 and c_2 are given by:

$$c_1(x = 0) = const \quad c_2(x = 1) = 0 \tag{10.35}$$

The two functions `guess` and `guess1` deliver initial guesses for both functions fulfilling the boundary conditions. The graphical output resulting from that algorithm is identical to Fig. 10.8.

It has been demonstrated that the procedure, using solutions of differential equations within the `fzero`-module, is applicable to parameter estimation in situations in which an explicit formula for the solution is not available. We chose a one-dimensional set-up as demonstration example and had to solve boundary value problems for ordinary differential equations. However, the same concept can be applied to more general set-ups, steady or unsteady, in one or more space dimensions. In general, the solutions of partial differential equations are required within the zero-search-algorithm.

References

Bauer P, Attinger S, Kinzelbach W (2001) Transport of a decay chain in homogeneous porous media: Analytical solutions. J Contaminant Transport 49:217–239

Grützmacher G (2006) Umweltbundesamt (German Federal Environment Agency), Personal communication

Guerrero JS, Skaggs TH, vam Genuchten MTh (2009) Analytical solutions for multi-species contaminant transport subject to sequential first-order decay reactions in finir media. Transport Porous Media 80:373–387

Pekdeger A (2006) Freie Universität Berlin, Germany, Personal communication

Rom N(2005) Welche Werkzeuge stellt FEMLAB für die Problemoptimierung bereit. *COMSOL News*(1): 15 (in German)

Yuan D, Kernan W (2007) Explicit solutions for exit-only radioactive decay chains. J Appl Phy 101:094907-1–12

Part II
Advanced Modeling Using MATLAB®

Chapter 11
Flow Modeling

Flow, of course, is a crucial topic in many environmental sciences. Flow is the carrier of advective transport (see Chap. 3). Biogeochemical species are transported by flow through environmental compartments and through systems of compartments. Often transport with the flow is the fastest process by which a species of potentially hazardous impact, starting from a source, reaches a sensitive region.

Let's take a repository for radioactive waste as an example. In several countries of the world final storage facilities are envisaged located in some nearly impermeable geological environments in the deep sub-surface. The main safety problem with these waste disposal sites is concerned with the identification of flow. Containments and barriers of any type are not able to shut off heat producing, acid, radioactive and/or toxic material for long time periods. There is the danger that in the long run hazardous substances find a subsurface flow path, which takes them up to the surface. Even if that are long distances and long time periods, the potential thread still remains, as those nuclides with long half-lives remain active for 1,000s of years. Though the transport along the flow path may take several 1,000s or 10,000s of years, this is a much faster process than any other one. Thus, it is important to understand and model flow paths and fields (Fig. 11.1).

Speaking of flow in environmental sciences not always means the same thing. There are various types of fluids and fluxes. In the hydrosphere, containing as different compartments as creeks, rivers, lakes and the sea with coastal waters, continental margins and the deep sea sub-compartments, water is the flow medium. Aquifers and the pore space of aquatic sediments also belong to the hydrosphere. Air is the flow medium of the atmosphere. In the unsaturated zone, between the earth surface and the groundwater table, water and air are both fluids, although with quite different characteristics. There is also flow in the biosphere, for example, when water is taken up by the roots and transferred to the green parts of plants above the earth surface.

According to the differences concerning the medium and the compartments and sub-compartments respectively, there is not the one and only approach to modeling flow phenomena. In the sequel, the term *free fluids* is used if the flow medium occupies the entire volume. A contrasting term is *porous media* flow, where the

E. Holzbecher, *Environmental Modeling*,
DOI 10.1007/978-3-642-22042-5_11, © Springer-Verlag Berlin Heidelberg 2012

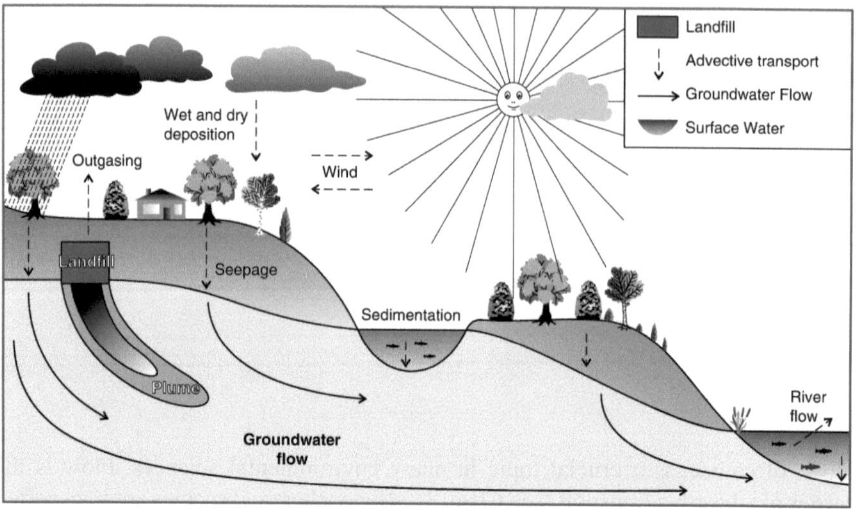

Fig. 11.1 Pollutant environmental pathways with relevant advective transport

fluid flow occurs within the pore space of a solid material, as in aquifers, aquatic sediments or the soil, the latter with seepage and air flow in the unsaturated or vadose zone.

11.1 The Navier-Stokes Equations for Free Fluids

Considering all the different phenomena of flow fields, it may seem to be amazing that one mathematical approach is well accepted as a fundamental description. It is valid for all types of free flow, either laminar or turbulent, in bounded or unbounded domains. The equations are a basis for many situations, although simplifications or extensions are necessary. The generally accepted *Navier*[1]*-Stokes*[2] *equations* can be noted as follows:

$$\rho \frac{\partial}{\partial t}\mathbf{v} + \rho(\mathbf{v} \bullet \nabla)\mathbf{v} - \rho\mathbf{f} + \nabla p + \eta\nabla^2\mathbf{v} = 0 \tag{11.1}$$

with fluid density ρ, velocity vector $\mathbf{v}$, volume force $\mathbf{f}$ and pressure p. p and $\mathbf{v}$ are the dependent variables. For a one-dimensional description the system (11.1) reduces

[1] Louis Marie Henri Navier (1785–1836), French mathematician and physicist.
[2] George Gabriel Stokes (1819–1903), Irish mathematician and physicist.

to a single equation, in two-dimensional space there are two differential equations given by system (11.1), and in three-dimensional space there are three.

Equations (11.1) are derived from the principle of momentum conservation, where momentum is defined as the product $\rho\mathbf{v}$. From that point of view the Navier-Stokes equations are for fluid mechanics what is Newton's Law for classical mechanics. The first two terms in (11.1) represent temporal change and advective transport of momentum. The $\rho\mathbf{f}$-term introduces outer forces as for example gravity. The connection with pressure is given by the ∇p-expression. The final term, including the viscosity η as parameter, represents the internal friction within the fluid. A more general formulation of the Navier-Stokes equations can additionally take the effect of compressibility into account (Guyon et al. 1997).

Detailed derivations of the Navier-Stokes equations can be found in textbooks on fluid mechanics; see for example Guyon et al. (1997). In the derivation of the formulation (11.1) it is assumed that the internal shear stress within the fluid is proportional to the change of the velocity in transverse direction. The dynamic viscosity η is the proportionality factor in that relationship, which can also be traced back to Newton. Water is a Newtonian fluid for which such a relation is valid, while different formulations result for non-Newtonian fluids. The change of water viscosity in the temperature range between 0°C and 50°C is depicted in Fig. 11.2.

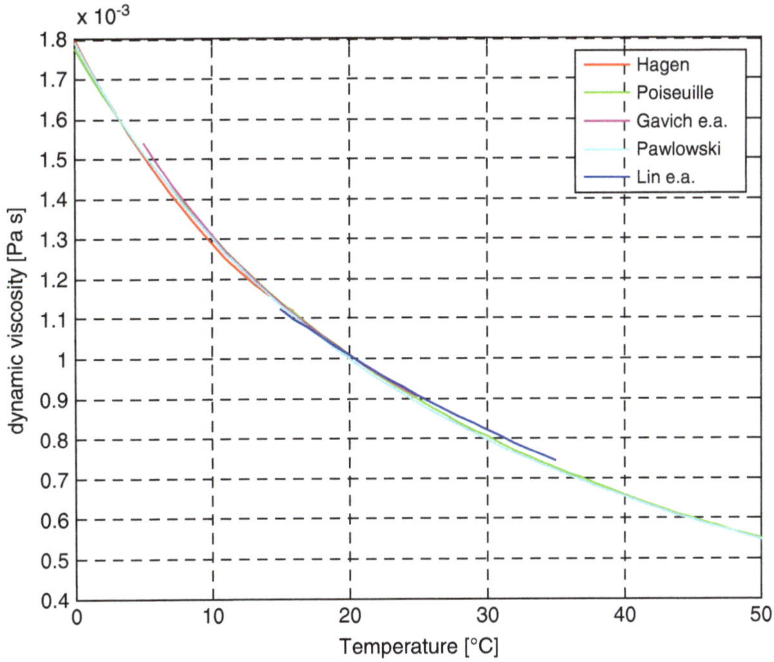

Fig. 11.2 Change of dynamic viscosity of water; according to different authors

Equations (11.1) are completed by the continuity equation

$$\frac{\partial}{\partial t}\rho - \nabla \bullet (\rho \mathbf{v}) = \rho Q \tag{11.2}$$

which represents the principle of mass conservation for the fluid. As outlined in Chap. 3, the derivation of (11.2) is analogous to the derivation of the mass conservation for species, utilizing the equation for fluid flux: $\mathbf{j} = \rho\mathbf{v}$. Q represents volume sinks and sources within the flow region. Not taken into account in both (11.1) and (11.2) is the case in which the fluid phase does not cover the entire space.

That can be included by an additional factor, which represents the share of the concerned phase on the entire volume. In the system of (11.1) and (11.2) the number of equations equals the number of unknown variables p and $\mathbf{v}$. From the mathematical aspect, the most problematic term in the equations is the second term of (11.1), which is nonlinear.

The plot is produced by a *'viscosity_dyn.m'*, included in the accompanying software.

There are few classical solutions for the complete set of Navier-Stokes (11.1) and (11.2). Analytical solutions are mostly restricted to special circumstances. One example is incompressible laminar flow through a pipe, see Sidebar 11.1. Incompressible flow concerns fluids with constant density, for which the continuity (11.2) simplifies to:

$$\nabla \bullet \mathbf{v} = Q \tag{11.3}$$

where the right hand side represents sources and sinks, measured as volumetric rate. In absence of sources and sinks the equation becomes identical to the condition for divergence-free vector fields:

$$\nabla \bullet \mathbf{v} = 0 \tag{11.4}$$

The *Reynolds*[3] *number Re* is defined by $Re = v_{\text{char}} \cdot H_{\text{char}}/\upsilon$, with a characteristic velocity v_{char}, a characteristic length H_{char} and kinematic viscosity $\upsilon = \eta/\rho$. In situations in which the Reynolds number is above that value, the flow regime becomes *turbulent*. In turbulent flow small disturbances are amplified, and the assumption of zero velocity components perpendicular to the main flow direction is not valid anymore. For turbulent flow there are no analytical solutions of the Navier-Stokes equations.

[3] Osborne Reynolds (1842–1912), English physicist.

Sidebar 11.1: One-Dimensional Laminar Flow (Hagen-Poiseuille)

One of the simplest examples for a solution of the Navier-Stokes equations is the two-dimensional horizontal flow between two plates, in a steady state situation with constant density ρ. Assume that the z-axis is directed perpendicular to the plates and the distance of the plates is given by Δz. The y-direction is neglected in the two-dimensional set-up. Moreover, the assumption of a vanishing velocity component v_z perpendicular to a pressure gradient in x-direction can be made. From (11.2) to (11.4) follows that

$$\frac{\partial^2}{\partial x} v_x = 0$$

i.e. the velocity is not changing in x-direction. If a constant pressure gradient is given in x-direction ($\Delta p/\Delta x$), the steady state version of the (11.1) reduces to equations for v_x:

$$\eta \frac{\partial^2}{\partial z^2} v_x + \frac{\Delta p}{\Delta x} = 0$$

The solution of the differential equation for v_x is a quadratic function of z. The two integration constants are derived from the condition that the velocity component vanishes at the plates at positions $z = \pm\Delta z/2$. The resulting solution can be expressed as:

$$v_x = v_{\max}\left(1 - \frac{4z^2}{\Delta z^2}\right)$$

The velocity distribution is visualized in Fig. 11.3. v_{max} is the maximum velocity at the halfway between the plates, given by the formula:

$$v_{\max} = -\frac{\Delta z^2}{8\eta}\frac{\Delta p}{\Delta x}$$

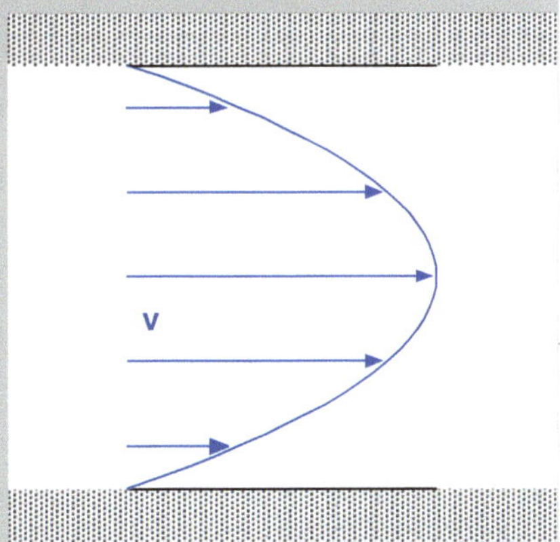

Fig. 11.3 Laminar flow between two plates

(*continued*)

The mean velocity is given by:

$$v_{\mathrm{mean}} = -\frac{\Delta z^2}{12\eta}\frac{\Delta p}{\Delta x}$$

and the flux per unit width by:

$$q = -\frac{\Delta z^3}{12\eta}\frac{\Delta p}{\Delta x}$$

The equation states a linear relationship between fluid flux q and the pressure gradient $\Delta p/\Delta x$. So far, one-dimensional flow between two plates in cartesian coordinates has been studied as most simple situation. one-dimensional flow within a pipe can be treated similarly using a two-dimensional coordinate system. Instead of the z-coordinate, the radial coordinate r has to be considered, and the differential equation for v_x in cylinder coordinates takes the form:

$$\eta\frac{1}{r}\frac{\partial}{\partial r}\left(r\frac{\partial}{\partial r}v_x\right) + \frac{\Delta p}{\Delta x} = 0$$

In analogy to the procedure demonstrated just before one obtains the solution:

$$v_x = v_{\max}\left(1 - \frac{r^2}{R^2}\right)$$

where R denotes the radius of the pipe. The paraboloid shape is depicted in Fig. 11.4.

$$v_{\max} = -\frac{R^2}{4\eta}\frac{\Delta p}{\Delta x}$$

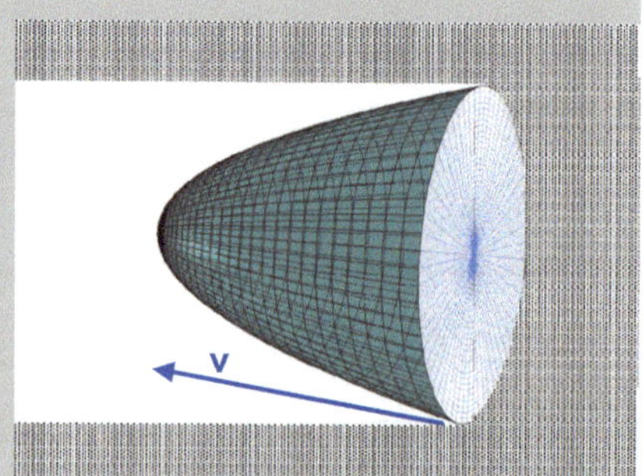

Fig. 11.4 Laminar flow of paraboloid shape in a pipe with circular cross-section

(continued)

The mean velocity is given by:

$$v_{\mathrm{mean}} = -\frac{R^2}{8\eta}\frac{\Delta p}{\Delta x}$$

Regarding the corresponding head h of the fluid as a measure for pressure, it is also possible to write:

$$v_{\mathrm{mean}} = -\frac{R^2}{8\nu}\frac{\partial h}{\partial x}$$

Taking into account that the total flux through the pipe is known, one obtains the classical result of Hagen[4] and Poiseuille[5]

$$Q = \int_0^R v_x(r) 2\pi r dr = \frac{\pi R^4}{8\eta}\frac{\Delta p}{\Delta x}$$

The formula was first experimentally developed by Hagen and by Poiseuille independently. According to the Hagen-Poiseuille formulae there is a linear relation between the flux Q and the pressure gradient $\Delta p/\Delta x$, and between the characterising velocities and the pressure gradient. Such a relationship is typical for situations in which the friction on solid walls is a dominant process. A linear relation between velocity and pressure, or flux and hydraulic head, holds not only in systems of pipes of small diameter but for porous media flow in general. Darcy's Law for porous media states exactly such a relation (see Chap. 11.3).

The validity of the given formulae is limited by the dimensionless Reynolds-number of $Re = 2300$, where the diameter is taken as characteristic length and the mean velocity as characteristic velocity.

Using MATLAB®, the focus will be on analytical solutions, which can be applied for special cases only. In the following subchapter cases will be considered in which internal friction can be neglected. Thereafter special systems are in the focus which are dominated by friction.

[4] Gotthilf Heinrich Ludwig Hagen (1797–1884), German engineer.

[5] Jean Louis Marie Poiseuille (1799–1869), French physician and physicist.

11.2 The Euler Equations and the Bernoulli Theorem

There are situations in which the friction can be neglected. For frictionless *ideal fluids*, i.e. fluids with zero viscosity, the *Euler*[6] *equations* are written in modern notation as:

$$\rho \frac{\partial}{\partial t}\mathbf{v} + \rho(\mathbf{v} \bullet \nabla)\mathbf{v} - \rho\mathbf{f} + \nabla p = 0 \tag{11.5}$$

which were published by Euler in 1750.

Utilizing potential theory for fluid mechanics is the classical approach ("classical hydrodynamics" according to Prandtl and Tietjens 1957), developed already by the Bernoulli's[7] and Euler. The works of Euler in Berlin and St. Petersburg not only mark the completion of classical fluid mechanics (Szabó 1987), but also the origin of an approach by which natural phenomena in the laboratory or in the field are described by differential equations and their solutions.

The general Navier-Stokes equations have few analytical solutions (for an example see Sidebar 11.1). Thus, they usually have to be solved by special software packages utilizing numerical methods, such as finite differences, finite elements, or finite volumes. The use of numerical methods by applying the `pdepe` command was already described in Part I of the book. However, `pdepe` can be used for 1D problems only. For higher dimensional problems the MATLAB® partial differential toolbox has to be applied, which is not described here. In the following chapters, it is the aim to examine the use of MATLAB® for potential flow.

Potential flow is an umbrella term for a technique to obtain analytical solutions. Analytical solutions are explicit formulae for the unknown variables, sometimes also referred to as *closed form solutions* (Narasimhan 1998). If a flow field is irrotational, i.e. if the condition[8]

---

[6] Leonard Euler (1707–1783), Swiss mathematician.

[7] Johann Bernoulli (1667–1748), Daniel Bernoulli (1700–1782), Swiss mathematicians.

[8] '$\times$' denotes the cross product for vectors, which for vectors $\mathbf{r}_1 = (x_1, y_1, z_1)^{\mathrm{T}}$ and $\mathbf{r}_2 = (x_2, y_2, z_2)^T$ is defined by:

$$\mathbf{r}_1 \times \mathbf{r}_2 = (x_2y_3 - x_3y_2, x_3y_1 - x_1y_3, x_1y_2 - x_2y_1)^T$$

If the nabla-operator is used as first vector and $\mathbf{v} = (v_x, v_y, v_z)^T$ as second, one obtains:

$$\nabla \times \mathbf{v} = \left(\frac{\partial}{\partial y}v_z - \frac{\partial}{\partial z}v_y, \frac{\partial}{\partial z}v_x - \frac{\partial}{\partial x}v_z, \frac{\partial}{\partial x}v_y - \frac{\partial}{\partial y}v_x\right)^T$$

For the special case of flow in the 2D (x,y)-plane follows the equation:

$$\nabla \times \mathbf{v} = \left(0, 0, \frac{\partial}{\partial x}v_y - \frac{\partial}{\partial y}v_x\right)^T$$

$$\nabla \times \mathbf{v} = 0 \tag{11.6}$$

is fulfilled at a time instant, this property remains valid further on. It can be shown that under condition (11.6) a potential φ exists, which is characterized by the property

$$\mathbf{v} = \nabla\varphi \tag{11.7}$$

and that it fulfills the Bernoulli theorem (for example: Gallavotti 2002):

$$\rho\frac{\partial\varphi}{\partial t} + \frac{\rho}{2}v^2 + \rho\phi + p = C \tag{11.8}$$

where ϕ denotes the potential of the force vector $\mathbf{f}$ (i.e. $\mathbf{f} = \nabla\phi$). C is a constant in a simply connected domain. If condition (11.6) is skipped, C is a constant for each streamline but not in the entire flow domain. For steady states the first term in (11.8) can be omitted. In fact, the Bernoulli theorem yields a relation between pressure and velocity. In following chapters methods will be explained how the potential can be determined, from which the velocity field is derived.

Sidebar 11.2 Open Channel Flow

Open channel flow is defined as flow in any situation in which a liquid has a free surface, such as in channels, rivers, streams, ditches, uncovered conduits and discharge from tailings ponds. There is open channel flow in closed channels, such as pipes, tunnels or adits, if the liquid does not fill the entire cross-section. Open channel flow is not under pressure, with gravity as the driving force.

Some characteristics of open channel flow can be derived from the Bernoulli theorem. For steady conditions according to (11.8) and (11.10), one can state that the left hand side of the Bernoulli (11.8)

$$\frac{v^2}{2g} + h\cos(\beta) = H_e$$

is a constant, which represents total energy in the dimension of height and is therefore denoted as H_e, the energy height (see also textbooks on fluid mechanics, for example: Schröder 1995). h denotes the height of the water column, i.e. the water level with reference to the zero level at the bottom of the flowing water body. Using the formula between mean flow velocity v, cross-sectional area A and flux $Q = Av$, one may write the equation in the form:

$$\frac{Q^2}{2gA(h)^2} + h\cos(\beta) = H_e$$

(*continued*)

If the flow is kept at a constant value, the question arises, which water levels result from the formula. In order to answer the question, the area has to be expressed in terms of water level. For a rectangular cross-section one obtains:

$$\frac{q^2}{2gh^2} + h\cos(\beta) = H_e$$

where q denotes the flow rate per unit width. The formula yields an energy height as function of water depth h. The situation can be visualized using MATLAB®.

The corresponding M-file is included in the accompanying software under the name *"OpenChannel.m"*

Figure 11.5 illustrates that there is a minimum energy height $H_{e,min}$. The water level, corresponding with $H_{e,min}$, is commonly referred to as critical height h_{crit}. For all possible levels $H_e > H_{e,min}$, there are two possible water table positions; one above h_{crit} and one below h_{crit}. Both values for the height of the water column are denoted as *conjugated heights*. In the former situation the velocity is lower than in the latter situation. That's why the first case is called *subcritical*, while the second case is called *supercritical*. Height and velocity at the critical state are given by:

$$h_{crit} = \sqrt[3]{q^2/g\cos(\beta)} \qquad v_{crit} = \sqrt{gh_{crit}\cos(\beta)}$$

which can be derived from the condition $\partial H_e/\partial h = 0$. For $\beta = 0$ the critical state can be related to the *Froude*[9] *number*

$$\mathrm{Fr} = \frac{v}{\sqrt{gh}}$$

Flow is subcritical flow for $\mathrm{Fr} < 1$ and supercritical for $\mathrm{Fr} > 1$. In open channels and regulated rivers changes from subcritical to supercritical or vice versa can be observed at locations where the channel characteristics change, as channel boundary roughness, channel bottom slope, lower boundary elevation, etc. (Rouse 1978). Most natural rivers are in the sub-critical regime in most parts (Olsen 2002), except in the vicinity of waterfalls, weirs or other structures.

(*continued*)

[9] William Froude (1810–1879) English engineer.

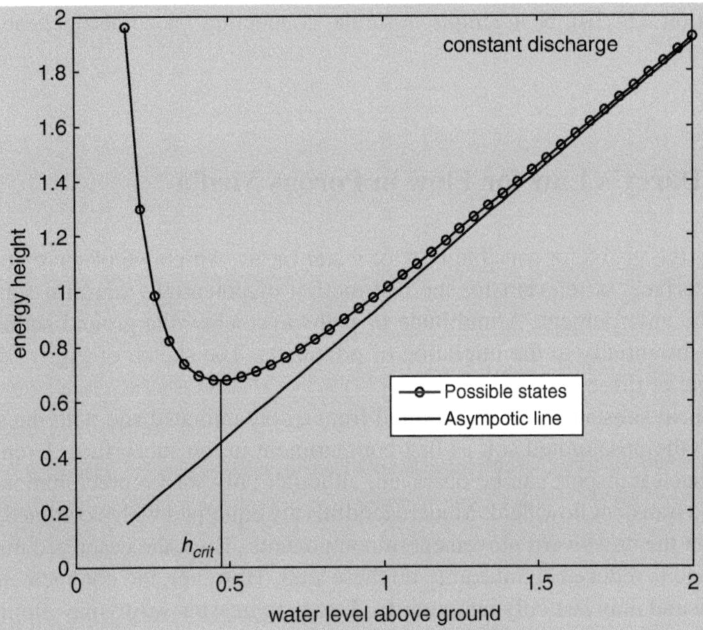

Fig. 11.5 Energy height and water level characteristic for open channel flow

Ideal fluids are quite rare, but they exist. Helium below a temperature of 2.17°K becomes a suprafluid without viscosity. However, there are situations with more common conditions, in which potential flow theory provides an approximate solution. At high velocities disturbances due to friction, for example at the boundary of an obstacle, cannot develop into the fluid, at least not within the short time of the passing fluid. Obstacles in a fast flowing fluid can thus be treated by potential flow. Also for fluids with high Reynolds numbers the Euler equations can be taken as an approximation if the flow is far away from any walls and not turbulent (Guyon et al. 1997).

Potential flow also plays an important role in hydraulic engineering (Schröder 1995). From (11.8) an explicit formula for total pressure p can be derived. For the steady state in the gravity field follows directly:

$$p = p_{ref} + \rho g z - \frac{\rho}{2} v^2 \tag{11.9}$$

where z denotes the opposite direction of gravity, g acceleration due to gravity, and p_{ref} a pressure reference value. If, for constant density, dynamic pressure $p_{dyn} = p - \rho g z$ is expressed as height of a fluid column (denoted as h), the result is:

$$h = h_{ref} - \frac{v^2}{2g} \tag{11.10}$$

Equation (11.10) is a simple formula connecting piezometric head h and velocity v.

11.3 Darcy's Law for Flow in Porous Media

Not only the visible or sensible flow of water or air, which we observe above the ground surface, is relevant for the distribution of potentially harmful substances within the environment. A multitude of pathways below the ground surface contribute substantially to the migration of pollutants. The sketch of Fig. 11.6 visualizes some of those.

Chemical substances are transported from a contaminated site with the seepage flow into the unsaturated soil as first compartment in the subsurface. Even in arid regions such transport can be observed, although only scarce precipitation events produce a transient flow field. Modern landfills are equipped with a confined bottom to prevent the downward movement of components. Thus, the described migration by seepage is reduced significantly on these sites. However, the confinement is not complete and may last only temporarily. Toxic, aggressive waste may diminish the sealing function over long time scales. Old landfills or contaminated sites can be sealed at the top by a cover which is more or less impermeable to water. This helps to reduce flow and advective transport effectively.

Contaminants can pass the unsaturated zone and enter the groundwater compartment. While in the vadose zone the dominant flow direction is vertically downward, in groundwater layers the horizontal velocity component usually dominates. The substances may thus be transported to more vulnerable regions. Where water is

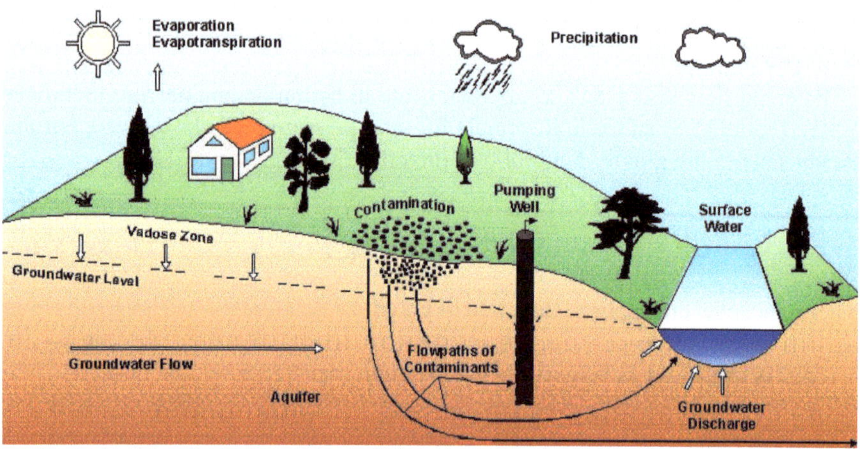

Fig. 11.6 Crucial sub-surface flow paths at waste disposal sites

pumped for drinking water or any other purposes, the contaminants can return back to the ground surface with the potential to produce hazards in the antroposphere.

The time scale for such a return can differ substantially. In the vicinity of surface water bodies the residence time in the subsurface compartments may be as short as several days or weeks. When deeper groundwater layers are involved, the time scale of residence is surely to be counted in years. In several regions fossil groundwater is pumped which was recharged several 1,000s of years before. In comparison to the ambient state without any anthropogenic influence, the natural residence time within the subsurface may be significantly shortened by pumping wells.

It was already mentioned in Sidebar 11.1 that in situations in which the wall friction is a relevant process a linear relationship between flux or velocity on one side and the pressure or head gradient on the other side can be expected. The relationship was derived from the Navier-Stokes (11.1) and (11.4) for one-dimensional pipe flow. A similar situation is given for fluid flow within the pore space of a porous matrix or porous medium. Pour water movements in sediments, seepage through the soil, or groundwater flow in aquifers are environmental systems that fall into that category.

The mentioned proportionality between flux and pressure drop for porous media is stated in *Darcy's Law*. In 1856, Henry Darcy[10] was the first who published such a proportionality law. He has performed a series of experiments in metal columns filled with sand. An experimental set-up, which in many parts resembles the original facility, is sketched in Fig. 11.7.

Driven by a pressure gradient water flows from the inlet of the column to the outlet. The pressure is kept constant if the two piezometer pipes are connected with water reservoirs of constant height. The height difference Δh between both reservoir levels is taken as a measure for the pressure difference. The finding from the Darcy experiment is stated mathematically as follows:

$$Q/A \propto \Delta h/L \tag{11.11}$$

where Q denotes the volumetric flow rate, A the cross-sectional area, L the length of the column and Δh the head gradient. In the 150 years that passed since the first publication, Darcy's law has been confirmed to be valid for a huge variety of porous media and for wide ranges of velocities and scales. In terms of the Reynolds-number the limiting value is $Re = 10$ (Bear 1972). For higher values of Re a generalized formulation has to be used, but in most subsurface aqueous environments the velocities are so low that Re stays below the threshold.

Nowadays it is common to formulate the law as an explicit equation for the Darcy velocity in an infinitesimal scale of the three-dimensional space:

$$\theta \mathbf{v} = -K_f \nabla h \tag{11.12}$$

[10] Henry Darcy (1803–1858), French engineer.

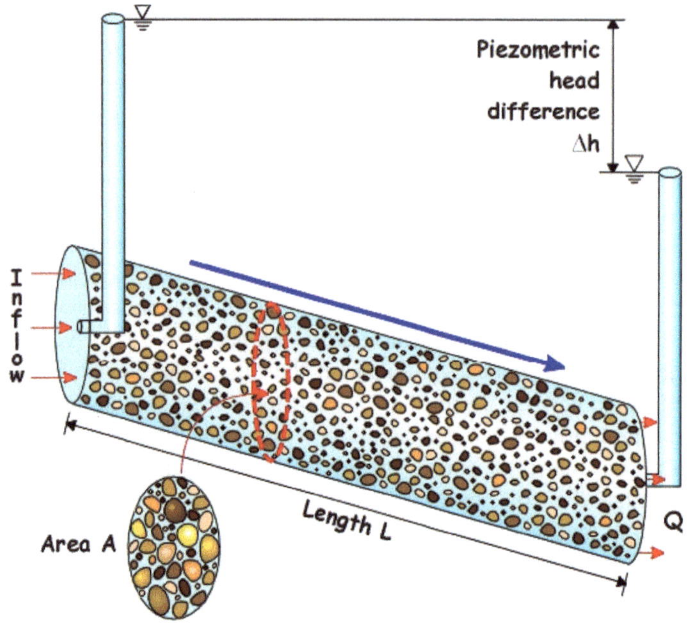

Fig. 11.7 Typical set-up of a Darcy-experiment; for examination of Darcy's Law and the determination of hydraulic conductivity

The left side of the equation denotes the *Darcy velocity*, which is derived from the real mean interstitial flow velocity by multiplication with the porosity θ. It is, in fact, the same variable as on the left side of (11.11); i.e. it is the velocity which represents the fluid flux. On the right side of equation (11.12) the head gradient ∇h is a generalization of the right side of (11.11) replacing the head difference per unit length. K_f is the proportionality factor, which is mostly referred to as hydraulic conductivity. It has the physical unit of velocity [L/T], [m/s] in MKS units. K_f depends on the properties of pore space, such as pore diameter and length, connectivity and porosity, or pore structure in general. The structure of the pores is far too complex to predict the K_f- value from microscopic properties. For porous pipe structures the hydraulic conductivity can be estimated on the basis of the Hagen-Poiseuille formula (see Sidebar 11.1).

The K_f- value also depends on the properties of the fluid which flows through the pore space. It is accepted that the dependencies can be separated by the approach:

$$K_f = \frac{k\rho g}{\eta} \tag{11.13}$$

where k represents the permeability (physical MKS unit: [m^2]) that depends on the porous medium, while density ρ and dynamic viscosity η are fluid properties.

Table 11.1 Classification and examples for hydraulic conductivities and permeabilities

K_f [m/s]	10^0	10^{-1}	10^{-2}	10^{-3}	10^{-4}	10^{-5}	10^{-6}	10^{-7}	10^{-8}	10^{-9}	10^{-10}	10^{-11}	10^{-12}
k [m^2]	10^{-7}	10^{-8}	10^{-9}	10^{-10}	10^{-11}	10^{-12}	10^{-13}	10^{-14}	10^{-15}	10^{-16}	10^{-17}	10^{-18}	10^{-19}
Pervious	■	■	■										
Semi–"				■	■	■	■	■					
Impervious									■	■	■	■	■
Gravel	■	■											
Sand			■	■	■								
Fine sand						■	■	■	■				
Peat					■	■							
Clay							■	■	■	■	■	■	■

$g = 9.81$ m^2/s is the acceleration due to gravity. Concerning the flow of water, it has to be taken into account that the dynamic viscosity changes by a factor of 2 between 0°C and 25°C. The permeability varies across a relatively wide range, which is shown in Table 11.1.

Note that with the corresponding version of the ∇-operator, the given formulation of Darcy's Law (11.12) is valid in 1D, 2D and 3D situations. two- and three-dimensional situations often make it necessary to distinguish between conductivities in different directions. In the mathematical formulation this can be taken into account by using a tensor (a matrix) $\mathbf{K_f}$ instead of a scalar K_f value, or a permeability tensor $\mathbf{k}$ instead of k. Some situations require an even more general formulation of Darcy's Law. For example for variable density flow (Holzbecher 1998) the head gradient on the right side of the equation has to be replaced by the gradient of the dynamic pressure:

$$\theta \mathbf{v} = -\frac{\mathbf{k}}{\mu} \nabla (p - \rho g z) \tag{11.14}$$

As additional variable besides velocities either hydraulic head h or pressure p appears. The concepts based on both variables are equivalent as long as there are no density gradients, which is a general assumption here. The mathematical treatment given here is similar to the derivations presented in textbooks on groundwater flow (Bear 1972; Bear and Verruijt 1987). In both formulations of Darcy's Law, (11.12) and (11.14), it is obvious that it is not the absolute value of pressure or head that determines the velocity. Flow, in its size and direction, is induced by the pressure or head gradient. For that reason it is irrelevant according to which reference value pressure and head are measured. From one application to the other the reference frame often is chosen very differently taking the specific circumstances in consideration.

In the mathematical formulation of porous media flow Darcy's Law replaces the momentum conservation (11.1). The conservation of mass principle for porous media can be formulated similarly to (11.3). The generalized formulation of the steady state is

$$\nabla \bullet \theta \mathbf{v} = Q \tag{11.15}$$

When in formula (11.15) the velocity $\mathbf{v}$ is replaced with the help of Darcy's Law (11.12), one obtains:

$$\nabla \bullet \mathbf{K_f} \nabla h = Q \tag{11.16}$$

When the conductivity as a material parameter for a porous medium is known, the differential (11.16) has a single unknown variable: h, the hydraulic or piezometric head. For given boundary conditions, the differential (11.16) has to be solved for h. A generalized version, which is very appropriate for 3D groundwater flow, results if formulation (11.14) is taken into the continuity equation:

$$\nabla \bullet \frac{\mathbf{k}}{\mu} \nabla (p - \rho g z) = Q \tag{11.17}$$

where pressure p is the dependent variable.

For unsteady flow the generalization of (11.16) is given by:

$$S \frac{\partial h}{\partial t} = \nabla \bullet \mathbf{K_f} \nabla h \tag{11.18}$$

where S is the storage coefficient, which is dimensionless. It denotes the volume of water released per unit area of aquifer and per unit drop in head. As far as a confined aquifer (see Chap. 12) is concerned, the storage coefficient is a function of the compressive qualities of water and matrix structures of the porous material. In 2D horizontal models for the unconfined aquifer (see Chap. 12) the storage coefficient is mainly determined by the change of the water column depth and thus may take much higher values than in the confined aquifer.

Codes for modeling groundwater flow apply numerical methods to solve (11.16) or (11.17) for the steady state, or (11.18) for the transient situation (Holzbecher 2002). In a homogeneous porous medium the (11.16) becomes equivalent to the Poisson equation. With the core version of MATLAB® numerical solutions can be obtained for one-dimensional cases, and analytical solutions for one- and two-dimensional cases. This will be presented in the next chapters.

11.4 Flow in Unsaturated Porous Media

One speaks of an unsaturated situation if there are two fluids filling the pore space of a porous medium, a gaseous and an aqueous phase. Soil is the most important environmental compartment, in which the unsaturated situation is met. The unsaturated layer between the land surface on the top and the groundwater table at the bottom is the unsaturated or vadose zone.

In order to model the more complex flow phenomena under unsaturated conditions, an extended formulation of Darcy's law is applied. The hydraulic conductivity for water decreases with the water content. Here the *volumetric water content* as a generalization of porosity is denoted by θ. Its maximum value is equal to the porosity of the porous medium. As an alternative parameter, the effective saturation S_e can be used. S_e takes values between 0 (gas phase only) and 1 (aqueous phase only):

$$S_e = \frac{\theta - \theta_r}{\theta_s - \theta_r} \tag{11.19}$$

where θ_r is the residual water content and θ_s the water content in the saturated situation.

Several mathematical relationships have been proposed for the dependency between hydraulic conductivity and water content. Mualem (1976) suggests a power law:

$$K(S_e) = K{S_e}^n \tag{11.20}$$

with a soil specific exponent n. The value for K on the right side of the equation is the conductivity of the saturated soil. For most soil types the exponent takes values between three and four (Brooks and Corey 1964). van Genuchten (1980) proposed a formula, which is frequently found in publications :

$$K(S_e) = K\sqrt{S_e}\left[1 - \left(1 - {S_e}^{1/m}\right)^m\right]^2 \tag{11.21}$$

The application in the vadose zone requires the hydraulic head to be split into a term representing the effect of total pressure p and buoyancy, as it was done in (11.17) already. With pressure head ψ as a measure of total pressure in a length unit (representing the height of a corresponding water column), the (11.18) for 1D gets the form:

$$\frac{\partial \theta}{\partial t} = \frac{\partial}{\partial z} K(\theta) \frac{\partial}{\partial z}(\psi - z) \tag{11.22}$$

On the left side of (11.22) storage of water is described by the change of water content.

The code for the retention curve visualization is included in the accompanying software under the name *"retention.m"*.

In the unsaturated zone the pressure head takes negative values. Sometimes the term *suction head* is used for the negative of pressure head. ψ becomes zero at the groundwater table. The saturation-suction relationship, often referred to as *retention curve*, is an empirical relationship, which has to be considered in soil modeling. Van Genuchten (1980) uses the mathematical form:

$$S_e = 1/(1 + |\alpha\psi|^n)^m \tag{11.23}$$

with parameters α and $n = 1 - 1/m$. The unit of α is [1/L]. Parameter m is identical to the one introduced in (11.21). Figure 11.8 depicts some example retention curves. There are various other formulations of the retention curve, which are not repeated here. A problem that is seldom tackled is the hysteresis of the retention curve, which means that the curve is not unique. In fact, experiments have shown that the saturation-suction curve for dewatering is often very different from the curve for re-wetting.

There are two possible ways to compute problems of unsaturated flow, based on (11.22) and the retention curve $\psi(\theta)$. Some authors prefer to use the retention curve to rewrite the term on the right side of (11.22) as function of θ. A MATLAB® implementation, using that approach, is presented by Hornberger and Wiberg (2005). The alternative approach is to rewrite the left side as follows:

$$\frac{\partial\theta}{\partial\psi}\frac{\partial\psi}{\partial t} = \frac{\partial}{\partial z}K(\psi)\frac{\partial}{\partial z}(\psi - z) \tag{11.24}$$

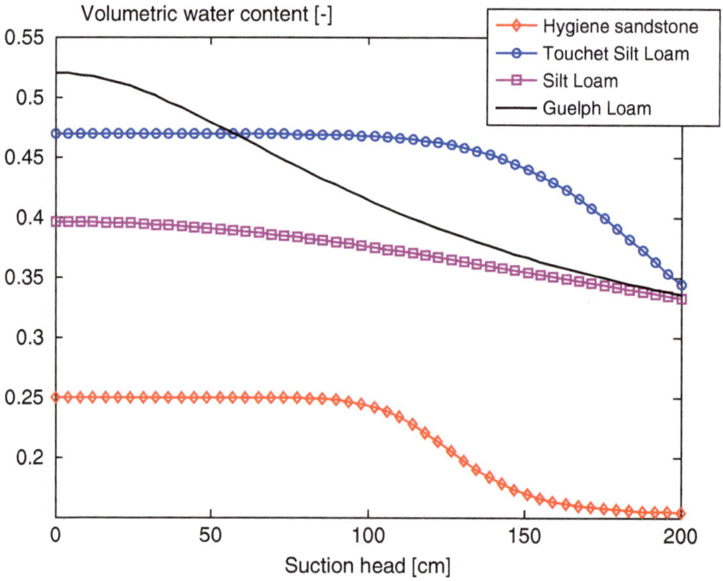

Fig. 11.8 Examples of retention curves, data from Hornberger and Wiberg (2005); produced using MATLAB®

The coefficient of the pressure derivative on the left side is evaluated based on the retention curve formula. For the van Genuchten formulation (11.23) one obtains:

$$\frac{\partial \theta}{\partial \psi} = \frac{(\theta_s - \theta_r) nm\alpha (-\alpha h)^{n-1}}{(1 + |\alpha \psi|^n)^{m+1}} \tag{11.25}$$

The differential (11.24) is the so-called *Richards*[11] *equation*, which can also be found in a slightly different notation:

$$\frac{\partial \theta}{\partial \psi}\frac{\partial \psi}{\partial t} = \frac{\partial}{\partial z} K(\psi)\left(\frac{\partial \psi}{\partial z} - 1\right) \tag{11.26}$$

The following M-file is an implementation of the Richards equation with the van Genuchten formulation for the suction-saturation and the conductivity-saturation relationships.

```
L = 200;                    % length [L]
s1 = 0.5;                   % infiltration velocity [L/T]
s2 = 0;                     % bottom suction head [L]
T = 4;                      % maximum time [T]
qr = 0.218;                 % residual water content
f = 0.52;                   % porosity
a = 0.0115;                 % van Genuchten parameter [1/L]
n = 2.03;                   % van Genuchten parameter
ks = 31.6;                  % saturated conductivity [L/T]

x = linspace(0,L,100);
t = linspace(0,T,25);

options = odeset('RelTol',1e-4,'AbsTol',1e-4,'NormControl','off',...
  'InitialStep',1e-7)
u=pdepe(0,@unsatpde,@unsatic,@unsatbc,x,t,options,... s1,s2,qr,f,a,n,ks);

figure;
title('Richards Equation Numerical Solution, computed with 100 mesh
  points');

subplot (1,3,1);
plot (x,u(1:length(t),:));
xlabel('Depth [L]');
ylabel('Pressure Head [L]');

subplot (1,3,2);
plot (x,u(1:length(t),:)-(x'*ones(1,length(t)))');
xlabel('Depth [L]');
ylabel('Hydraulic Head [L]');

for j=1:length(t)
    for i=1:length(x)
        [q(j,i),k(j,i),c(j,i)]=sedprop(u(j,i),qr,f,a,n,ks);
    end
end

subplot (1,3,3);
plot (x,q(1:length(t),:)*100)
xlabel('Depth [L]');
ylabel('Water Content [%]');
```

[11] Lorenzo Adolph Richards (1904–1993), US-American soil physicist.

```
% -------------------------------------------------------------------
function [c,f,s] = unsatpde(x,t,u,DuDx,s1,s2,qr,f,a,n,ks)
[q,k,c] = sedprop(u,qr,f,a,n,ks);
f = k.*DuDx-k;
s = 0;
% -------------------------------------------------------------------
function u0 = unsatic(x,s1,s2,qr,f,a,n,ks)
u0 = -200+x;
if x < 10 u0 = -0.5; end

% -------------------------------------------------------------------
function [pl,ql,pr,qr] = unsatbc(xl,ul,xr,ur,t,s1,s2,qr,f,a,n,ks)
pl = s1;
ql = 1;
pr = ur(1)-s2;
qr = 0;
%------------------- soil hydraulic properties -------------------
function [q,k,c] = sedprop(u,qr,f,a,n,ks)
m = 1-1/n;
if u >= 0
    c=1e-20;
    k=ks;
    q=f;
else
    q=qr+(f-qr)*(1+(-a*u)^n)^-m;
    c=((f-qr)*n*m*a*(-a*u)^(n-1))/((1+(-a*u)^n)^(m+1))+1.e-20;
    k=ks*((q-qr)/(f-qr))^0.5*(1-(1-((q-qr)/(f-qr))^(1/m))^m)^2;
end
```

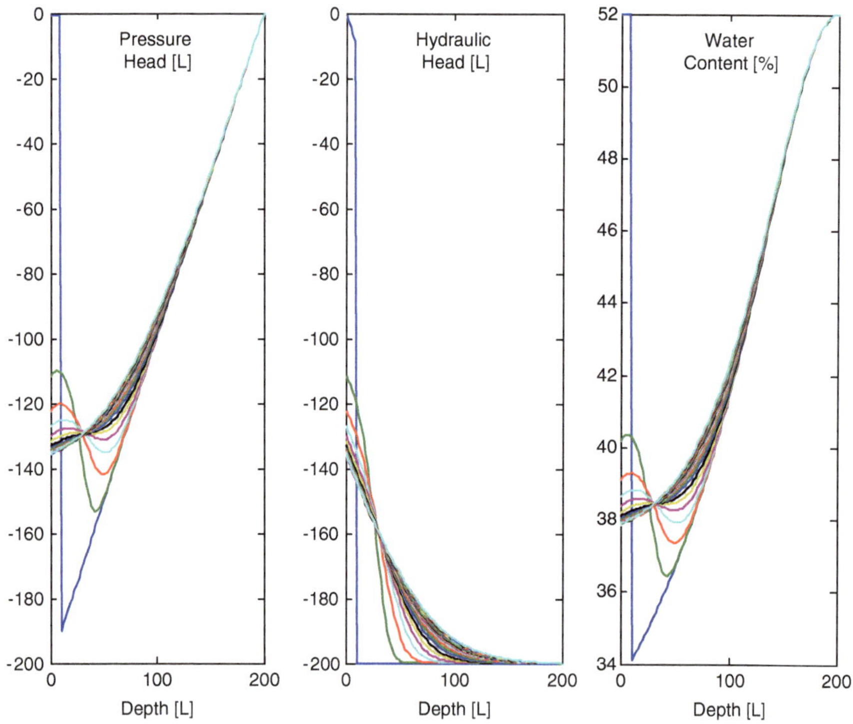

Fig. 11.9 Solution of Richards equation for infiltration within the soil compartment

The complete code is included in the accompanying software under the name *"richards.m"*.

The output for the example data-set is reproduced in Fig. 11.9. The initial profile is linear in the major deeper part of the soil column. In the upper 10 cm, the linear profile is disturbed by a layer with high pressure head and high water content. There is constant inflow specified as input condition at the top of the column, and a zero pressure head 2 m below the top, representing the groundwater table.

The simulation shows the gradual development towards a steady state with constant infiltration and a gradual increase of volumetric water content from 38% towards its maximum of 52%. Parameters for soil properties were taken from Hornberger and Wiberg (2005). The physical units for all data are a combination from length in [cm] and time in [h].

References

Bear J (1972) Flow through porous media. Elsevier, New York, p 764

Bear J, Verruijt A (1987) Modeling groundwater flow and pollution. D. Reidel Publ, Dordrecht, p 414

Brooks RH, Corey AT (1964) Hydraulic properties of porous media, Colorado State University. Hydraulic Papers No 3. Ford Collins, p 27

Gallavotti G (2002) Foundations of fluid dynamics. Springer, Berlin, p 513

Guyon E, Hulin JP, Petit L (1997) Hydrodynamik. Vieweg, Braunschweig, p 466, in German

Holzbecher E (1998) Modeling density-driven flow in porous media. Springer, Heidelberg

Holzbecher E (2002) Groundwater modeling – simulation of groundwater flow and pollution. FiatLux Publ, Fremont, e-book

Hornberger G, Wiberg P (2005) Numerical methods in hydrological sciences. Am. Geophys. Union, Washington DC, e-book

Mualem Y (1976) A new model for predicting the hydraulic conductivity of unsaturated porous media. Water Res Res 12(3):513–522

Narasimhan TN (1998) Hydraulic characterization of aquifers, reservoir rocks and soils: A history of ideas. Water Res Res 34(1):33–46

Olsen NRB (2002) Hydroinformatics, fluvial hydraulics and limnology. The Norwegian Univ. of Sci. and Techn, Trondheim, p 113

Prandtl L, Tietjens OG (1957) Fundamentals of hydro- and aeromechanics. Dover, New York, p 270

Rouse H (1978) Elementary mechanics of fluids. Dover, New York, p 376, reprint

Schröder RCM (1995) Technische hydraulik. In: Mehlhorn G (ed) Der ingenieurbau: Grundwissen 2, hydrotechnik – geotechnik. Ernst & Sohn, Berlin, pp 123–214, in German

Szabó I (1987) Geschichte der mechanischen prinzipien. Birkhäuser, Basel, p 571, in German

van Genuchten MTh (1980) A closed-form equation for predicting the hydraulic conductivity of unsaturated soils. Soil Science Soc of America 44(5):892–898

Chapter 12
Groundwater Drawdown by Pumping

Aquifers are a valuable source of water. Groundwater is available and used in many parts of the world for industrial and municipal purposes and for public water supply. In several regions and urban centers the percentage of groundwater on public water supply reaches 100%. Although the chemistry of subsurface water may be very different, groundwater quality can fulfil highest standards nevertheless.

Groundwater is pumped from single wells or galleries of several wells. In the vicinity of the wells the water table may decrease, depending on the type of the aquifer. In all cases the piezometric head decreases, which is explained in more details below.

Environmental studies in connection with groundwater withdrawal are necessary for several reasons. The maximum yield, which can be extracted on a sustainable basis, is of high concern for the well operating agency. The drawdown of the water table itself may also be of ecological importance, as eco-systems in the catchment of the well can be affected. Wetlands for example are vulnerable systems, which react quite sensitive to changes of the sub-surface or surface water table.

Water quality is another important topic for water withdrawal systems. If the quality of pumped water is not sufficient, knowledge about the well catchment and the flowpaths may enable counter-measures in order to avoid or reduce the migration of polluted water towards the pumping facilities. Recharge wells may prevent such migration if operated at an appropriate location and an appropriate recharge rate.

In this chapter we examine the change of piezometric head h in the vicinity of a single pumping well. There are analytical solutions for h as function of distance from the well centre r, which can be computed easily using MATLAB®. In all cases other causes for groundwater flow (for example base flow) are neglected. More complex situations are treated in the following two chapters.

12.1 Confined Aquifer

A confined aquifer is a permeable groundwater layer between two impermeable layers (aquitards), as shown in Fig. 12.1. In idealized situations, which are treated in this chapter, groundwater flows in a permeable layer, the aquifer, from all sides

E. Holzbecher, *Environmental Modeling*,
DOI 10.1007/978-3-642-22042-5_12,

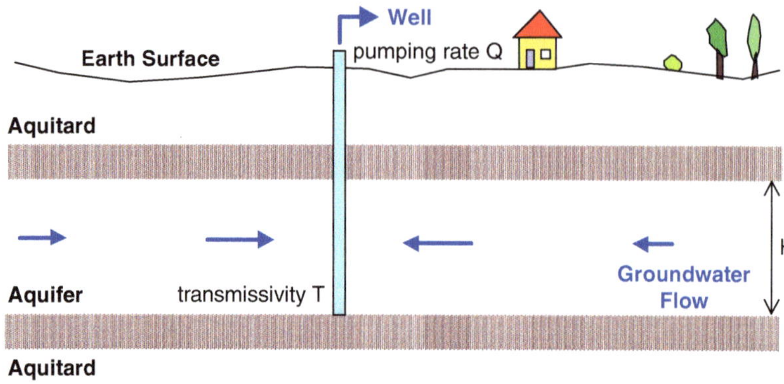

Fig. 12.1 Schematic cross-sectional view of a well pumping from a confined aquifer

radially towards an installed pumping well. It is assumed that the situation is totally equal in all radial directions, which allows the use of the radius r as the space variable. It is also assumed that there are no differences in vertical direction: the well is screened across the entire aquifer and there are no differences concerning the hydraulic properties within the permeable layer. The aquifer remains water saturated, i.e. there are no parts that fall dry due to pumping.

In the idealized situation shown in Fig. 12.1, the aquifer is characterized by a thickness H [m] and a transmissivity T [m^2/s]. In the transmissivity parameter the hydraulic conductivity K of the porous material and the thickness of the aquifer H are represented:

$$T = K \cdot H \tag{12.1}$$

T increases with thickness; T is higher for more permeable aquifers. It is assumed that the well withdraws water at a constant rate Q [m^3/s], which allows the description of the steady state groundwater flow. The relevant variable for the analysis of groundwater flow is the piezometric head h, which changes with the distance r from the well position. Piezometric head is the key variable for flow (see Darcy's Law, Chap. 11), quantifying the height of the water table above some reference level measured by a piezometer. A piezometer is a pipe that is open at both ends, and reaches into the aquifer with the lower end). h decreases if the well is approached and can be calculated by using the formula of Thiem (1906):

$$h(r) = h_0 + \frac{Q}{2\pi T} \log\left(\frac{r}{r_0}\right) \tag{12.2}$$

with:

h_0 piezometric head above base at radius r_0 [m]
Q pumping rate [m^3/s]
T transmissivity of the aquifer [m^2/s]
r_0 radius [m]

A derivation of the formula is found in Sidebar 12.1. Here we describe a short command sequence by which the piezometric head values in the vicinity of the well are calculated. At first, the values of the input parameters have to be given:

```
h0 = 5;

T = 5.e-6;
Q = 1.e-4;
r0 = 0.1;
```

Afterwards the radius vector `r` is specified. For each radius within the vector the lowering of the piezometric head h is calculated:

```
r = [0.1:0.1:20];
```

The following command initiates the computation of the desired values according to formula (12.2):

```
h = h0 + (Q/(2*pi*T))*log(r/r0);
```

In MATLAB®, `log` is the natural logarithm to the basis e. In the vector `h` we now find all piezometric heads for the radii, given in the radius vector `r`.

For

```
plot (r,h);
ylabel ('piezometric head [m]'); xlabel ('distance [m]');
```

The result is shown in Fig. 12.2.

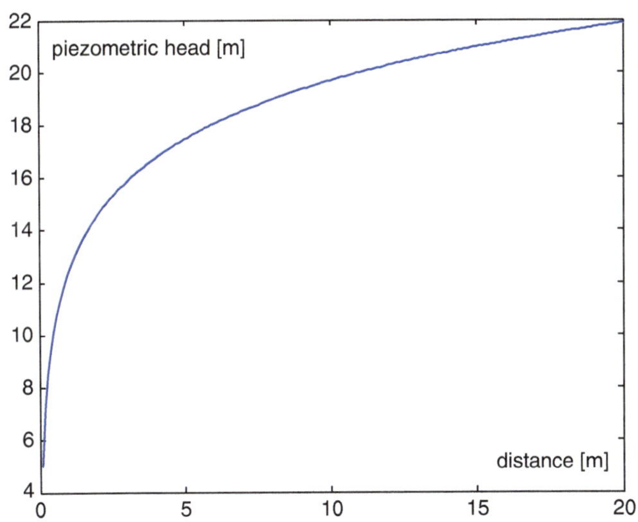

Fig. 12.2 Drawdown of groundwater piezometric head in a confined aquifer due to pumping (Thiem formula)

12.2 Unconfined Aquifer

In contrast to a confined aquifer, an unconfined aquifer (also called phreatic aquifer) is not limited by an impermeable layer from above. The upper boundary of an unconfined aquifer is given by the groundwater table. Between the groundwater table and the earth surface the unsaturated zone is located, where the pore space within the porous material is filled with water and air. Within the aquifer it is only water that flows in the pore space. The situation is schematically depicted in Fig. 12.2.

If measured in reference to the aquifer base, the variable h is the water saturated thickness of the aquifer, which is the distance between the position of the groundwater table and the base of the aquifer below. In contrast to the confined aquifer, in the unconfined aquifer piezometric head h represents the position of the groundwater table (Fig. 12.3).

The following formula adapts the Thiem (12.2) to the situation of an unconfined aquifer. It is derived in Sidebar 12.1 and delivers piezometric head h in the distance r from a well:

$$h^2(r) = h_0{}^2 + \frac{Q}{\pi K} \log\left(\frac{r}{r_0}\right) \tag{12.3}$$

with:

h_0 water level in well [m]
Q pumping rate [m^3/s]
K aquifer hydraulic conductivity [m/s]
r_0 well radius [m]

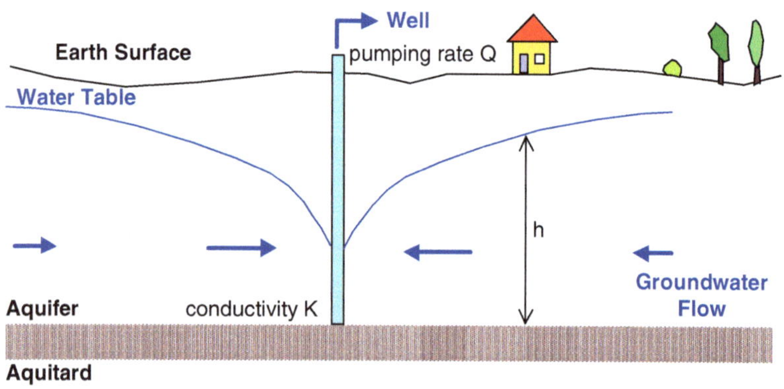

Fig. 12.3 Schematic cross-sectional view of a well pumping from an unconfined aquifer

Sidebar 12.1: Derivation of Thiem's Equations for Confined and Unconfined Aquifers

In the confined aquifer horizontal flow towards a well in a steady state needs to fulfil the volume conservation equation:

$$2\pi r H v_r = Q$$

for all radii r with radius-dependent velocity v_r, aquifer depth H and pumping rate Q. According to Darcy's Law holds:

$$v_r = K\frac{\partial h}{\partial r}$$

Both equations together deliver a differential equation for $h(r)$:

$$r\frac{\partial h}{\partial r} = \frac{Q}{2\pi T}$$

with $T = KH$. As the right hand side is a constant, the differential equation can also be written as follows:

$$\frac{\partial}{\partial r}\left(r\frac{\partial h}{\partial r}\right) = 0$$

In order to obtain a solution formula, we proceed with a reformulation of the equation:

$$\frac{\partial h}{\partial r} = \frac{Q}{2\pi T}\frac{1}{r}$$

The solution can simply be obtained by integration:

$$h = \frac{Q}{2\pi T}\log(r) + C$$

with integration constant C. If the head h_0 at a position r_0 is given, the integration constant can be determined:

$$C = h_0 - \frac{Q}{2\pi T}\log(r_0)$$

The formula (12.2) given above results.

In the unconfined situation one starts analogously with the volume conservation principle:

(continued)

$$2\pi r h v_r = Q$$

Instead of the total height of the groundwater layer, the height of the water table h above the base has to be considered. Using Darcy's Law, as stated above, yields:

$$rh\frac{\partial h}{\partial r} = \frac{Q}{2\pi K}$$

or

$$r\frac{\partial h^2}{\partial r} = \frac{Q}{\pi K}$$

with the general solution:

$$h^2 = \frac{Q}{\pi K}\log(r) + C$$

As above, the knowledge of a pair (r_0,h_0) helps to determine the integration constant C and the formula (12.3) results.

h changes with the radius r, as already explained above. We compute the changing values in a short command sequence. In the MATLAB® command window specify the new input parameter first:

```
K = 1.e-4;
```

The following line initiates the computation of the vector of piezometric heads:

```
h = sqrt(h0*h0 + (Q/(pi*K))*log(r/r0));
```

`sqrt` denotes the squareroot. With the next command the results are shown as a green broken ine:

```
plot (r,h,'--g');
```

Exercise. Change the value for K and compare drawdowns of piezometric head in a single figure! Use the command

```
hold on;
```

to keep the graphic, and the command

```
legend (,K=1.e-4');
```

to show a legend. The legend, like other additions to or corrections of the graphic, could also be added from the figure editor itself. A title is plotted by using the `title` command:

```
title ('Thiem unconfined');
```

12.3 Half-confined Aquifer

The situation in a half-confined aquifer is depicted in Fig. 12.4: an aquifer is overlain by a half-permeable layer. Thus, the pumped water partially originates from the aquifer itself, partially from the overlying strata, which is connected through the half-permeable layer.

For a thick half confined aquifer, de Glee (1930)[1] derived a formula describing the drawdown s of piezometric head at a distance r from a well:

$$s(r) = \frac{Q}{2\pi T} K_0\left(\frac{r}{\sqrt{Tc}}\right) \tag{12.4}$$

with:

Q pumping rate [m^3/s]
T transmissivity [m^2/s]
c resistance of half-permeable layer [s]
K_0 modified Bessel function 2. type 0. order

In MATLAB®, put in the new parameter c:

```
c = 1.e7;
```

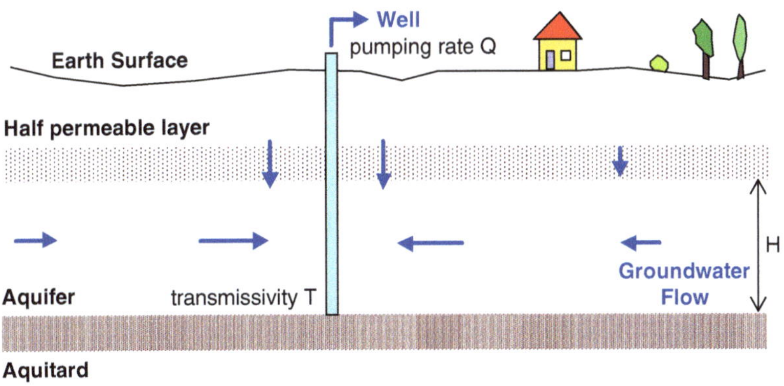

Fig. 12.4 Schematic cross-sectional view of a well pumping from a half-confined aquifer

[1] Gerrit Jan de Glee (1897–1975), Dutch hydrologist.

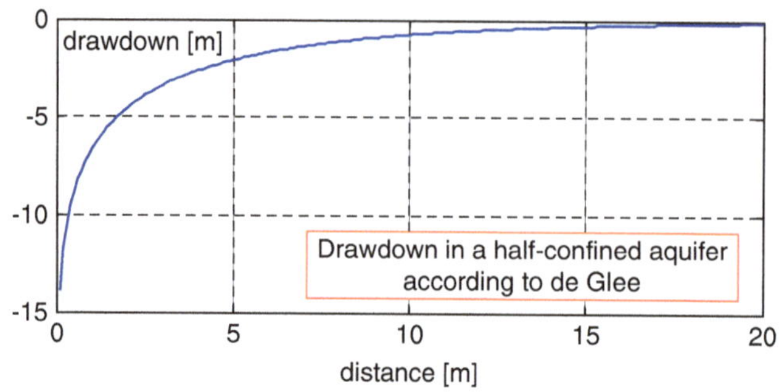

Fig. 12.5 Drawdown of groundwater piezometric head in a half-confined aquifer due to pumping, according to de Glee (1930)

and calculate the vector of drawdowns according to:

```
s = (Q/(2*pi*T))*besselk(0,r/sqrt(T*c));
```

The correct variant of the Bessel function (here `besselk`) and the parameters are found in MATLAB-help. The graphical representation of the results (Fig. 12.5) is shown by using the `plot` command:

```
plot (r,-s);
```

In Fig. 12.6 drawdowns in a confined, an unconfined and a half-confined aquifers are compared. The drawdown for the half-confined situation lies between the drawdown for the confined and the unconfined aquifers. The user may easily find parameter values for which that reasonable result is not true. The reason for the apparent incompatibility is that all three formulae are valid under different conditions. The formula of de Glee is derived for the half-space below the half-permeable layer, i.e. under the assumption that the aquifer is too extended, making the value of its thickness irrelevant.

The complete code is included in the accompanying software under the name '*welldrawdown.m*'

12.4 Unsteady Drawdown and Well Function

In a confined aquifer the drawdown of the piezometic head s is given by the formula of Theis (1983)[2]. s is a function of the distance from the well r and the time t after the start of pumping:

[2] Charles Vernon Theis (1900–1987), US-American hydrogeologist.

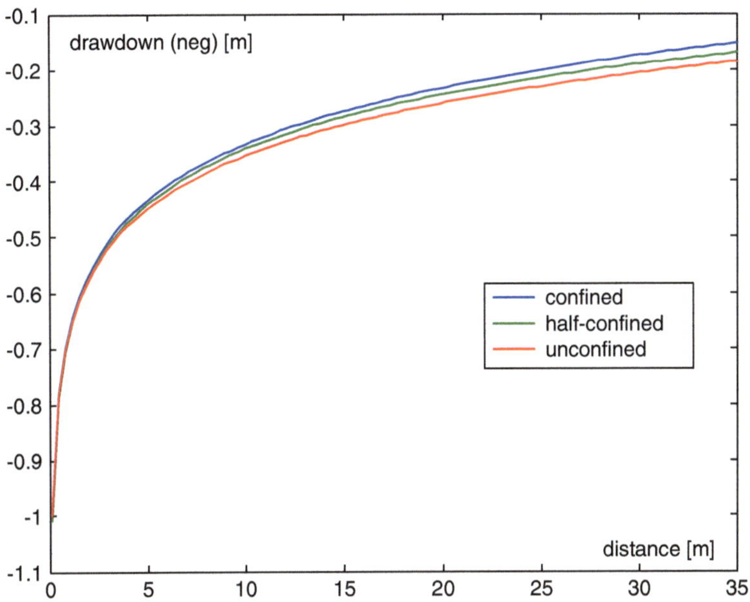

Fig. 12.6 Steady drawdown of groundwater piezometric head in a confined, a half-confined and an unconfined aquifer

$$s(r) = \frac{Q}{4\pi T} W\left(\frac{Sr^2}{4Tt}\right) \tag{12.5}$$

with:

Q pumping rate [m^3/s]
T aquifer transmissivity [m^2/s]
S storage coefficient of the aquifer [1]

In the formula appears the function W, with an argument which is usually abbreviated as u. $W(u)$ is so important for well drawdown that it is named *well function* in the corresponding literature. In the mathematical literature, the same function is called the *exponential integral* and is mostly referred to as $E_1(u)$. The name is explained by the definition of $E_1(u)$:

$$W(u) = E_1(u) = \int_u^\infty \frac{\exp(-\varsigma)}{\varsigma} d\varsigma \tag{12.6}$$

In MATLAB® the well function can be found under its mathematical notation. It is called by `expint`. In order to compute the formula given above, specify the *storage parameter* as new parameter:

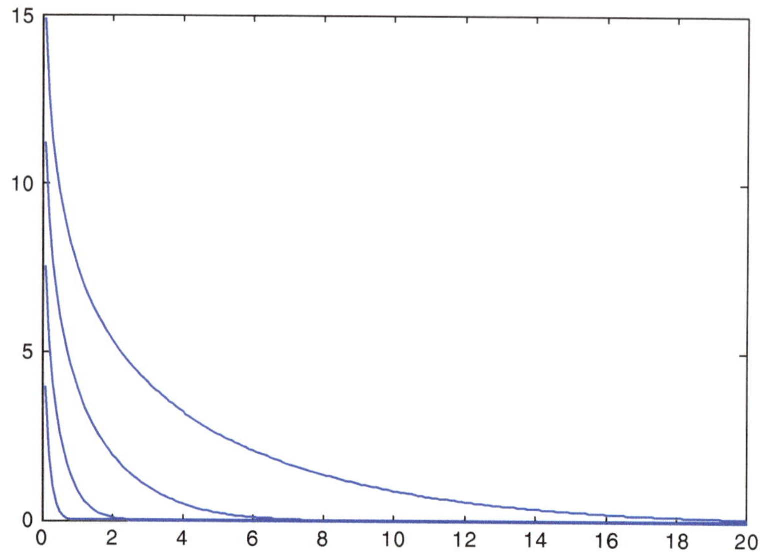

Fig. 12.7 Unsteady drawdown of groundwater piezometric head in a confined aquifer due to pumping according to Theis; parameters given in text, for times t = 10^3, 10^4, 10^5 and 10^6

```
S = 0.1;
```

Then specify the time for which drawdown is to be computed:

```
t = [1000000];
```

Calculate and plot the results by the following commands:

```
s = (Q/(4*pi*T))*expint(S*r.*r/(4*T*t));
plot (r,-s);
```

Figure 12.7 depicts the resultinmg drawdowns at four different time instants.

12.5 Automatic Transmissivity Estimation

The formulae given in the previous sub-chapters can not only be used for the computation of the groundwater drawdown and lowering of piezometric head but also for parameter estimation. In so called *pumping tests* water is pumped from one well, while drawdown is observed in some surrounding boreholes or piezometers. The result is a series of drawdown values; an example data set is given in Table 12.1. The conductivity or transmissivity of the aquifer is to be determined, i.e. we have a task of inverse modeling as already introduced in Chap. 10.

Table 12.1 Pumping test example data-set (after: Krusemann and de Ridder 1973)

Time at r = 30 [min]	Drawdown r = 30 [m]	Time at r = 90 [min]	Drawdown r = 90 [m]	Time at r = 215 [min]	Drawdown r = 215 [m]
0	0	0	0	0	0
0.1	0.04	1.5	0.015	66	0.089
0.25	0.08	2.0	0.021	127	0.138
0.5	0.13	2.16	0.023	185	0.165
0.7	0.18	2.66	0.044	251	0.186
1.0	0.23	3.0	0.054		
1.4	0.28	3.5	0.075		
1.9	0.33	4.0	0.090		
2.33	0.36	4.33	0.104		
2.8	0.39	5.5	0.133		
3.36	0.42	6.0	0.153		
4.0	0.45	7.5	0.178		
5.35	0.50	9.0	0.206		
6.8	0.54	13.0	0.250		
8.3	0.57	15.0	0.275		
8.7	0.58	18.0	0.305		
10.0	0.60	25.0	0.348		
13.1	0.64	30.0	0.364		

In simple cases parameter estimation can be performed manually, i.e. the concerned parameter is adjusted until a reasonable coincidence between observed and calculated values is obtained. Here we follow the procedure, described in Chap. 12.5, performing automatic parameter estimation using MATLAB®. The procedure is demonstrated for the Thiem formula (12.2), i.e. for the determination of the transmissivity of a confined aquifer. The example is based on a data set given by Krusemann and de Ridder (1991)), measured for a pumping test at the 'Oude Korendijk', the Netherlands. Values for steady state drawdown were obtained at four positions in different distances from the well. In MATLAB®, distances and drawdowns are specified in vectors:

```
rfit = [0.8 30 90 215];
sfit = [2.236 1.088 0.716 0.25];
```

Next pumping rate [m^3/d] and reach of the well are given, as well as an initial guess for the transmissivity [m^2/d]:

```
Q = 788;
reach = 500;
T = 700;
```

The estimation is performed by utilizing the MATLAB® zero-search function `fzero`:

```
T = fzero(@myfun,T);
```

with an appropriate function `myfun`. The function is derived from the residual condition

$$\|res\| = \sqrt{\sum \left(h(r_{fit}) - \left(h_0 - s_{fit}\right)\right)^2} \text{ is minimal} \tag{12.7}$$

When the reach of the well with condition $h_0 = 0$ is considered, condition (12.7) is equivalent to finding the minimum of the objective function

$$e(T) = \sum \left(h(r_{fit}) + s_{fit}\right)^2 \tag{12.8}$$

This has the following necessary condition:

$$\frac{\partial e}{\partial T} = 2\sum \left(h(r_{fit}) + s_{fit}\right)\frac{\partial h}{\partial T}(r_{fit}) = 0 \tag{12.9}$$

Using the Thiem formula, the derivative can be written as:

$$\frac{\partial h}{\partial T} = -\frac{Q}{2\pi T^2}\log\left(\frac{r}{r_0}\right) = -\frac{h}{T} \tag{12.10}$$

and thus the condition can be reformulated as follows:

$$\sum \left(h(r_{fit}) + s_{fit}\right)\frac{h(r_{fit})}{T} = 0 \tag{12.11}$$

The vector notation is:

$$\frac{1}{T}\left(\mathrm{h}(r_{fit}) + s_{fit}\right)h(r_{fit})^T = 0 \tag{12.12}$$

It is convenient to use the function in an M-file, which should look similar to:

```
function f = myfun(T);
global rfit sfit reach Q

% calculate Thiem solution
h = Q*log(rfit/reach)/T/2/pi;

% specify function f to vanish
f = (h+sfit)*h'/T;
```

The result for the example data set is: $T = 352$ m^2/d, which is obtained after few iterations within the MATLAB® `fzero`–module. Figure 12.8 depicts the

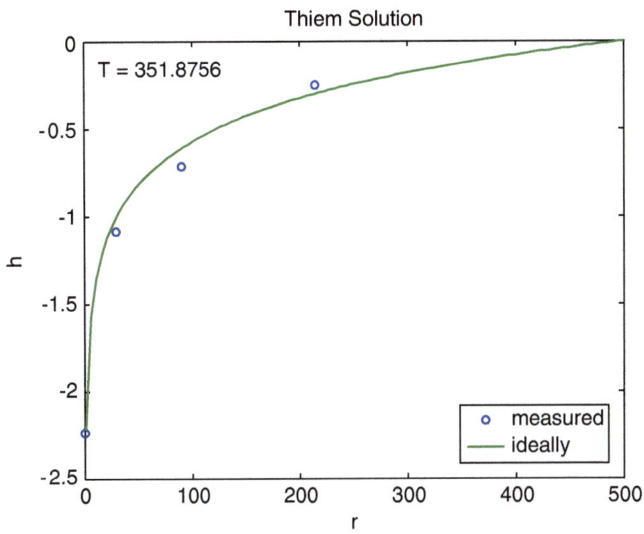

Fig. 12.8 Automatic transmissivity estimation in MATLAB® based on Theis solution

hypothetical drawdown for the ideal case, calculated by MATLAB®, and the measured drawdowns.

The MATLAB® module for automatic transmissivity estimation based on Theis steady-state solution can be found under the name *'thiem_test.m'*. The described results can be obtained by setting the `test`–parameter in the input section of the M-file to 1.

The example demonstrates a procedure for the determination of transmissivity in a confined aquifer using the Thiem formula (12.2). The method can be performed similarly for the other formulae given in this chapter. One can determine hydraulic conductivity in an unconfined aquifer with the help of formula (12.3). One can estimate transmissivity and/or resistance of the half-permeable layer for a half-confined aquifer using the de Glee formula (12.4), and one may obtain the transmissivity and storativity of a confined aquifer using the Theis formula (12.5).

Automatic Transmissivity Estimation Exercise

Write an M-file, similar to the example given above, and perform an automatic parameter estimation for an unsteady pumping test using the Theis formula (12.5). As two parameters have to be estimated, use a structure of two functions, where the second function for the estimation of the transmissivity is called within the first for the estimation of the storage coefficient.

As an exercise, use the data set from Table 12.1. There are three observation points in the distances $r = 30$ m, 60 m and 215 m from the well. In the columns of the table corresponding times and drawdowns are given for the various measurements. The example data-set was taken from Krusemann and de Ridder (1973).

Hint: use the derivative of the well function

$$\frac{\partial W(u)}{\partial u} = \frac{\exp(-u)}{u} \tag{12.13}$$

to show the following two equalities:

$$\frac{\partial W}{\partial T} = \frac{\partial W}{\partial u} \cdot \frac{\partial u}{\partial T} = \frac{-\exp(-u)}{T} \tag{12.14}$$

and

$$\frac{\partial W}{\partial S} = \frac{\partial W}{\partial u} \cdot \frac{\partial u}{\partial S} = \frac{\exp(-u)}{S} \tag{12.15}$$

Formulae (12.4) and (12.5) have to be used in the automatic parameter estimation procedure that is based on derivatives and is demonstrated above and in Chap. 10.4.

References

de Glee GJ (1930) Over grondwaterstromingen bij wateronttrekking door middel van putten. J. Waltman, Delft, p 175, in Dutch

Krusemann GP, de Ridder NA (1973) Untersuchung und Anwendung von Pumpversuchen. Verlagsges R Müller, Köln, p 191, in German

Krusemann GP, de Ridder NA (1991) Analysis and evaluation of pumping test data. Intern. Inst. for Land Reclamation and Improvement (ILRI), Wageningen, p 377

Theis CV (1983) The relation between the lowering of the piezometric surface and the rate and duration of discharge of a well using ground-water storage. In: Freeze RA, Back W (eds) Physical hydrogeology. Hutchinson Ross Publ, Stroudsburg, pp 141–146

Thiem G (1906) Hydrologische methoden. J.M.Gebhardt, Leipzig, p 56, in German

Chapter 13
Aquifer Baseflow and 2D Meshing

13.1 1D Analysis

In general a flow field, as introduced in Chap. 11, may be different at every location and, for transient flow, at every time instant. In contrast, one-dimensional base flow represents a constant vector independent of time and space. Such an idealized situation is seldom met in environmental compartments, but it may serve as an approximate description of field situations. An example could be the groundwater flow between two parallel channels which have a constant but different water level. However, in parts of a regional watershed often a constant flow field is assumed as a simplification of the real situation. The 1D assumption is also often justified for experimental set-ups, for example in column experiments. The simplest flow pattern for a fluid set-up between two plates is also 1D.

For a basic description, we assume that the x-axis is chosen in flow direction. The flux per unit width results as product of the height h of the water column and the velocity u:

$$q = h \cdot u \tag{13.1}$$

The unit of q is [L^2/T]. When the fluid fills only part of the space, i.e. in porous media or in multi-phase situations, u is the product of the real velocity v multiplied with the share θ of the porespace or the corresponding phase on the total volume:

$$u = \theta \cdot v \tag{13.2}$$

In porous media, u is the Darcy-velocity or filter velocity. A slightly more generalized situation is given if water height and velocity are allowed to change along the flowpath (in x-direction):

$$q = h(x)u(x) \tag{13.3}$$

E. Holzbecher, *Environmental Modeling*,
DOI 10.1007/978-3-642-22042-5_13, © Springer-Verlag Berlin Heidelberg 2012

with:

q water flux per unit width [m^2/s];
h height of the water column [m];
v velocity [m/s].

For a confined aquifer (see Chap. 12) the height remains constant: h is equal to the thickness of the aquifer H. In porous media Darcy's Law is valid, which can be formulated as

$$u(x) = -K\frac{\partial h}{\partial x} \tag{13.4}$$

with: K hydraulic conductivity [m/s]

Replacing u in (13.3) by the formula (13.4) delivers:

$$q = -T\frac{\partial h}{\partial x} \tag{13.5}$$

as transmissivity is the product of hydraulic conductivity and aquifer height: $T = K{\cdot}H$. Equation (13.5) is a differential equation for the function $h(x)$. For constant transmissivity T the equation is easy to solve:

$$h(x) = -\frac{q}{T}x + h_0 \tag{13.6}$$

For the unconfined aquifer (see Chap. 12) the starting point is not (13.5) but the following:

$$q = -Kh\frac{\partial h}{\partial x} \tag{13.7}$$

The differential equation (13.7) can also be written as:

$$q = -\frac{1}{2}K\frac{\partial h^2}{\partial x} \tag{13.8}$$

which has the solution:

$$h^2(x) = -\frac{2q}{K}x + h_0^2 \tag{13.9}$$

Another relevant term is the discharge potential φ [m^3/s] with derivative $-q$ as defining condition. The discharge potential cannot be measured directly but is introduced, because it fits into the theoretical framework. φ plays an important role in the following chapters. Note that the findings for the potential are valid, whether an aquifer is confined or unconfined. However, the transition from potential φ to piezometric head h depends on the conditions of the aquifer.

The transformation formulae are derived from the equivalent defining equation:

$$\varphi(x) = -\int q dx \tag{13.10}$$

and further:

$$\varphi(x) = \begin{cases} KH \int \frac{\partial h}{\partial x} dx & \text{for the confined aquifer} \\ K \int h \frac{\partial h}{\partial x} dx = \frac{K}{2} \int \frac{\partial h^2}{\partial x} dx & \text{for the unconfined aquifer} \end{cases} \tag{13.11}$$

Thus follows finally:

$$\varphi(x) = \begin{cases} T(h(x) - h_0) & \text{for the confined aquifer} \\ \frac{K}{2}\left(h^2(x) - h_0^2\right) & \text{for the unconfined aquifer} \end{cases} \tag{13.12}$$

Note that the relation $T = KH$ was applied. The potential is not unique. The addition of a constant does not change the potential property.

13.2 1D Implementation

For modelling with MATLAB® set input parameters first:

```
h0 = 10;
Qx0 = 5.e-5;
K = 0.0001;
```

Then define the vector `x` with distances at which h is to be calculated:

```
x = [1:1:100];
```

In the following line the vector with piezometric heads for the confined aquifer is calculated utilizing formula (13.6):

```
x = [1:1:100];
```

The calculation of the unconfined situation, according to formula (13.9), is similar:

```
hu = sqrt (h0*h0 + (2*Qx0/K)*x);
```

With the following commands heads in confined and unconfined aquifers are compared:

```
plot (x,hc,'-b',x,hu,'--g');
legend ('confined','unconfined',2)
```

Exercise 13.1. Change q to a more realistic value that is 2 orders of magnitude lower

Quite often the discharge q is not known. Instead, the groundwater level h_1 in a certain distance L is known from measurements. The derivation of the solution, as shown above, delivers the formulae:

$$h(x) = \begin{cases} h_0 + \dfrac{h_1 - h_0}{L}x & \text{for the confined aquifer} \\ \sqrt{h_0^2 + \dfrac{h_1^2 - h_0^2}{L}x} & \text{for the unconfined aquifer} \end{cases} \tag{13.13}$$

Obviously, h depends only on the observed values h_0 and h_1 and the length L. h is independent of the material parameter K and the aquifer depth H.

Exercise 13.2. Compare the confined and the unconfined situation in a graph, as it is shown in Fig. 13.1!

The solution is obtained by using the following commands:

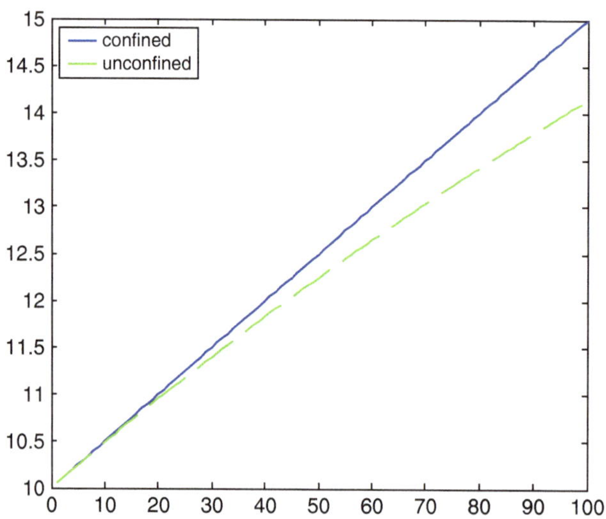

Fig. 13.1 Piezometric head for a confined and an unconfined aquifer with identical hydraulic properties

```
L=100;
h1 = 12;
hc = h0 + ((h1-h0)/L)*x;
hu = sqrt(h0*h0 + ((h1*h1-h0*h0)/L)*x);
plot (x,hc,'--g',x,hu,'-b');
legend ('confined','unconfined',2)
```

13.3 2D Implementation

Of course most flow fields in environmental systems are higher-dimensional. In this chapter we are concerned with 2D flow fields, which are more complex than 1D fields, but in general still simpler than 3D flows. The first task is to represent a 1D flow field in 2D.

In order to start with a 2D description, one needs to know how to represent meshes. In MATLAB®, the easiest way to obtain a mesh is the `meshgrid` command. It is best explained by an example:

```
[x,y] = meshgrid ([0:10:100],[10:2:20])

x =
     0    10    20    30    40    50    60    70    80    90   100
     0    10    20    30    40    50    60    70    80    90   100
     0    10    20    30    40    50    60    70    80    90   100
     0    10    20    30    40    50    60    70    80    90   100
     0    10    20    30    40    50    60    70    80    90   100
     0    10    20    30    40    50    60    70    80    90   100
y =
    10    10    10    10    10    10    10    10    10    10    10
    12    12    12    12    12    12    12    12    12    12    12
    14    14    14    14    14    14    14    14    14    14    14
    16    16    16    16    16    16    16    16    16    16    16
    18    18    18    18    18    18    18    18    18    18    18
    20    20    20    20    20    20    20    20    20    20    20
```

A rectangular mesh, a grid, is produced for which the x-coordinates are given by the first vector in the brackets and the y-coordinates by the second vector in the brackets. If the command is called with a single vector as parameter, that one is used for both x- and y-direction. The `meshgrid` command produces two matrices, one containing the x-coordinates, and one containing the y-coordinates. In the example the matrices are denoted by `x` and `y` also. Using the commands

```
hc = h0 + (Qx0/(K*h0))*x;
hu = sqrt (h0*h0 + (2*Qx0/K)*x);
```

the piezometric heads for the confined and for the unconfined aquifers are calculated for all grid points. Both `hc` and `hu` are also matrices here which have the same dimension as `x` and `y`. The fields are plotted as surfaces by using the command:

```
surf (x,y,hc);
hold on
surf (x,y,hu);
```

Here the mesh is rectangular and regular, but the `surf`-command can also be used for irregular meshes.

The next step is to extend the model for situations in which the aquifer is partially confined and partially unconfined. The potential φ (in the M-file: `phi`) is first evaluated without a constant. Then an appropriate constant `phi0` is added. The constant is computed differently for the confined and for the unconfined situation. The confined situation is given if `h0`, the head at mesh position (1,1), exceeds the aquifer thickness `H`. In order to use this simple criterion, it is required that the hydraulic head is measured to zero level at the aquifer base (Fig. 13.2).

The formulae (13.12) are applied. The potential for the confined situation gets another constant, which is $-\frac{1}{2}KH^2$. Inserting this constant the potential becomes a continuous function at locations with $h = H$, where the conditions change from confined to unconfined or vice versa. The potential value at these 'critical points' is given by:

$$\varphi_{crit} = \frac{1}{2}KH^2 + \varphi_0 \tag{13.14}$$

The formulae for the head thus become:

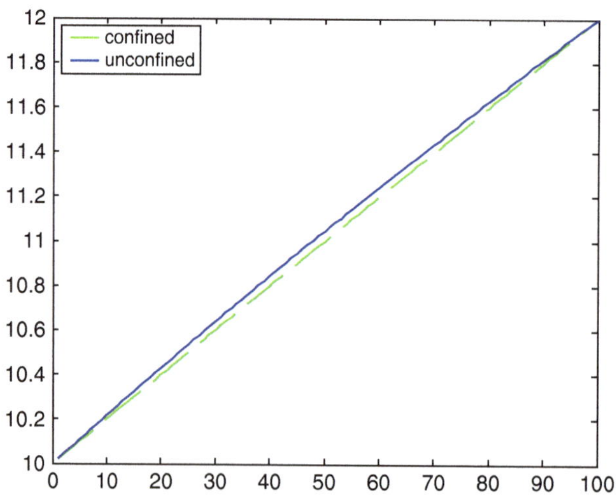

Fig. 13.2 The difference of piezometric head between two specified levels for a confined and an unconfined aquifer is obviously marginal

$$h(x) = \begin{cases} \dfrac{H}{2} + \dfrac{\varphi(x)}{KH} & \text{for the confined aquifer} \\ \sqrt{\dfrac{2}{K}\varphi(x)} & \text{for the unconfined aquifer} \end{cases} \tag{13.15}$$

The entire M-file is given by:

```
Qy0 = -1.e-6; Qx0 = 0.0;
K = 0.001;
h0 = 9.99;
H = 10;

[x,y] = meshgrid ([0:2:200],[0:2:200]);
phi = -Qx0*x - Qy0*y;
if (h0>=H)
   phi0 = -phi(1,1) + K*H*h0 - 0.5*K*H*H;     % confined
else
   phi0 = -phi(1,1) + 0.5*K*h0*h0;            % unconfined
end
phi = phi + phi0;
phicrit = 0.5*K*H*H + phi0; % margin between confined and unconfined
confined = (phi>=phicrit);
h = confined.*(0.5*H+(1/K/H)*phi)+~confined.*sqrt((2/K)*phi);
surf (x,y,h);
```

In the final part of the M-file we use a matrix containing the values 0 and 1, depending on the hydraulic situation of the aquifer at the corresponding mesh position. The array confined contains a 1 if the aquifer is confined, and a 0 if it is

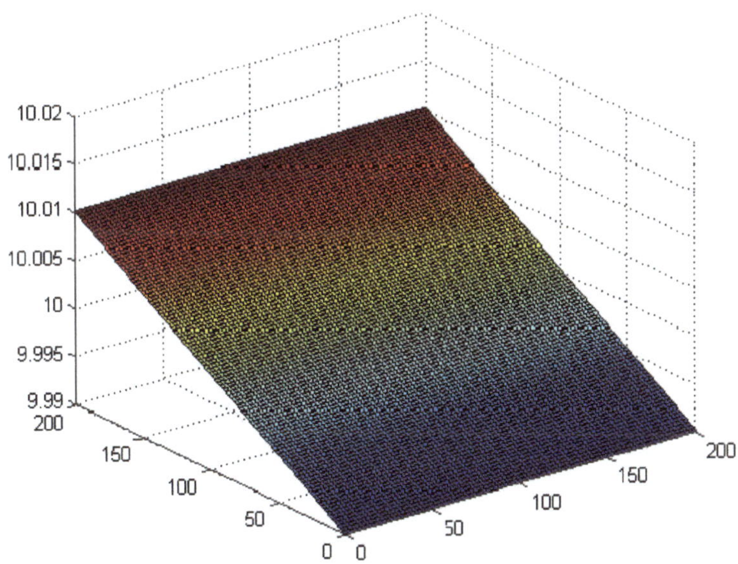

Fig. 13.3 Head distribution for the reference input data

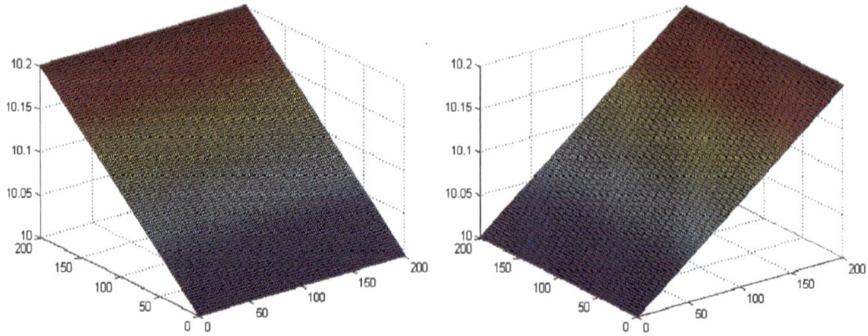

Fig. 13.4 Two examples for baseflow for confined aquifer (with reference head 10 m, $K = 10^{-4}$ m/s); *left*: in x-direction ($Q_{x0} = -10^{-6}$, $Q_{y0} = 0$), *right*: in y-direction ($Q_{x0} = 0$, $Q_{y0} = -10^{-6}$)

unconfined. The matrix is created simply by the formula `phi>=phicrit`. In the following statement the matrix `h` is calculated using the formula for the confined or for the unconfined situation, depending on the situation at each mesh node. For that purpose the `confined` and the `~confined` arrays are used, with the negation operator `~` that switches between 0 and 1 at each location.

The following figure depicts the output of the M-file for the reference input data-set (Fig. 13.3).

Exercise 13.3. Change `Qx0` and `Qy0`! Two examples are given below in Fig. 13.4.

13.4 Meshs and Grids

Some commands concerning MATLAB® 2D graphics have already been used in previous chapters and subchapters (see Chap. 4 for `surf` and `contourf`). MATLAB® 2D graphics, justifying the name of the software, is based on matrices, i.e. 2D arrays. The values within a matrix can be visualized by several graphics commands. The user may try the following simple sequence:

```
A = rand(10);
surf (A);
```

which produces a surface plot of a random matrix (Fig. 13.5):

Other commands that work on single matrices are: `plot (A)`, `contour (A)`, `contourf (A)`, `mesh (A)` and `waterfall (A)`. In these introductory examples the x-y-positions in the graph are simply the indices corresponding to rows and columns of the matrix. In most 2D graphics such a simple mesh of integers is not appropriate. Thus the graphical representation starts with the computation of the mesh.

The basic mesh generating `meshgrid` command was already used in this Chapter before. Using `meshgrid` in 2D, two 1D vectors are transformed into two 2D arrays; the latter represent the x- and y- coordinates of the mesh nodes. As already

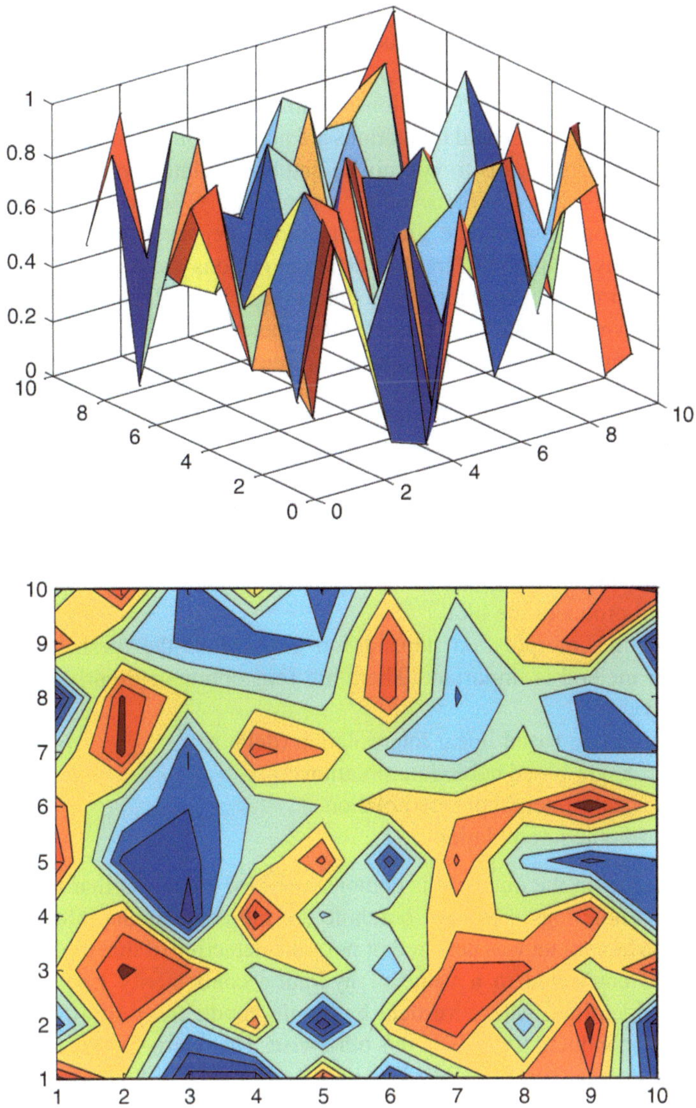

Fig. 13.5 Surface plot and filled contours for a matrix

demonstrated, the command can also be used for the generation of 3D meshes. If the input vectors are equidistant, a *regular mesh* is produced, in which all areal *elements* of the mesh are of same size. If the vectors are not equidistant, the resulting mesh is *irregular*.

MATLAB® contour commands are based on a rectangular grid. The contouring algorithm is quite simple. For a given contour level v the positions between two neighboring grid points are computed based on linear interpolation if the contour level lies between the variable values of these two positions x_i and x_j, i.e. if $u(x_i) < v < u(x_j)$ or $u(x_i) > v > u(x_j)$ hold. In a second step the algorithm draws lines connecting the calculated positions within the same block.

Contours can also be created if positions are not distributed on a rectangular grid. An intermediate operation is to interpolate and extrapolate given data to a rectangular grid. This is done by the `griddata` command. The command has several parameters and options and the graphical output, i.e. the exact locations of the contour lines may be quite different, if different options are used.

We demonstrate the procedure for a dataset that is available in three columns: x-coordinates, y-coordinates and values, all gathered in an array `X`. The following command sequence can be used to produce a figure of filled contours:

```
[x,y] = meshgrid(linspace(xmin,xmax,100),linspace(ymin,ymax,100));
z = griddata (X(:,1),X(:,2),X(:,3),x,y);
contourf (x,y,z);
```

`xmin`, `ymin`, `xmax` and `ymax` are the limits of the x- and y-region. By the first command the mesh is calculated and stored in the variables `x` and `y`. The second command initiates the interpolation. Interpolated data are stored in variable `z`. The last command produces the plot.

Of course the interpolation depends on gridsize of the mesh. But also the interpolation method is important. There are four options currently implemented in MATLAB®:

- 'Linear': triangle-based linear interpolation (default)
- 'Cubic': triangle-based cubic interpolation
- 'Nearest': nearest neighbor interpolation
- 'v4': biharmonic interpolation

The influence on the outcome is demonstrated for a data-set that was extracted from 100 borehole logs. For each borehole the file contains the information, if a certain subsurface strata was detected or not. The detection of that highly impermeable strata is connected to a value of hydraulic conductivity (see Chap. 11.3). Detection is identified with a vertical hydraulic conductivity of 10^{-8} m/s, while there is a high conductivity of 10^{-3} m/s otherwise. The interpolations using the four different methods available in MATLAB®, lead to very different results that are presented in Fig. 13.6.

Results from linear and cubic interpolation look most similar at first sight, but a closer look reveals that the cubic method delivers negative values at several places of the map. Like other material properties and like concentrations, conductivities can not take negative values and the result of cubic interpolation is thus not appropriate for processing in an environmental model.

Biharmonic spline is a method, for which the curvature of the approximated surface plays an important role (Sandwell 1987). As a result local peaks in the data set have an influence on a larger region in their vicinity. The method is not appropriate in this example, as the regions with negative values are even more extended than for cubic interpolation.

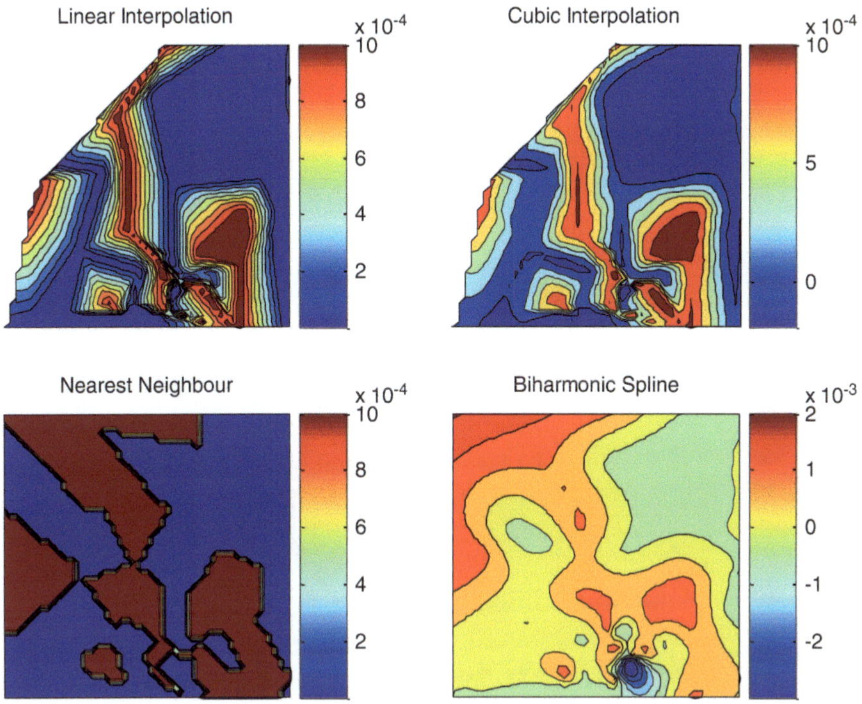

Fig. 13.6 Comparison of interpolation methods

The 'nearest neighbor' method delivers a completely different picture, as the variable takes only two values. There is no transition zone. Concerning its geological significance the result seems to be the closer to reality, as there are no locations with intermediate values: the impermeable layer is either present or not.

There are also differences between the four methods concerning extrapolation. Linear and cubic interpolation are based on triangulation of the model region and thus deliver no values outside the subregion, for which data are available. Outside that region `NaN` is given by the algorithm, i.e. there are no values extrapolated. Consequently in these regions, near the upper and right boundary, no colours are depicted in the plot. The other two methods deliver values for the entire grid, although extrapolation may not be justified. By using the 'nearest neighbor' method, the user has to set NaN's manually before plotting, if extrapolation is not wanted.

References

Sandwell DT (1987) Biharmonic spline interpolation of Geos-3 and Seasat altimeter data. Geophys Res Lett 14(2):139–142

Chapter 14
Potential and Flow Visualization

14.1 Definition and First Examples

Potential and streamfunction are mathematical functions that cannot be observed directly in the real world, but which turn out to be extremely powerful concerning the calculation and visualization of 2D flow fields. There are applications for all types of fluids, for free flow of gases and liquids, as well as for porous media flow. Electro- and magnetodynamics are other scientific fields where potential theory is applied extensively.

The notation potential refers to a function φ, from which a flow field is derived by the gradient of φ. φ is a velocity potential if:

$$\mathbf{v} = \pm\nabla\varphi \tag{14.1}$$

For steady incompressible fluids (see Chap. 2), for which the continuity equation $\nabla \cdot \mathbf{v} = 0$ is valid, follows the *potential equation* or *Laplace*[1] *equation*[2]:

$$\nabla^2\varphi = 0, \quad \text{in 2D: } \frac{\partial^2\varphi(x,y)}{\partial x^2} + \frac{\partial^2\varphi(x,y)}{\partial y^2} = 0$$
$$\text{in 3D: } \frac{\partial^2\varphi(x,y,z)}{\partial x^2} + \frac{\partial^2\varphi(x,y,z)}{\partial y^2} + \frac{\partial^2\varphi(x,y,z)}{\partial z^2} = 0 \tag{14.2}$$

The short form, using the ∇-operator, is valid for 2D and 3D cases. In fluid dynamics the potential φ has the physical unit of [m^3/s]. The name potential is connected with the property that at each location of the model region the flux or velocity vector can be derived from the gradient of the potential:

[1] Pierre-Simon Laplace (1749–1827), French mathematician and astronomer.

[2] The formulation $\Delta\varphi = 0$ can be found frequently, which makes sense, as the Laplace operator Δ is formally defined as $\Delta := \nabla \bullet \nabla$.

E. Holzbecher, *Environmental Modeling*,
DOI 10.1007/978-3-642-22042-5_14,

$$\mathbf{v} = \nabla\varphi \tag{14.3}$$

There is an entire mathematical discipline dealing with solutions of the 2D potential equation. Subject of *complex analysis* are *harmonic functions* that are solutions of the 2D Laplace equation. In this chapter we deal with 1D parallel flow that is represented by the potential

$$\varphi(x,y) = \varphi_0 + \varphi_x x + \varphi_y y \tag{14.4}$$

Additionally there are sources and sinks in the infinitely extended space, which are represented by the potential:

$$\varphi(x,y) = \frac{Q}{2\pi}\log(|\mathbf{r} - \mathbf{r}_0|) \tag{14.5}$$

where φ_x, φ_y and φ_0 are constant numbers. Q denotes the source- or sink-rate, $\mathbf{r}_0$ the location of the source or sink in 2D space and $\mathbf{r} = (x,y)$ the vector towards the current location. According to vector analysis, $\mathbf{r} - \mathbf{r}_0$ is the vector connecting source/sink location with the current position. $|\mathbf{r} - \mathbf{r}_0|$ is the length of the connecting vector, equal to $\sqrt{(x-x_0)^2 + (y-y_0)^2}$.

In the following we examine potentials emerging from the superposition of formulae (14.4) and (14.5). According to the *principle of superposition*, the sum of the functions is a solution of the Laplace equation too. The principle is a trivial consequence from the fact that the potential equation is linear. In MATLAB® such functions can be visualized easily as demonstrated by the following command sequence.

```
xmin = -1; xmax = 1;                         % x-coordinates
ymin = 0; ymax = 2;                          % y-coordinates
x0 = 0; y0 = .905;                           % source/sink location
Qx0 = 0.1; Qy0 = 0;                          % baseflow components
Q = 1;
                                         % source/sink rate
% mesh generation
xvec = linspace(xmin,xmax,100);
yvec = linspace(ymin,ymax,100);
[x,y] = meshgrid (xvec,yvec);                % create mesh

% processing
r = sqrt((x-x0).^2+(y-y0).^2);               % distances to well
phi = -Qx0*x - Qy0*y + (Q/(2*pi))*log(r);    % potential

%post-processing
surf (x,y,phi);                              % surface plot
```

With the first five instruction lines the parameter values are specified. In the next step, 100 equidistant positions at the intervals on the x- and y-axis are computed before the mesh is constructed and stored in the `x` and `y` arrays, using MATLAB® `meshgrid`. Next, the array `r` is computed, which contains the distances to the

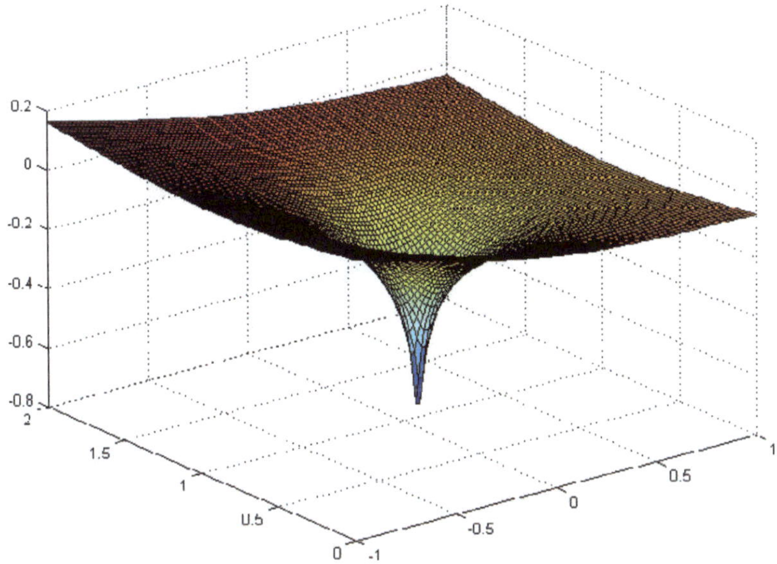

Fig. 14.1 Surface plot for a potential function; derived from superposition of 1D baseflow and flow towards a sink

source/sink position for all mesh-points. The potential array `phi` is evaluated in the following command and finally plotted as surface plot.

The result of the entire series of commands is given in Fig. 14.1. The outer gradient, due to baseflow, is clearly visible as well as the dramatic drawdown where the sink is approached. As the M-file is written, a positive value for `Q` produces a sink, while a negative value results in a source. The user may check the direction and gradient of baseflow by variation of the variables `Qx0` and `Qy0`. Note that the given logarithm has a single singularity at the source/ sink position with $\mathbf{r} = \mathbf{r}_0$.

In 3D the point source/sink solution, corresponding to formula (14.5), is given by

$$\varphi(x, y, z) = -\frac{Q}{4\pi} \frac{1}{|\mathbf{r} - \mathbf{r}_0|} \tag{14.6}$$

In MATLAB® the 3D potential is calculated by the following command sequence. The resulting plot is depicted in Fig. 14.2.

```
Q = 1;
Qx0 = 1;
i = linspace(1,3,50)
[x,y,z] = meshgrid (i,i,i)
r = sqrt((x-2.05).^2+(y-2.05).^2+(z-2.05).^2);
xslice = [1.5;1.9;2.3];
yslice = [3];
zslice = [1.5;2.2];
phi = -Qx0*x-Q/4/pi/r;
slice (x,y,z,phi,xslice,yslice,zslice)
```

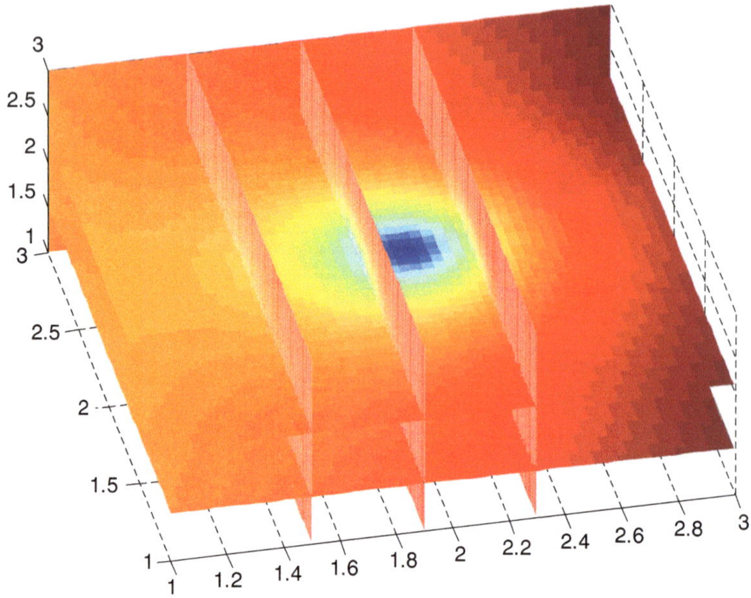

Fig. 14.2 Potential flow in 3D

14.2 Potential and Real World Variables

From the Euler equations (see Chap. 11) for irrotational potential flow the following formula can be derived (Guyon et al. 1997):

$$\rho \frac{\partial \varphi}{\partial t} + \rho \frac{v^2}{2} + p + \rho \varphi_f = C \tag{14.7}$$

where C is a constant for the entire domain. φ_f denotes the potential of an outer force. Equation (14.7) resembles the Bernoulli theorem (see Chap. 11), which holds for all solutions of the Euler equations but with a constant C on streamlines only. It is a formula connecting the potential and real world variables p and v. If the potential is known, the velocity can be obtained by formula (14.1), and (14.7) becomes an equation for the pressure p as only unknown variable. Some conditions concerning ρ and φ are required additionally, which was already discussed in detail by Prandtl and Tietjens (1934).

In 2D porous media flow it is usual to use the discharge vector $\mathbf{q}$ as the negative gradient of the discharge potential φ:

$$\mathbf{q} = -\nabla \varphi, \quad \text{in 2D:} \begin{pmatrix} q_x(x,y) \\ q_y(x,y) \end{pmatrix} = \begin{pmatrix} -\dfrac{\partial \varphi(x,y)}{\partial x} \\ -\dfrac{\partial \varphi(x,y)}{\partial y} \end{pmatrix} \tag{14.8}$$

The discharge vector has the dimension [m^2/s] and denotes the volume of water flowing in the x-y-plane per unit space of z-direction. Discharge vector and Darcy-velocity $\mathbf{u}$ [m/s] are related according to the formula:

$$\mathbf{q}(x,y) = \begin{cases} H \cdot \mathbf{u}(x,y) & \text{for the confined aquifer} \\ h(x,y) \cdot \mathbf{u}(x,y) & \text{for the unconfined aquifer} \end{cases} \tag{14.9}$$

with H thickness of the confined aquifer [m] and h height of watertable above the base of the unconfined aquifer [m].

For the real interstitial velocity $\mathbf{v}$ [m/s] holds:

$$\mathbf{q}(x,y) = \begin{cases} H \cdot \theta \cdot \mathbf{v}(x,y) & \text{for the confined aquifer} \\ h(x,y) \cdot \theta \cdot \mathbf{v}(x,y) & \text{for the unconfined aquifer} \end{cases} \tag{14.10}$$

with porosity θ. Using Darcy's Law $\mathbf{u} = -K \cdot \nabla h$ (with hydraulic conductivity K), a formula for the calculation of the potential φ from the piezometric head h can be given:

$$\varphi(x,y) = \begin{cases} K \cdot H \cdot h(x,y) + C_c & \text{for the confined aquifer} \\ \frac{1}{2}K \cdot h(x,y)^2 + C_u & \text{for the unconfined aquifer} \end{cases} \tag{14.11}$$

h [m] is the piezometric head above the base, both for the confined and the unconfined case. In the unconfined aquifer, h corresponds to the position of the groundwater table. C_u and C_c are constants that are irrelevant for the flow field: when the potential is differentiated, the two constants vanish. However, C_u and C_c are relevant for the relation between h and φ. Details are given below.

The condition that the head has no jump, where the aquifer changes from confined to unconfined state, yields a condition for C_u und C_c. If both formulae for the marginal condition $h = H$ are evaluated, both potential values are equal under the condition:

$$C_c = C_u - \frac{1}{2}KH^2 \tag{14.12}$$

One obtains a continuous potential $\varphi(x,y)$, with which it is possible to describe aquifers being partly confined and partly unconfined. Altogether one may thus write:

$$\varphi(x,y) = \begin{cases} K \cdot H \cdot h(x,y) - \frac{1}{2}K \cdot H^2 + \varphi_0 & \text{for the confined aquifer} \\ \frac{1}{2}K \cdot h(x,y)^2 + \varphi_0 & \text{for the unconfined aquifer} \end{cases} \tag{14.13}$$

where the notation φ_0 is used instead of C_u. The transition between confined and unconfined situation is given for the (critical) potential value:

$$\varphi_{crit} = \frac{1}{2}KH^2 + \varphi_0 \tag{14.14}$$

In order to compute piezometric head from the potential, which is obtained as solution of the potential equation, the (14.13) have to be resolved for h. The result is:

$$h(x,y) = \begin{cases} \left(\varphi(x,y) + \frac{1}{2}K \cdot H^2 - \varphi_0\right)/(KH) & \text{for the confined aquifer} \\ \sqrt{2(\varphi(x,y) - \varphi_0)/K} & \text{for the unconfined aquifer} \end{cases} \tag{14.15}$$

Equations (11.16) and (11.17) are valid for porous media flow independent of the number of spatial coordinates, i.e. in 1D, 2D and 3D. Under certain conditions these may be reformulated as potential equations. From (11.6) can be derived that in case of constant conductivity K the potential equation is valid for $\varphi = K \cdot h$ in model regions without sinks or sources. One may also use the pressure formulation (11.7) to obtain a connection between potential and pressure:

$$\varphi = \frac{\mathbf{k}}{\mu}\nabla(p - \rho g z) \tag{14.16}$$

14.3 Example: Groundwater Baseflow and Well

In the following example the formulae, derived for aquifers in the previous subchapter, are applied to a system of wells in an aquifer. Input values are typical for hydro geological set-ups. There are aquifer thickness, hydraulic conductivity and baseflow in both coordinate directions. Moreover, there is a reference value for piezometric head, which is to be valid at a very specific position of the model region.

Well coordinates and pumping rate are to be specified as well as the extension of the model region:

```
% Baseflow
H = 5.;                  % thickness [L]
h0 = 5.5;                % reference piezometric head [L]
K = 5.e-5;               % hydraulic conductivity [L/T]
Qx0 = 1.e-6;             % baseflow in x-direction [L^2/T]
Qy0 = 0;                 % baseflow in y-direction [L^2/T]
% Well
x0 = 100;                % x-coordinate well position [L]
y0 = 0;                  % y-coordinate well position [L]
Q = 1.e-4;               % pumping / recharge rate [L^3/T]
% Mesh
xmin = 0; xmax = 200;% min./max. x-position of mesh [L]
ymin = -100; ymax = 100;% min./ max. y-position of mesh [L]

% Reference point position in mesh
iref = 1; jref = 1;

xvec = linspace(xmin,xmax,100);
yvec = linspace(ymin,ymax,100);
[x,y] = meshgrid (xvec,yvec);                          % mesh
r = sqrt((x-x0).*(x-x0)+(y-y0).*(y-y0));               % distances to well
phi = -Qx0*x + Qy0*y + (Q/(2*pi))*log(r);              % potential
phi0 = -phi(iref,jref) + K*H*h0 - 0.5*K*H*H;...
                                          % reference potential
hc = 0.5*H+(1/K/H)*(phi+phi0);                         % confined
surf (x,y,hc);                                         % surface plot
```

In the execution part the mesh is computed first in the arrays `x` and `y`. The vector of distances from the well position is stored in `r`. Then the potential is calculated in `phi`. The reference potential is determined to produce the reference head value `h0` at the (i_{ref},j_{ref}) position of the mesh. For the chosen input values, the position of the reference head has the coordinates (x_{min},y_{min}). The user has to choose different values if the reference value is to be valid at another location of the model region. The array of piezometric heads `hc` for the confined aquifer is computed from the potential using (14.15) with a value for reference potential `phi0` calculated before. The final command initiates the figure plot.

The last commands are valid for the confined aquifer. For the unconfined aquifer the command sequence needs to be changed only slightly. The potential remains the same. Only the part in which the head is computed from the potential has to be altered. The command for `phi0` is concerned, as well as the computation of the head `hu` for the unconfined aquifer. One may use the following commands

```
phi0 = -phi(iref,jref) + 0.5*K*h0*h0;          % reference potential
hu = sqrt ((2/K)*(phi+phi0));                  % unconfined
contourf (x,y,hu);
```

in order to produce a contour plot, as shown in Fig. 14.3. The reference head value needs to be below the aquifer thickness; `H = 6` was used in the computation.

The M-file can be extended to account for situations in which the aquifer is partially confined and partially unconfined in the model region. To do that, the M-file must be changed in the part where the head is calculated:

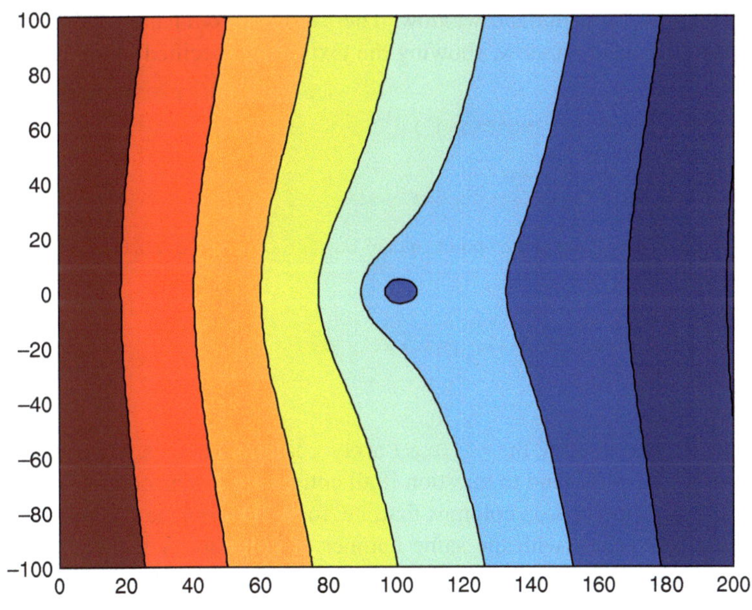

Fig. 14.3 Drawdown in an aquifer in the vicinity of a pumping well

```
if h0 > H
    phi0 = -phi(iref,jref) + K*H*h0 - 0.5*K*H*H;
else
    phi0 = -phi(iref,jref) + 0.5*K*h0*h0; % reference potential
end
hc = 0.5*H+(1/K/H)*(phi+phi0);              % head confined
hu = sqrt ((2/K)*(phi+phi0));               % head unconfined
phicrit = phi0 + 0.5*K*H*H;           % transition confined / unconfined
confined = (phi>=phicrit);            % confined / unconfined indicator
h = confined.*hc+~confined.*hu;             % head
```

The reference value `phi0` is computed in the first five lines. If the specified reference head `h0` exceeds the aquifer depth `H`, the aquifer is confined, otherwise it is unconfined. For those situations the reference potential is computed using different formulae, which are obtained by solving (14.13) for φ_0.

Then the heads `hc` and `hu` for the confined and the unconfined situation are both calculated. `phicrit` is the critical potential value at which the aquifer switches between confined to unconfined state, according to formula (14.14). The `confined` array contains a 1 at every entry corresponding to a mesh node, where the aquifer is confined, and a 0 at every entry that corresponds with a mesh node where the aquifer is unconfined. In the final command the real 2D head array is computed. Each entry related to a location in the confined part takes the `hc` value, and the `hu` value if that is related to a location in the unconfined part of the aquifer. '~' is the negation operator, switching an 1-entry to 0 and vice versa.

It is useful to indicate to the user the state of the aquifer within the model-region. In the demonstration example below, messages concerning the state of the aquifer are displayed in the command window. The `display` command produces a message in the command window, showing the text string specified in the brackets.

```
if all(all(confined))
    display ('aquifer confined');
else
    if all(all(~confined))
        display ('aquifer unconfined');
    else
        display ('aquifer partially confined and unconfined');
    end
end
if any(any(h<0))
    display ('aquifer falls partially dry');
    h = max(0, h);
end
```

For a confined aquifer the `confined` array contains only 1s. Here we use the MATLAB® `all` command to question if all entries are 1. The command has to be used twice, as it operates on columns first, i.e. for a 2D array the command `all(A)` produces a line vector with the same number of columns as `A`. The column has a 1-entry for each column containing non-zero elements only. The second call gathers all columns. Some other MATLAB® commands operate on the same idea.

The following double `all` command on the `~confined` array delivers a 1 if the aquifer is unconfined. For the remaining situations, in which the aquifer is neither confined nor unconfined, the corresponding message is given in the nested `else` block.

The `any` command is applied in the last four lines. It asks if the condition is true for any entry of the matrix `phi<0`. Any user in doubt may examine that for each matrix `A` the expression `A<0` is a matrix too! If the head value, representing the height of the piezometer level table above the aquifer base, becomes less or equal zero, the aquifer is fallen dry. Is that true in any part of the chosen region, the corresponding message is given in the command window. The final command changes all negative head values (as they are impossible) to zero.

The commands outlined above are part of the *"gw_flow.m"* file, which is included in the accompanying software. The groundwater baseflow and well example is extended in Chap. 15.

Exercise 14.1. Vary parameters and examine in which parts the aquifer is confined and unconfined, and when the aquifer falls dry!

14.4 MATLAB® 2D Graphics

Based on mesh data several output commands for 2D and 3D meshes are available in MATLAB®. Some of these (`contourf` and `surf`) have already been introduced in the previous sub-chapters. They will be demonstrated here by extending the M-file developed in this chapter until this point. First a graphic option is introduced in the specification part of the M-file:

```
% Graphical output options
gsurfh = 1;            % piezometric head surface plot
```

The user controls the graphical output by the `gsurfh` parameter. At the end of the previous version of the M-file add the plot commands:

```
%----------------------------------graphical output----------------
if gsurfh
    figure; surf (x,y,h);                                  % surface
end
```

If the graphics control parameter is nonzero, the figure editor is opened and a surface of the head values is plotted.

As another extension of the M-file, two further graphics options are added:

```
gcontf = 20;            % no. filled contour lines (=0: none)
gquiv = 1;              % arrow field plot
```

which are connected to additional output commands at the end of the file:

```
figure;
if gcontf                                          % filled contours
    contourf (x,y,h,gcontf,'w');
    colormap (winter);
    colorbar; hold on;
end
if gquiv
    quiver (x,y,u,v,'y'); hold on;                 % arrow field
end
```

With the `gcontf` parameter there is the option to plot filled contours of head values. If such a plot is not wanted, set the parameter to zero. `gcontf` not only serves as a switch for contour plots; it also contains the number of contours to be plotted. The contours' color is white (formal parameter `'w'`).

The MATLAB® `colormap` command determines the use of colors in the plot. The default colormap is `jet`, which has all colours of the rainbow. Here we select the `winter` colormap, in which the colors blue and green are preferred. The user may select other predefined colormaps. It is most convenient to explore colormaps from the figure editor. Use the 'Colormap…' sub-menu entry of the 'Edit' main entry in the figure editor. Figure 14.4 depicts the colormap editor and the choice of pre-defined colormaps.

The `colorbar` command displays a colorbar next to the figure, on the right. The colorbar relates the colors to numbers, e.g. to head values in the given example.

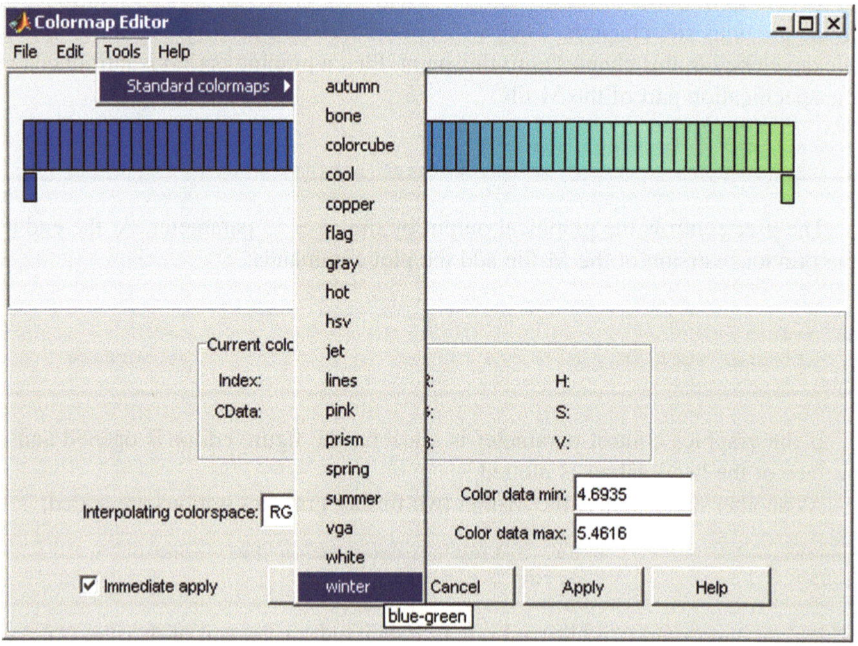

Fig. 14.4 Colormap editor

Minimum and maximum values of colorbar are calculated automatically by the software.

The `gquiv` switch is related to a `quiver` call. For a given velocity field the command depicts an arrow field. Input values are, aside from the mesh coordinates `x` and `y`, the velocity components `u` and `v`. The latter are obtained from the potential by the `gradient` command:

```
[u,v] = gradient (-phi);
```

which corresponds to (14.8). In the example the arrows are plotted in yellow (formal parameter `'y'`). Note that `quiver` can only be used for 2D flow fields and equidistant meshes.

Figure 14.5 depicts the output of the groundwater flow example, as far as developed to this point. The plot was obtained after zooming in the figure editor. Flow towards the well is clearly visible. Visible also is the increase of velocity if the well is approached.

Finally another graphic option is introduced. The `streamline` command produces flowpaths, which are also based on the velocity field. Flowpaths trace the flow of a particle within the flow field; for that reason the method is also called *particle tracing*. Flowpaths are identical to streamlines in steady state flow fields and identical to contours of the streamfunction; see Chap. 15 for differences and advantages/disadvantages compared to streamfunction plots.

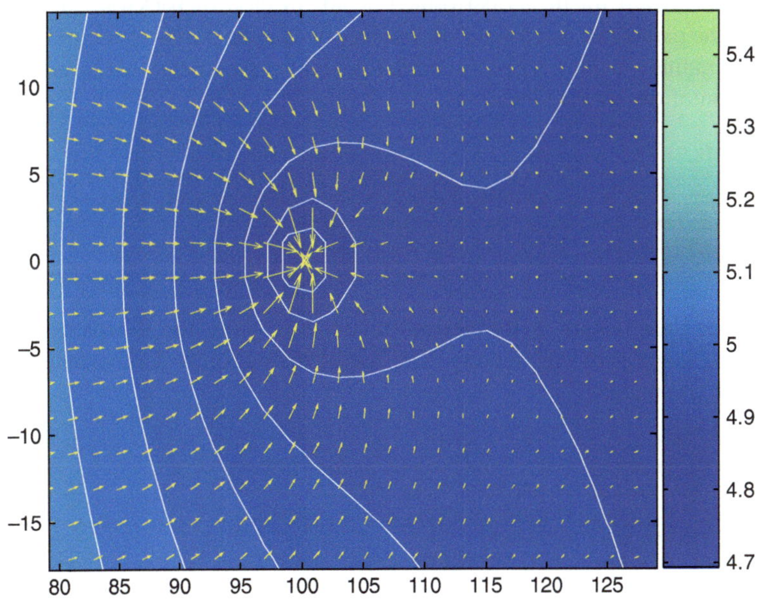

Fig. 14.5 Example graphic showing contours, colorbar and arrow field

We add a `gflowp_fit` switch as another option where graphic options are specified. At the end we add the following commands:

```
if gflowp_fit                                                % flowpaths
    xstart = []; ystart = [];
    for i = 1:100
        if v(1,i) > 0 xstart = [xstart xvec(i)];...
                ystart = [ystart yvec(1)]; end
        if v(100,i) < 0 xstart = [xstart xvec(i)];...
                ystart = [ystart yvec(100)]; end
        if u(i,1) > 0 xstart = [xstart xvec(1)];...
                ystart = [ystart yvec(i)]; end
        if
                ystart = [ystart yvec(i)]; end
    end
    h = streamline (x,y,u,v,xstart,ystart);
    set (h,'Color','red');
end
```

The `streamline` command, almost at the end of the last command listing, obtains mesh and velocity field information like the `quiver` command. There are two new vectors to be added, which have to contain the starting values for the flowpaths. With the `for` loop the starting positions for the flowpaths are determined. Each mesh node at the boundary becomes a start position if the flow velocity is directed into the model region.

A nice way to determine the catchment of a well is to choose startpositions near the well position and trace the flowpath backward in time. In the code we introduce another option parameter `gflowp_bit`. The value of the parameter determines whether backward tracing is performed at all. In case of backward tracing, the value of the parameter determines the number of starting positions around the well. The well radius `R` is introduced as another new input parameter that is used for the calculation of the starting locations.

```
if gflowp_bit
    xstart = x0 + R*cos(2*pi*[1:1:gflowp_bit]/gflowp_bit);
    ystart = y0 + R*sin(2*pi*[1:1:gflowp_bit]/gflowp_bit);
    h = streamline (x,y,-u,-v,xstart,ystart);
    set (h,'Color','y');
end
```

Another nice feature is the option to use dots along the flowpaths in order to indicate the size of the velocity. The use of the option is demonstrated on the following commands:

```
if gflowp_dot
    [verts averts] = streamslice(x,y,u,v,gflowp_dot);
    sc = 10/mean(mean(sqrt(u.*u+v.*v)));
    iverts = interpstreamspeed(x,y,u,v,verts,sc);
    h = streamline (iverts);
    set (h,'Marker','.','Color','y')
end
```

As a switch, the variable `gflowp_dot` is introduced. The `streamslice` command delivers flowpaths for the given mesh and velocity field similar to the `streamline` command. The information is gathered in two data structures. The structure `verts` contains coordinates for all streamlines. If `averts` is not returned, arrows indicate the direction along the flowpaths instead of dots. `gflowp_dot` has the additional function to determine the streamline density: a doubled value produces approximately twice as much streamlines. `interpstreamspeed` evaluates the speed along the streamlines, which is scaled by the `sc` parameter. `sc` is the inverse of the mean velocity, multiplied by 10. `mean` initiates the calculation of the mean value and has to applied twice, as we have arrays of dimension 2. Some manual adjustment is surely necessary here: the author found it often appropriate to use the mean velocity multiplied by a factor of 10. The last two commands have already been explained: streamlines are finally plotted in yellow colour, using dots as time markers.

The result for a well with pumping rate $Q = 2{\cdot}10^{-4}$ m^3/s is depicted in Fig. 14.6. The pumping rate is $2{\cdot}10^{-4}$ m^3/s.

Exercise 14.2. Confirm the following formulae for a control sample by choosing an appropriate graphical output!

1. The distance between the well location and the stagnation point downstream from the well is given by:

$$d = \frac{Q}{2\pi Q_{x0}} \tag{14.17}$$

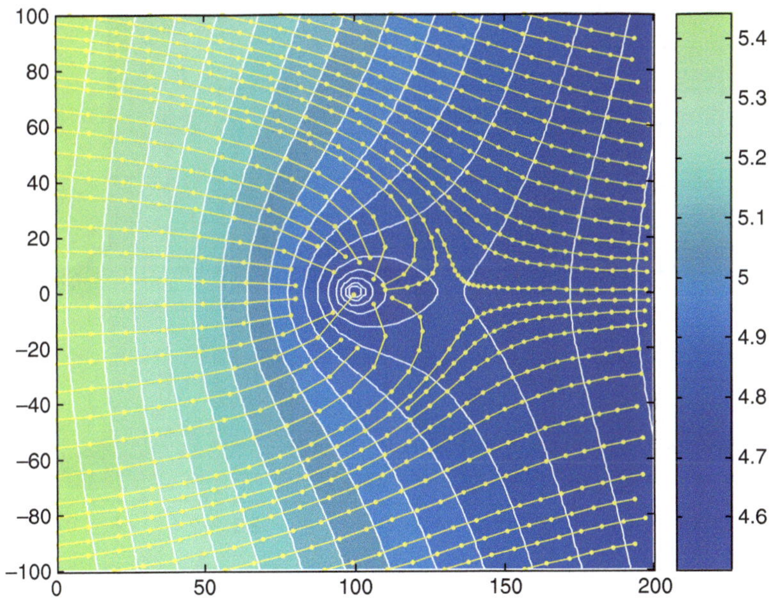

Fig. 14.6 Flow towards a well, depicted by filled head contours and flowpaths with speed dots

2. The width of the well catchment in far upstream distance is:

$$b = \frac{Q}{Q_{x0}} \tag{14.18}$$

14.5 MATLAB® 3D Graphics

Potential flow can also be computed in 3D. The potential for a sink or a source is given by the formula:

$$\varphi(x, y, z) = -\frac{Q}{4\pi} \frac{\log}{|\mathbf{r} - \mathbf{r}_0|} \tag{14.19}$$

instead of formula (14.5). This can be combined with baseflow in x-direction. An example M-file for a sink or source at position (2,2,2) is shown below. Slices `xslice`, `yslice` and `zslice` are defined for all three different space directions, which are then used within the `slice` command. Streamlines are drawn using `streamline`.

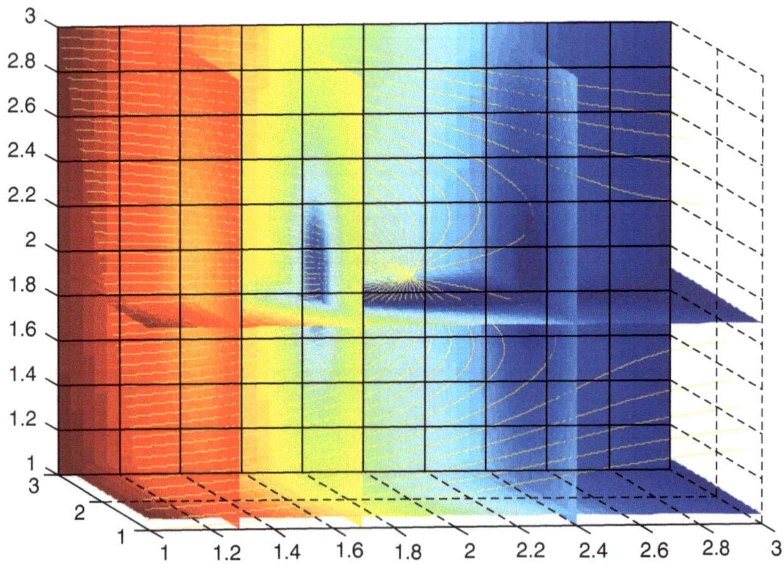

Fig. 14.7 3D potential flow, computed and visualized using MATLAB®

```
Q = 1;                         % source/ sink rate [L^3/T]
Qx0 = 0.2;                     % baseflow [L^2/T]
i = linspace(1,3,50);
[x,y,z] = meshgrid (i,i,i);
r = sqrt((x-2).^2+(y-2).^2+(z-2).^2);
xslice = [1.3;1.7;2.4];
yslice = [3];
zslice = [1.1;1.9];
phi = -Qx0*x-Q/4/pi/r;
slice (x,y,z,phi,xslice,yslice,zslice); hold on;
[u,v,w] = gradient (-phi);
h = streamline (x,y,z,u,v,w,ones(1,50),i,i);
set (h,'Color','y')
```

or

```
coneplot (x,y,z,u,v,w,)
```

The M-file is included in the accompanying software under the name '*3D_flow.m*'.

Figure 14.7 shows the flow field produced by the M-file after some manual changes using the MATLAB® Figure editor. The color range for the color map was restricted and lines on slices were omitted.

References

Guyon E, Hulin JP, Petit L (1997) Hydrodynamik. Vieweg, Braunschweig, p 466, in German
Prandtl L, Tietjens OG (1934) Fundamentals of hydro- and aeromechanics. Dover, New York, p 270

Chapter 15
Streamfunction and Complex Potential

15.1 Streamfunction

The *streamfunction* is another mathematical construct that is of high importance for models using analytical solutions. Together with the potential the streamfunction enables the visualization of flow patterns that can hardly be produced by other methods. In 2D the streamfunction Ψ is defined by the equations:

$$q_x = -\frac{\partial \Psi}{\partial y} \quad q_y = \frac{\partial \Psi}{\partial x} \tag{15.1}$$

The derivatives of the streamfunction are the components of the discharge vector. In contrast to the potential, the negative of the y-derivative delivers the x-component of discharge, and the x-derivative delivers the y-component of discharge. From the defining equations follows that the streamfunction also fulfils the potential equation:

$$\begin{aligned} \frac{\partial^2 \Psi}{\partial x^2} + \frac{\partial^2 \Psi}{\partial y^2} &= \frac{\partial}{\partial x}\frac{\partial \Psi}{\partial x} + \frac{\partial}{\partial y}\frac{\partial \Psi}{\partial y} = \frac{\partial}{\partial x} q_y - \frac{\partial}{\partial y} q_x \\ &= \frac{\partial}{\partial x} K \frac{\partial h}{\partial y} - \frac{\partial}{\partial y} K \frac{\partial h}{\partial x} = K\left(\frac{\partial^2 h}{\partial x \partial y} - \frac{\partial^2 h}{\partial y \partial x}\right) = 0 \end{aligned} \tag{15.2}$$

Streamlines are characterized by the property that the tangentials are perpendicular to the contours of the potential or the head. The mathematical proof utilizes the connection between streamfunction and discharge vector (15.1) as well as the connection between potential and discharge vector (14.3):

$$q_x = \frac{\partial \varphi}{\partial x} \qquad q_y = \frac{\partial \varphi}{\partial y} \tag{15.3}$$

E. Holzbecher, *Environmental Modeling*,
DOI 10.1007/978-3-642-22042-5_15, © Springer-Verlag Berlin Heidelberg 2012

The condition for orthogonality is then obtained through:

$$\frac{\partial\varphi}{\partial x}\frac{\partial\Psi}{\partial x}-\frac{\partial\varphi}{\partial y}\frac{\partial\Psi}{\partial y}=-q_xq_y+q_xq_y=0 \tag{15.4}$$

There are explicit formulae for the streamfunction for special flow elements, for example:

$$\Psi(x,y)=\begin{cases}-Q_{x0}y+Q_{y0}x & \text{for baseflow}\\ \dfrac{Q}{2\pi}\vartheta(x,y) & \text{for a well}\end{cases} \tag{15.5}$$

with ϑ = angle of the connecting vector between position (x,y) and well location.

The physical unit of the streamfunction is [m^3/s], i.e. the dimension of volume flux. This is easiest explained regarding baseflow in x-direction. Q_{x0} is the volume flux per unit width $\Delta y = 1$. According to the formula (15.5) holds: $\Psi(x,y) = -Q_{x0}y$, representing is the flux between the x-axis and its parallel in distance y. It follows that between two horizontal lines at locations y_1 and y_2 and streamfunction values Ψ_1 and Ψ_2 the flux is given by

$$\Delta\Psi = \Psi_1 - \Psi_2 \tag{15.6}$$

It is a general property of the streamfunction that for any two locations in the model region the flux between these positions is given by the difference of their streamfunction values. This characteristic property of the streamfunction follows directly from (15.1) by integration along curves within the model region. The restricting condition is that the model region has to be simply connected (Needham 1997), which roughly means that it has no holes.

Imagine two arbitrary locations. If the streamfunction takes the same value at these positions, the volumetric flux between these locations is zero, which results as a special case from the above given property of the streamfunction. The property is obvious if the connecting line is the streamline itself, but it holds for every connecting line. Then, positive fluxes across the line are exactly outweighted by negative fluxes. In a more general sense one may consider arbitrary closed curves in the model region. Take an arbitrary point on this curve, representing both start and end point of the curve. The streamfunction property states that the integral of fluxes across this line is zero (because the streamfunction value at start and end is the same). Physically speaking, negative and positive fluxes outweigh each other exactly.

Because of the mentioned property, the streamfunction is a very appropriate measure for fluxes. In order to demonstrate the application of the streamfunction, we extend the M-file '*gw_flow.m*', which was developed in Chap. 14. That is done in three steps:

1. In the execution part of the M-file add a command by which the streamfunction is computed under the variable `psi`:

```
psi = -Qx0*y + Qy0*x + (Q/(pi+pi))*atan2((y-y0),(x-x0));
```

The command is an implementation of the two formulae given in (15.5). Following the principle of superposition (see Chap. 14), the formulae have to be added to account for a situation with baseflow and a single well. The angle ϑ is calculated by the arcustangens function. In MATLAB® there are two ways of computing the arcustangens. Here we chose not to use the standard `atan` command, which delivers angles in the interval between $-\frac{\pi}{2}$ and $\frac{\pi}{2}$. The `atan2` command is chosen, because angles in the range between $-\pi$ and π are computed. We will come back to that point below.

2. Add another plot option in the input part of the M-file:

```
gstream = 10;          % streamfunction plot
```

3. Plot baseflow with the `contour` command (without filling colors):

```
if gstream
    contour (x,y,psi,gstream,'w');
end
```

The `gstream` parameter is not only a switch for streamline graphics, but also determines the number of contours to be depicted. According to the property of the streamfucntion mentioned above, there is no flux between two locations on the same contour. This means that the Ψ-contours are *streamlines*, a term already met in Chap. 14 in connection with the MATLAB® `streamline` command. The streamfunction offers an alternative method for plotting streamlines.

Figure 15.1 shows the output of the M-file if both the streamline and the streamfunction contours are active. Streamfunction contours are depicted in black colour, lines from `streamline` in yellow, using the `gflowp_dot` option, introduced in Chap. 14. Obviously the flow field representations coincide[1]. But streamline distributions are obviously different. Using `streamline`, the lines are determined by starting points; the MATLAB® algorithm thus is a flowpath algorithm by which a flowpath is traced forward or backward in time. Another word for such a procedure is *particle tracking*. In 2D these are numerical solutions of the differential equations:

[1] Of course, there should be no cross-overs between the streamline patterns if both representations are correct. The user may play around and find out that under certain conditions there are cross-overs. One important example is given, when the mesh spacing in x- and y-direction is not equal, as the output of the streamline command is not a correct representation of the flow field.

$$\frac{\partial x}{\partial t} = v_x \ \frac{\partial y}{\partial t} = v_y \tag{15.7}$$

Streamlines, obtained as streamfunction contours, originate from a very different idea. When Ψ-contour levels are selected equidistantly, the same volume of fluid flows between neighbouring streamlines. Between two locations on neighbouring streamlines there is the same volumetric flux, independent of the locations, i.e. $\Psi_1 - \Psi_2$. If the streamfunction is plotted using equidistant values of streamfunction levels (as in the example above), the same amount of fluid flows between two neighbouring streamlines – that holds for all streamlines in the plot. Thus, streamlines provide a very nice visualization of velocities: where streamlines are dense, the velocities are high; where streamlines are far apart, velocities are low. The streamline density illustrates the velocity of a flow field.

Figure 15.1 shows another advantage of the streamfunction: flowpaths are easily followed into regions with high velocities, here into the immediate vicinity of the well. As illustrated in the figure, the particle tracking algorithms have difficulties in the vicinity of the well.

An advantage of the MATLAB® particle tracking is that markers can be used as indicators of velocity size. A disadvantage of the streamfunction approach is that *cuts* may appear. In Fig. 15.1 there is a cut, depicted as a thick black line between the well and the left side boundary. Such cuts appear where the stream function has a singularity. Using the `atan2` function in the M-file the cuts appear if the angle is $\vartheta = \pi$, which is 180° from the positive x-axis. Along the negative x-axis the arcustangens function has a jump, which becomes visible in the contour plot as a cut. If the `atan` function is called in the M-file, the flow pattern is still represented correctly but with cuts from the well singularity in vertical direction.

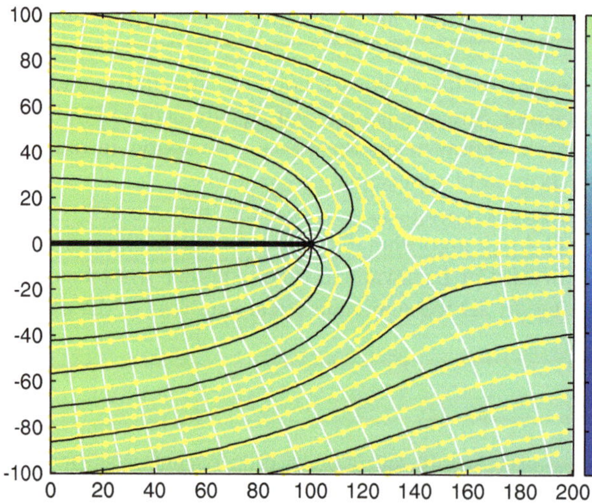

Fig. 15.1 Streamfunction contours and flowpaths for a single well in uniform baseflow

Exercise. The formula concerning the width of the upstream catchment of a well, given in Chap. 14, can easily be checked by the streamfunction property mentioned above. Hint: in the far distance from the well only the baseflow determines the flow pattern. Between two locations perpendicular to baseflow with distance b the water flux is $Q_{x0}b$.

15.2 The Principle of Superposition

The principle of superposition has been introduced in Chap. 14 and proved to be a powerful method to obtain solutions of the flow potential. We continue with further demonstrations, as the principle can also be applied to the streamfunction. For that purpose we extend the groundwater M-file, which was developed in the previous sections. The M-file is extended to account for several wells (the former version can be used for a single well only). Wells, i.e. their positions within the model region and their pumping rates, are introduced in the input part of the M-file and replace the concerned commands by the following sequence:

```
% Wells
xwell = [100 100 150];                % x-coordinates well position [m]
ywell = [0 50 100];                   % y-coordinates well position [m]
Qwell = 1.e-4*[1 1 1];                % pumping / recharge rates [m^3/s]
```

The sequence gives an example for the introduction of three wells with equal pumping rates. In the same manner the user may enter as many wells as she/he likes. There may be pumping and recharge wells in the same well gallery, each working at its own rate. The vectors, specified in the three lines, must have the same length.

In order to take several wells into account in the potential and streamfunction calculation, we use a `for` loop as follows:

```
phi = -Qx0*x - Qy0*y;
psi = -Qx0*y + Qy0*x;
for i = 1:size(xwell,2)
     r = sqrt((x-xwell(i)).*(x-xwell(i))+(y-ywell(i)).*(y-ywell(i)));
     phi = phi + (Qwell(i)/(2*pi))*log(r);    % potential
     psi = psi + (Qwell(i)/(2*pi))*atan2((y-ywell(i)),(x-xwell(i)));
end
```

While the index variable `i` takes values starting from 1 to the number of wells, the loop runs through all the wells. In each run through the loop, the analytic solution for the well is calculated and added to the amount calculated before, according to the principle of superposition. An example output is provided in Fig. 15.2.

The *'cplxPot.m'* M-file, included in the accompanying software, where the complex potential is applied, is an extension of *'gw_flow.m'*, in which several wells can be taken into account.

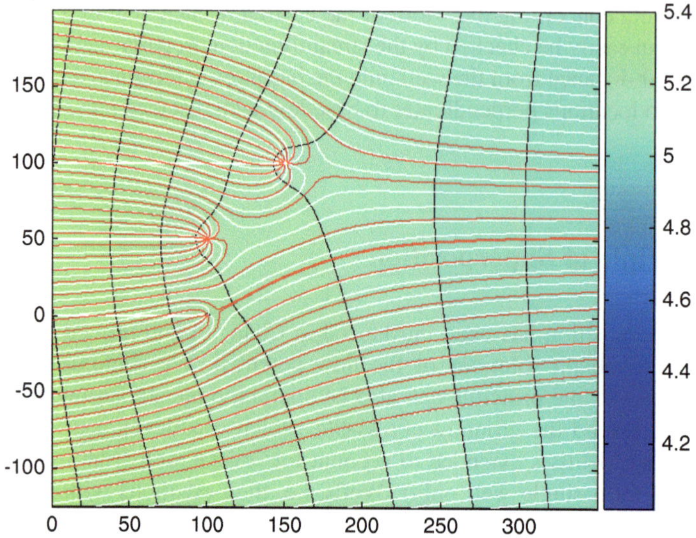

Fig. 15.2 Groundwater flow towards three wells (is potential, streamlines and flowpaths)

15.2.1 The Doublette

A problem which is suitable for the application of the program concerns a *doublette* system. Such a system consists of a pumping well and a recharge well. More complex installations with several recharge and discharge wells are not treated here, although they can be simulated easily using the presented procedure. Such systems are installed for several environmental purposes, for example, to extract freshwater and simultaneously dispose waste water.

Doublettes are a typical set-up of geothermal technology. In geothermal facilities hot water is pumped, usually from deep geological formations. In most applications the pumped water has a high mineral content and can neither be used for other purposes except for the use of heat, nor dumped into surface water bodies. Environmental regulations often require the fluid to be suppressed back into the subsurface region, from where it originates. The question of cold water breakthrough, which can be answered by modelling, is important as it defines the lifetime of the facility.

Also for the clean-up of contaminated groundwater doublette well systems are common. The pumping well takes the polluted fluid to a treatment station. After purification the treated water is brought back into the aquifer by the recharge well.

Exercise 15.1. Model a doublette system with recharge and pumping well 100 m apart from each other. Model a recharge and a discharge well with equal pumping rates and assume no baseflow. How do isopotentials and streamfunctions look like?

As shown in Fig. 15.3, the doublette flow pattern is symmetric in several respects. One symmetry axis connects both wells and is a streamline. The other

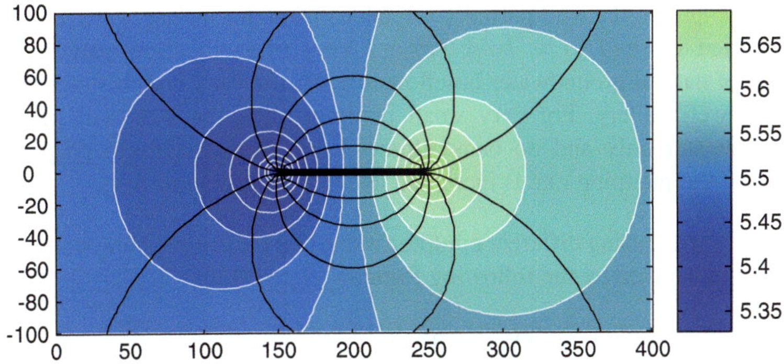

Fig. 15.3 Doublette system (iso potential and streamlines)

symmetry line, which is an isopotential line, is perpendicular to the connecting line. Note that the depicted is potentials are not completely identical on the left and right side of the figure, which stems from the fact that the internally determined potential levels are not symmetrical around the mean value. It is also easy to see that all streamlines and is potentials are circles. If that is not true on the user's computer display, it probably depends on the scaling that may not be the same for both length axes. Moreover, streamlines and is potentials meet at right angles. All these properties are well known from classical theory, but it is nice to see them confirmed using a small M-file.

15.2.2 *Mirror Wells*

The modeller may utilize non-existing wells in order to simulate special boundary conditions. For that purpose *virtual wells* are introduced that do not exist in reality. The previous example showed the simple doublette system with a constant potential line as a symmetry axis between both wells. The modeller may thus take advantage of that property and use such a doublette system to produce a constant potential boundary where it is needed. In fact, for each real well a virtual well needs to be located at the mirror position with respect to the is-potential line. For real pumping wells the *mirror well* should be a recharge well and vice versa.

A typical application is given for a well or a well gallery in the vicinity of a river or lake. The spatial gradient within the surface water body is usually very small and can be neglected. If both water bodies, groundwater and surface water, are well connected, the shoreline becomes a constant head boundary for the groundwater modeller. Constant head corresponds with constant potential. The formula for the analytical solution of such a situation is identical to the doublette solution. However, in this case the pattern needs to be computed only for that side of the bank where ground surface and well are located. The 'virtual side' should be omitted.

The water pumped in the well partially originates from the surface water body. This part is often referred to as *bank filtrate*. Very often the flow pattern is a combination of natural groundwater baseflow towards the bank and the imposed regime by the well gallery. For very small pumping rates the well withdraws ambient groundwater only and no bank filtrate. Bank filtrate starts to play an increasing roll if the pumping rate is increased above a critical value S_{cript}.

Exercise 15.2. Verify, using different pumping rates, that the critical pumping rate for bank filtration is given by the following formula

$$Q_{crit} = Q_{x0}\pi x_0 \tag{15.8}$$

where A_x determines base flow towards the is potential boundary and a_x the distance between well and boundary. Use a plot of the flow pattern to decide whether there is bank filtration or not!

Exercise 15.3. Confirm for some sample runs the dependence of groundwater and surface water withdrawal from pumping rate.

Figure 15.4 visualises the exact formula for the share of water originating from the surface water body (bank filtrate) and ambient groundwater. The formula is given by:

$$\frac{Q - \Delta\Psi}{Q} = \frac{2}{\pi}\left(\arctan\left(\sqrt{\frac{Q}{Q_{crit}} - 1}\right) - \frac{Q_{crit}}{Q}\sqrt{\frac{Q}{Q_{crit}} - 1}\right) \tag{15.9}$$

Hint: utilize that in streamfunction plots for equidistant levels the same amount of water flows between two contour lines. In the situation shown in Fig. 15.5 the percentage of bank filtrate is 4/11 $\approx$ 36%. The entire well discharge is divided into 11 equal parts, four of them originating from the bank on the left side of the figure.

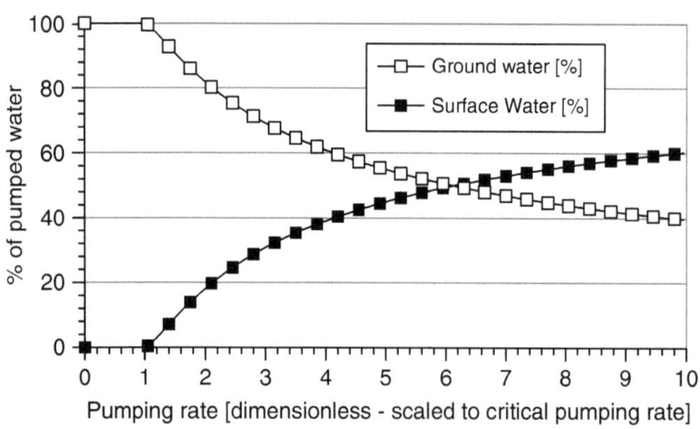

Fig. 15.4 Percentage of groundwater and bank filtrate withdrawal for a single well, depending on pumping rate

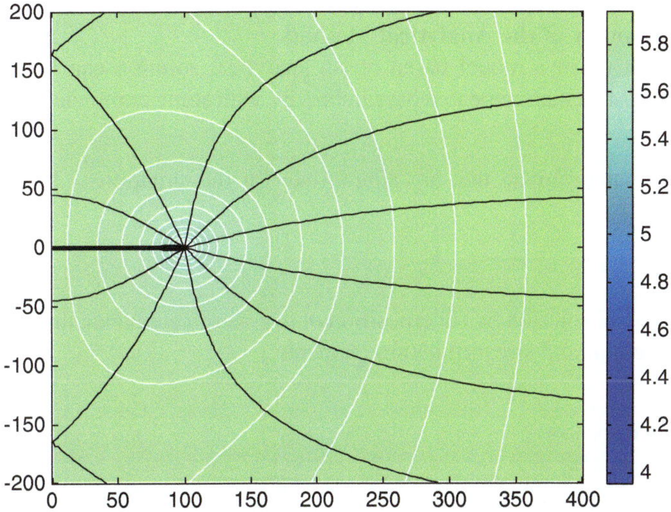

Fig. 15.5 Bank filtration example plot (for $a_x = 100$ m, $A_x = -10^{-6}$ m^2/s, $Q = 0.001164$ m^3/s)

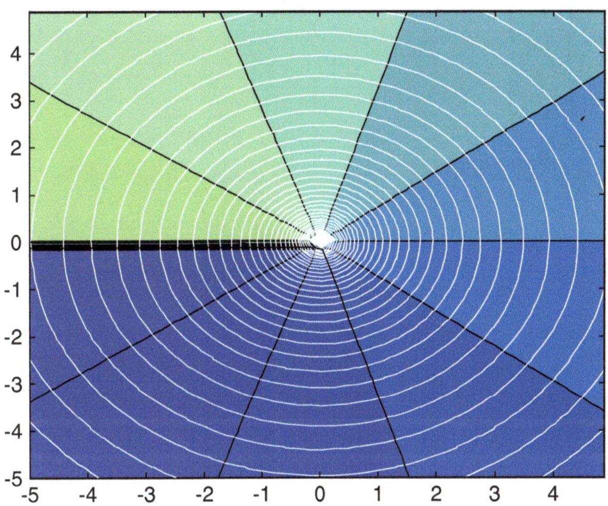

Fig. 15.6 Image of vortex; filled contours for potential and white contour lines for streamfunction

Those, who do not believe that the procedure works, may evaluate formula (15.9) to obtain the same result.

Figure 15.6 shows a vortex with circulating flow as another flow pattern which can be modelled by using the analytical method. The detailed steps of the method are summarized in Sidebar 15.1.

Sidebar 15.1: Summary of the Analytical Method

In summary, the set-up of a model based on an analytical solution can be divided into 4 steps, which are repeated here in brevity with some comments:

Step 1

Choose the analytical formulae that are appropriate for the situation to be modelled.

See Table 15.1.

Step 2

Compute the analytical solution by superposition of the chosen elements; that means: add all potential and streamfunction elements!

Step 3

Numerical Postprocessing

Compute hydraulic heads

$$h(x,y) = \begin{cases} \left(\varphi(x,y) + \frac{1}{2}K \cdot H^2 - \varphi_0\right)/(KH) & \text{for the confined aquifer} \\ \sqrt{2(\varphi(x,y) - \varphi_0)/K} & \text{for the unconfined aquifer} \end{cases}$$

Compute flux vector components

$$\left\{ \begin{aligned} q_x(x,y) &= -\frac{\partial \varphi(x,y)}{\partial x} \\ q_y(x,y) &= -\frac{\partial \varphi(x,y)}{\partial y} \end{aligned} \right\} \text{ or } \left\{ \begin{aligned} q_x(x,y) &= -\frac{\partial \Psi(x,y)}{\partial y} \\ q_y(x,y) &= \frac{\partial \Psi(x,y)}{\partial x} \end{aligned} \right\}$$

Compute velocity vector components

$$v_x(x,y) = \begin{cases} q_x(x,y)/H & \text{for the confined aquifer} \\ q_x(x,y)/h(x,y) & \text{for the unconfined aquifer} \end{cases} \quad \text{and}$$

$$v_y(x,y) = \begin{cases} q_y(x,y)/H & \text{for the confined aquifer} \\ q_y(x,y)/h(x,y) & \text{for the unconfined aquifer} \end{cases}$$

The last two formulae are to be processed for groundwater flow based on the discharge potential. For free fluids the potential has to be transformed in terms of pressure (see Chap. 14.2).

Step 4

Graphical Postprocessing

Plot (optionally):

- Hydraulic heads
- Velocity vectors
- Streamfunction contours
- Path lines

Table 15.1 Analytical elements for some fundamental flow patterns

Element	(Real) Potential φ	Streamfunction Ψ
Baseflow[2]	$-Q_{x0}x - Q_{y0}y$	$-Q_{x0}y + Q_{y0}x$
Well[3]	$\frac{Q}{2\pi}\log\left(\sqrt{(x-x_{well})^2+(y-y_{well})^2}\right)$	$\frac{Q}{2\pi}\arctan\left(\frac{y-y_{well}}{x-x_{well}}\right)$
Vortex[4]	$\frac{A}{\pi}\arctan\left(\frac{y-y_0}{x-x_0}\right)$	$\frac{A}{\pi}\log\left(\sqrt{(x-x_0)^2+(y-y_0)^2}\right)$
Di-pole[5]	$\frac{s}{2\pi}\frac{\cos\left(\arctan\left(\frac{y-y_0}{x-x_0}\right)-\beta\right)}{\sqrt{(x-x_0)^2+(y-y_0)^2}}$	$\frac{s}{2\pi}\frac{\sin\left(\arctan\left(\frac{y-y_0}{x-x_0}\right)-\beta\right)}{\sqrt{(x-x_0)^2+(y-y_0)^2}}$

15.3 Complex Analysis and Complex Potential

In MATLAB® both potential and streamfunction can be visualized more easily if one adopts the notation of the 2D plane using *complex numbers*. The simplest command sequence to produce a potential for a sink or source, are:

```
z=cplxgrid(30)
cplxmap(z,log(z))
```

In Fig. 15.7 the real part of the logarithm is plotted. Compare with Fig. 14.1.

Instead of the computation of the real potential and the streamfunction, the imaginary potential is evaluated. Real potential φ, streamfunction Ψ and imaginary potential Φ are connected by the formula:

$$\Phi = \varphi + i\Psi \tag{15.10}$$

i denotes the square root of -1, the imaginary unit. In MATLAB® specify the imaginary number `z` by:

```
z = i

z =          0 + 1.0000i
```

or

```
z = -1+2i

z =   -1.0000 + 2.0000i
```

In the latter example the *real part* of z is -1, and the *imaginary part* is 2. In MATLAB@ use the commands `real` and `imag`:

[2] With x-component Q_{x0} and y-component Q_{y0}.

[3] At position (x_{well}, y_{well}) with pumping rate Q.

[4] At position (x_0, y_0) with strength A.

[5] At position (x_0, y_0) with strength s and angle β.

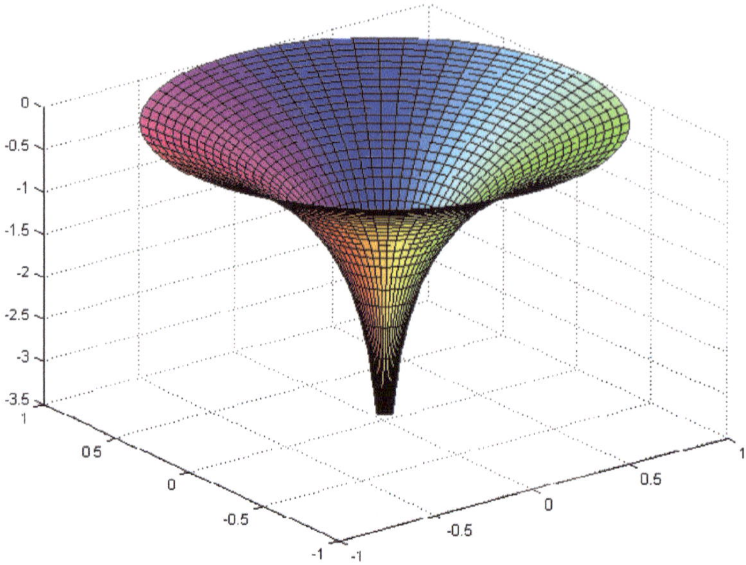

Fig. 15.7 Surface plot of the complex logarithm

```
real(z)

ans =
     -1

imag(z)

ans =
     2
```

Complex numbers are just another more general data type which MATLAB® can handle. Thus in MATLAB® there are arrays, vectors, matrices and functions of complex numbers, and the user may use those almost as she/he is accustomed to with real numbers. Any complex number can alternatively be characterized by an absolute value $r = \sqrt{x^2 + y^2}$ and an angle θ:

$$\mathbf{z} = re^{i\theta} \tag{15.11}$$

The corresponding MATLAB® command are:

```
abs(z)

ans =
     2.2361

angle(z)

ans =
     2.0344
```

The output for the angle lies between -π and π.

The conjugate complex is denoted by an over bar and defined as follows:

$$\bar{z} = x - iy \tag{15.12}$$

In MATLAB the conjugate complex of a complex number z is calculated by using the `conj` function:

```
conj(z)
```

Often it is easier to implement the imaginary potential instead of the separate computation of the real potential and the streamfunction (see examples in Table 15.2). When the complex potential is computed, (real) potential and streamfunction can be obtained as real and imaginary part of the imaginary potential:

$$\varphi = \mathrm{Re}(\Phi)\ \Psi = \mathrm{Im}(\Phi) \tag{15.13}$$

There are more analytical elements than those for baseflow and wells. Formulae for line sinks or line sources, for di-poles, for vortices and so forth can be found in the textbooks.

Table 15.2 Complex potentials for various flow patterns; basic flow patterns with parameters as in Table 15.1; with $\mathbf{Z} = \dfrac{z - \frac{1}{2}(\mathbf{z}_1 + \mathbf{z}_2)}{\frac{1}{2}(\mathbf{z}_2 - \mathbf{z}_1)}$

Element	(Imaginary) Potential Φ
Baseflow	$-\overline{\mathbf{Q}}_0\mathbf{z}$
Well	$\dfrac{Q}{2\pi}\log(\mathbf{z} - \mathbf{z}_{well})$
Vortex	$\dfrac{A}{\pi i}\log(\mathbf{z} - \mathbf{z}_0)$
Di-pole	$\dfrac{s}{2\pi}\dfrac{\exp(i\beta)}{\mathbf{z} - \mathbf{z}_0}$
Line-sink[6]	$\dfrac{\sigma L}{4\pi}\left\{(\mathbf{Z}+1)\log(\mathbf{Z}+1) - (\mathbf{Z}-1)\log(\mathbf{Z}-1) + 2\log\left[\frac{1}{2}(\mathbf{z}_2 - \mathbf{z}_1)\right] - 2\right\}$

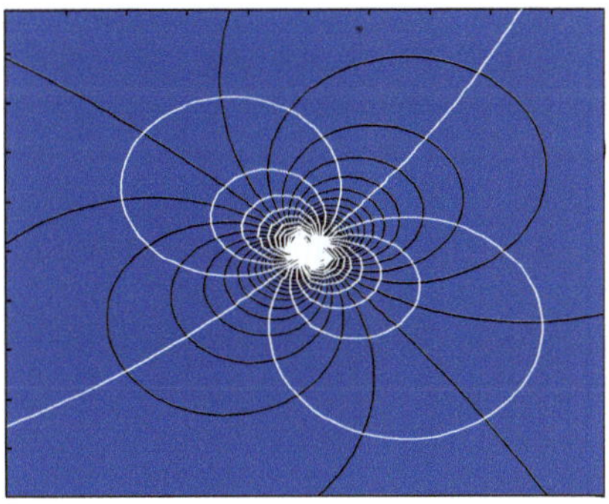

Fig. 15.8 Dipole pattern (streamlines white, potential contours black)

[6] Between positions $\mathbf{z}_1$ and $\mathbf{z}_2$with strength σ.

The M-file for the dipole is included in the accompanying software as *'dipole.m'* file. The figure (15.8) was produced by calling the function from the *'AnElements.m'* file with appropriate parameters. The *'AnElements.m'* file has options to superpose baseflow, di-pole, sources, sinks and vortices.

For the easy computation of complex potentials the *'AnElements.m'* file can be used. In the following example we demonstrate the superposition of a di-pole and a baseflow solution:

```
Q0 = .5;              % baseflow
s = -10;              % dipole strength
beta = 0;             % dipole angle
z0 = 0;               % position of well, vortex or di-pole

%---------------------------------------------------mesh--------
xmin = -5;            % minimum x-position of mesh [m]
xmax = 5;             % maximum x-position of mesh [m]
dx = .21;             % grid spacing in x-direction [m]
ymin = -5;            % minimum y-position of mesh [m]
ymax = 5;             % maximum y-position of mesh[m]
dy = .22;             % grid spacing in y-direction [m]

%------------------------------start positions (for streamline)--
zstart = xmin + i*(ymin+(ymax-ymin)*[0.1:0.1:1]);
zstart = [zstart xmin-i*0.001]

%----------------------------------------------execution---------
[x,y] = meshgrid ([xmin:dx:xmax],[ymin:dy:ymax]);    % mesh
z = x+i*y;

Phi = -Q0'*z;     % baseflow
if (s ~= 0) Phi = Phi + (s/(pi+pi))*exp(i*beta)./(z-z0); end %
%-------------------------------------------------output---------
colormap(bone);
contourf (x,y,real(Phi),70);
hold on;
[u,v] = gradient (-real(Phi));
h = streamline (x,y,u,v,real(zstart),imag(zstart));
set(h,'Color','w');
```

The M-file performs the major tasks of input, initialization, execution and output in consecutive blocks. Isopotential lines are visualized by filled contours, the streamlines by the `streamline` command. A streamline pattern could also be produced by contours for the imaginary part of the complex potential. However, by using the option to specify starting points one obtains a better representation of the circular limit streamlines.

Figure 15.9 results, which can be found in different application fields. The white lines can be interpreted as flow lines around a spherical obstacle in the 2D plane (Prandtl and Tietjens 1934). Black lines are the is-potential lines in that case. In groundwater modeling the same superposition is interpreted differently. The black

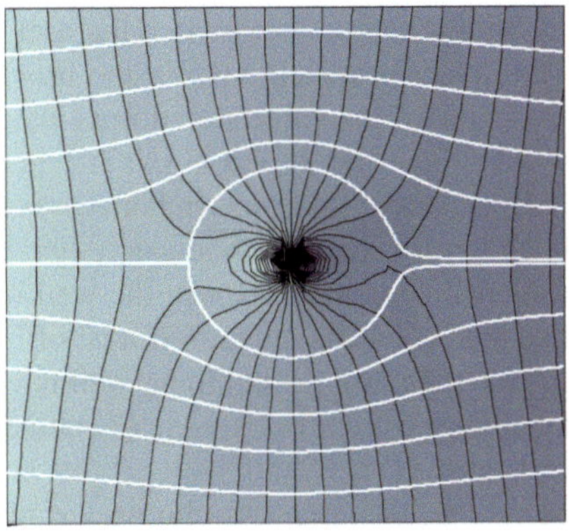

Fig. 15.9 Flow round an obstacle; groundwater flow towards a lake

lines represent streamlines. Here, baseflow is in vertical direction towards a circular lake, where a constant potential (water table) is given (Strack 1989).

White lines are the potential contours. Streamlines are depicted in black. The interior of the lake, within the white circle, has to be neglected in the groundwater flow problem.

The user may investigate the influence of the solution parameters, here `Q0`, `s` and `z0`.

15.4 Example: Vortices or Wells Systems

As an example for the various applications of analytical solutions for the complex potential an M-file is developed, by which systems of several vortices or wells can be evaluated and illustrated. Moreover, various types of boundaries can be considered, here as example along the x- or y-axes. The input section of the file is as follows:

```
wellvort = 1;                      % wells (1)/ vortices (0)  switch
mirrorx = 1;                       % symmetry property for x-axis
mirrory = 0;                       % symmetry property for y-axis
mirror = 0;                        % symmetry / antisymmetry switch
xspace = linspace(-100,100,200);   % x-values mesh
yspace = linspace(-100,100,200);   % y-values mesh
N = 3;                             % no. of wells/ vortices (random)
                                   %   for manual input set N = 0
zloc = [-20 40] + i*[40 60];       % well / vortex positions
s = [5 10];                        % pumping rates/ circulation rates
N0 = 30;                           % no. of filled contours
M0 = 8;                            % no. starting points around each well
                                     / vortex
gquiv = 1;                         % arrow field on/off
```

With the `wellvort` switch the user decides if wells or vortices are considered. The `mirrorx` switch enables the user to make the x-axis a symmetry boundary; analogously, the value stored in `mirrory` decides about the y-axis. It is possible to decide about the x- and y-axis independently: either option may be 'on' or 'off' independent of the other. In both cases the symmetry may result in an impermeable boundary or an open boundary. For the latter alternative the `mirror` switch is responsible: if it is 'on' (1), the well or vortex on the other side of the symmetry line is of identical strength; if it is 'off' (0), the mirror object has opposite strength. In the latter case the pumping well is vis-à-vis of a recharge well, and a clockwise circulating vortex is vis-à-vis of another one circulating counter-clockwise.

Under `xspace` and `yspace` the x- and y-values of the mesh are specified. Moreover, the user can choose the number of vortices or wells. There are two options: randomly distributed wells and manual input. For the first option the user has to choose a value `N` greater than 0. Otherwise the locations of the wells/vortices are specified in the `zloc` vector and their strengths in the `s` vector. The positions are given as complex numbers.

There are some further options. The number of start positions around the well is required. For well systems the starting positions concern streamlines, for vortices iso-potentials. There is the number of filled contour lines that represent streamlines for vortices plots and iso-potentials for well galleries. Finally, there is the option to switch on / off the output of arrow fields.

The next commands compute some auxiliary variables, which are needed in the following. `xrange` and `yrange` represent the length of the model region on both coordinate axes. `x` and `y` are the mesh variables in real numbers, `z` in complex numbers. `zloc` is a vector containing randomly determined positions of the vortices and wells. `s` is a vector containing the strength of all, also randomly determined. With the final command the minimum absolute value of all the vortices or wells strengths is computed and divided by the integer `K`.

```
xrange = xspace(end) - xspace(1);
yrange = yspace(end) - yspace(1);
[x,y] = meshgrid (xspace,yspace); z = x +i*y;
if N>0
   zloc = xspace(1)+xrange*rand(1,N)+i*(yspace(1)+yrange*rand(1,N))
    s = -10+round(20*rand(1,N)) % strength steps between -10 and 10
else
    N = size(s,2);
end
```

Note that the contents of the two vectors `zloc` and `s` are displayed in the command window. The user is thus able to find the exact data for the output in case she/he is interested to reproduce interesting results. In the next command lines virtual wells or image vortices are introduced depending on the options set in the specification part of the M-file. The position given by `conj(zloc)` is the mirror location of a well with respect to the *x*-axis, the *conjugate complex*. It is easy to check that the commands in the nested `if`-blocks fulfill the task, which is explained above.

```
if mirrorx
    zloc = [zloc conj(zloc)];
    if mirror
        s = [s s];
    else
        s = [s -s];
    end
    N = N+N;
end
if mirrory
    zloc = [zloc -real(zloc)+i*imag(zloc)];
    if mirror
        s = [s s];
    else
        s = [s -s];
    end
    N = N+N;
end
f = Phi(z,zloc,s);
```

In the next command, at the end of the last listing, the complex potential is evaluated. The function `Phi` appears at the end of the M-file:

```
function f = Phi (z,zloc,s)
f = 0*z;
for k = 1:size(s,2)
    f = f + s(k)*log(z-zloc(k))/2/pi;
end
```

The formula for the complex logarithm, as given in Table 15.2, is evaluated in every run through the `for`-loop and added to the former output. Note that all computations are performed in the space of complex numbers. The result of `log` is a complex number. All further commands are based on the complex potential, which is available under the variable `f` after execution.

At first, velocity vectors are computed from the real potential, which is contained in the real parts of the complex potential. The following four commands construct the legend, stating that streamlines are illustrated by white lines, and isopotential contours by blue lines. The last command selects the colormap.

```
[u,v] = gradient (-real(f));
whitebg([.753,0.753,0.753]);                       % legend
plot (xspace(1:2),yspace(1)*[1 1],'w'); hold on;
plot (xspace(1:2),yspace(1)*[1 1],'b');
legend ('streamlines','isopotentials')
colormap('jet');
```

In the next commands starting positions for the streamline command are calculated. Around each well and virtual well (vortex and image vortex) `M0` positions are chosen equidistantly.

```
xstart = []; ystart = [];
for k = 1:size(s,2)
    for j = 1:M0
        xstart = [xstart real(zloc(k))+xrange*cos(2*pi*j/M0)/1000];
        ystart = [ystart imag(zloc(k))+yrange*sin(2*pi*j/M0)/1000];
    end
end
xspace = linspace (xspace(1),xspace(end),20);
yspace = linspace (yspace(1),yspace(end),20);
```

With the final command new mesh-spacing vectors are computed. With the latter the arrow plot is performed. For fine meshes the arrows become unidentifiable, and thus it is better to visualize them on a coarser grid.

In the last command block a figure is composed out of filled contours, contours and arrows. This has to be done separately for a well gallery on one side and a vortices system on the other side. For the well option, the real part of the complex potential represents the real potential and the contours are given in blue color. The streamline command is started twice, once forward and once backward. This is realized by changing the sign in the velocity component. The user may increase the performance of the execution by choosing the correct set of start values for each `streamline` command. In the final `if` block the arow field is plotted. The course mesh is computed first and stored in variable `z`. Then the velocity field is re-calculated, based on another evaluation of the complex potential, before the `quiver` command plots arrows in the figure.

```
if wellvort
    contourf (x,y,real(f),N0,'b'); hold on;
    hh = streamline (x,y,u,v,xstart,ystart);
    set (hh,'Color','w');
    hh = streamline (x,y,-u,-v,xstart,ystart);
    set (hh,'Color','w');
    if gquiv
        [x,y] = meshgrid (xspace,yspace); z = x +i*y;
        [u,v] = gradient (-real(Phi(z,zloc,s)));
        quiver (x,y,u,v,gquiv,'w');
    end
```

For the vortex option the same tasks are performed but with slightly different details. The filled contour plot gets white isolines, as these represent streamlines in this case. The lines, plotted with the `streamline` command, are is-potentials in the vortex case and are thus plotted in blue. For the arrow field, it has to be taken into account that the real part of the complex potential represents the streamfunction, and thus the velocity components have to be computed following the (15.1).

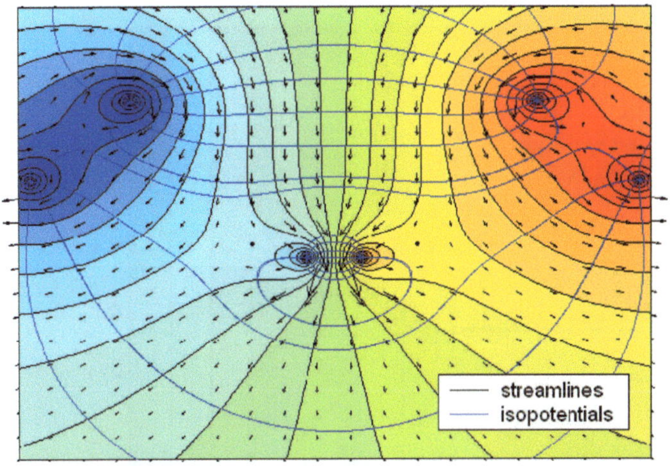

Fig. 15.10 Vortex plot example, as obtained using the *'wellvortex.m'* file; change of white color to black color by copy option in the figure editor

```
else
    contourf (x,y,real(f),N0,'w'); hold on;
    hh = streamline (x,y,u,v,xstart,ystart);
    set (hh,'Color','b');
    hh = streamline (x,y,-u,-v,xstart,ystart);
    set (hh,'Color','b');
    if gquiv
        [x,y] = meshgrid (xspace,yspace); z = x +i*y;
        [v,u] = gradient (-real(Phi(z,zloc,s))); u = -u;
        quiver (x,y,u,v,gquiv,'w');
    end
end
```

The M-file described in this sub-section is included in the accompanying software as *'wellvortex.m'* file.

An example output of the M-file is given in Fig. 15.10.

Exercise 15.4. Examine, by use of the M-file, which option combination of the `wellvort` and `mirror` switches leads to an impermeable or to an is-potential boundary?

15.5 Example: Thin Objects in Potential Flow

Another example for the application of the complex potential concerns thin objects in a baseflow field. A highly permeable or impermeable object is approached by a potential flow field with an angle α. One may imagine a situation in sub-surface

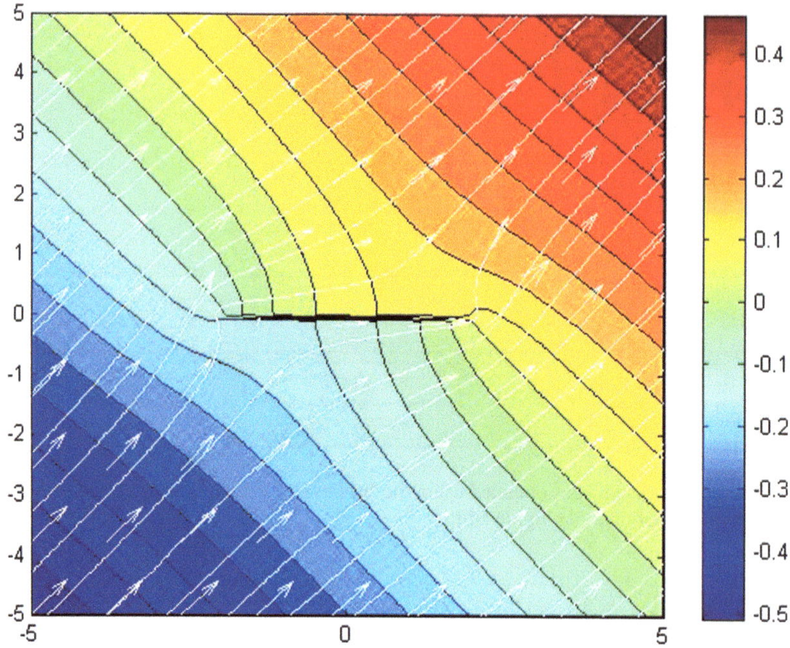

Fig. 15.11 Impermeable object in a flow field entering with an angle of 45°: isopotentials (*black*), streamlines (*white*), velocity field (*arrows*)

flow with a permeable fracture or an impermeable obstacle as well as in the atmosphere with a solid obstacle.

For the following it is assumed that the thickness of the object can be neglected, so that it becomes a line object in the 2D model region. For the impermeable case a solution has been given in form of a complex potential by Churchill and Brown (1984):

$$\Phi = v_\infty\left(z\cos(\alpha) - \mathrm{i}\sqrt{z^2 - a^2}\sin(\alpha)\right) \tag{15.14}$$

where v_∞ denotes the absolute value of velocity at infinity. a is the half length of the line object. In Fig. 15.11 we show the result for $a = 2$ and $\alpha = 45°$. Closer evaluation shows that the real part potential has a jump across the line object.

With a rotation of the angle the complementary solution, in which real and imaginary part of the solution of formula (15.14) are exchanged, can be used also for highly permeable objects. This is considered in the following program. Variable `perm` is to be used to switch between the two situations. In that case the streamfunction has a jump at the line object, representing the flux through the permeable formation.

```
perm = 0;               % =1: permeable fracture, =0 impermeable obstacle
velo = 1;               % velocity at infinity
alpha = pi/4;           % baseflow angle
a=2;                    % half-length of fracture / obstacle
xmin = -5; xmax = 5; ymin = -5; ymax = 5;
N = 101;

a = a*a;
xvec = linspace(xmin,xmax,N);
yvec = linspace(ymin,ymax,N);
[x,y] = meshgrid (xvec,yvec);                                  % mesh
z = x+i*y;
if perm
    Phi = velo*(z*cos(alpha)-i*sqrt(z.*z-a).*sin(alpha+(x<0)*pi+pi));
    contourf (x,y,imag(Phi),20);
    hold on; colorbar;
    contour (x,y,real(Phi),20,'w');
    [u,v] = gradient (imag(Phi));
else
    Phi = velo*(z*cos(alpha)-i*sqrt(z.*z-a).*sin(alpha+(x<0)*pi));
    contourf (x,y,real(Phi),20);
    hold on; colorbar;
    contour (x,y,imag(Phi),20,'w');
    [u,v] = gradient (real(Phi));
end
qx = ceil(size(x)/9); qy = ceil(size(y)/9);
quiver (x(1:qx:end,1:qy:end),y(1:qx:end,1:qy:end),...
    u(1:qx:end,1:qy:end),v(1:qx:end,1:qy:end),0.5,'w');
```

The M-file described in this sub-section is included in the accompanying software as *'fracture.m'* file.

References

Churchill RV, Brown JW (1984) Complex variables and applications. McGraw-Hill, New York
Needham T (1997) Visual complex analysis. Oxford University Press, New York
Prandtl L, Tietjens OG (1934) Fundamentals of hydro- and aeromechanics. Dover, New York, p 270
Strack ODL (1989) Groundwater mechanics. Prentice Hall, Englewood Cliffs, p 732

Chapter 16
2D and 3D Transport Solutions (Gaussian Puffs and Plumes)

16.1 Introduction

In the Chaps. 4–7 several solutions of the transport equation in a single space dimension (1D) have been presented. Analytical solutions were presented including various kinds of different processes. The models are valid under restricted conditions. In this chapter we show that analytical solutions are also available for higher dimensional problems.

The 1D formula can be applied to various situations in almost all environmental compartments. Such models are especially popular in air pollution modeling, for example in the modeling the local concentration distribution due to emissions from stacks, but they can also be applied to problems of point pollution in rivers, channels, lakes, reservoirs and the sea, in ground- and soil water.

Most solutions are derived from the 1D solution for a plume, which is given by the *normal distribution* function $f(x)$:

$$f(x) = \frac{1}{\sqrt{2\pi}\sigma} \exp\left(-\frac{1}{2}\left(\frac{x-\mu}{\sigma}\right)^2 \right) \tag{16.1}$$

The normal distribution has most applications in statistics. μ is the mean value of the distribution and σ the standard deviation. As the normal distribution is also referred to as Gaussian[1] distribution, these methods are also named *Gaussian plumes* (in case of steady state) or *Gaussian puffs* (for the transient case). One may find the term *Gaussian models* too.

The normal distribution $f(x)$ with $\mu = 0$ and $\sigma = \sqrt{2Dt}$ is the solution of the transport equation

$$\frac{\partial c}{\partial t} = \frac{\partial}{\partial x} D \frac{\partial c}{\partial x} \tag{16.2}$$

[1] Carl Friedrich Gauss (1777–1855), German mathematician.

E. Holzbecher, *Environmental Modeling*,
DOI 10.1007/978-3-642-22042-5_16,

which accounts for diffusion only. The initial condition for this solution is the peak with infinite concentration at $x = 0$. This mathematical construction is termed the *δ-function*. More precisely the δ-function is defined by the two properties:

$$\delta(x) = \begin{cases} \infty & \text{for } x = 0 \\ 0 & \text{for } x \neq 0 \end{cases} \quad \text{and} \quad \int_{-\infty}^{\infty} \delta(x)dx = 1 \tag{16.3}$$

The normal distribution converges to zero for $x \to \pm\infty$. This is a suitable boundary condition for an excess concentration, i.e. when the background value is subtracted from the measured value.

The solution (16.1) can be extended to the situation where the plume is transported within a flow field. If the fluid moves with velocity v in x-direction, the solution is:

$$c(x,t) = \frac{M}{\sqrt{4\pi t}\sqrt{D}} \exp\left(-\frac{(x - vt)^2}{4tD} \right) \tag{16.4}$$

M denotes the total mass per unit area in the fluid system. The concentration c is a solution of the transport equation, which accounts for diffusion and advection:

$$\frac{\partial c}{\partial t} = \frac{\partial}{\partial x} D \frac{\partial c}{\partial x} - v \frac{\partial c}{\partial x} \tag{16.5}$$

The solution (16.4) can be implemented easily in MATLAB®. The following command sequence is an example:

```
Dx = 0.000625;                  % diffusivity
v = 0.1;                        % velocity
M = 1;                          % mass
xmin = -0.05; xmax = 2.15;      % x-axis interval
t = [1:4:20];                   % time
%-----------------------------execution-------------------------
x = linspace (xmin,xmax,100);
c = [];
for i = 1:size(t,2)
    xx = x - v*t(i);
    c = [c; (M/sqrt(4*pi*Dx*t(i)))*ones(1,size(x)).*...
        exp(-(xx.*xx)/(4*Dx*t(i)))];
end
%--------------------------------output--------------------------
plot (c'); hold on; xlabel ('space');
ylabel ('concentration'); title ('1D Gaussian puff');
```

The space interval is divided into 100 parts. Within each run through the `for`-loop, which represents the time instants, the solution is evaluated for the entire space vector `x`. The resulting vector of concentrations is appended as a new line in the concentration matrix `c`. Note that using the `plot` command in the final plotting

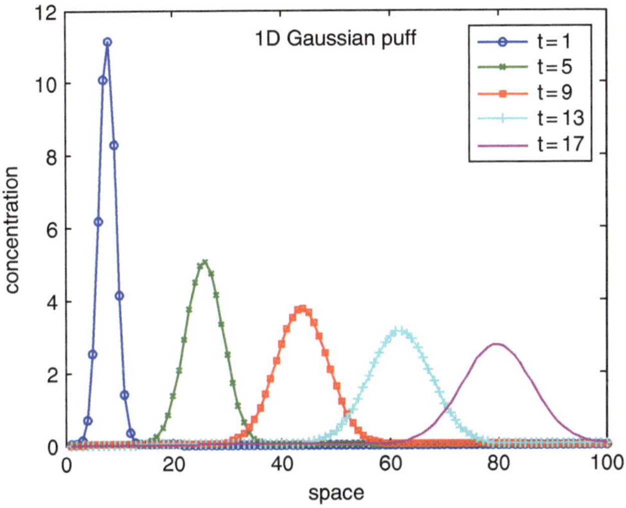

Fig. 16.1 1D Transport solution for instantaneous source for five time instants

section the column indices of the matrix appear on the x-axis. Similar constructions were already used in Chap. 4 for simulations of concentration fronts.

Figure 16.1 depicts the output, slightly modified in the MATLAB® figure editor. The concentration distribution for five different time instants is plotted. Obviously, the plume moves downstream while the shape of the bell function changes. The peak concentration decreases with time, and the space interval, in which elevated concentrations show up, widens.

Most command sequences, presented on the following pages, are gathered in the '*GaussianPuff.m*' file, which is included in the accompanying software. Between various solutions the user may choose using option parameters in the input section of the file.

The presented formula (16.4) has been applied in almost all branches of environmental sciences and only few examples can be listed. Based on (16.4), Maloszewski et al. (1994) explain tracer experiments in karstic aquifers, Sukhodolov et al. (1997) deal with dispersion in a lowland river, Wang and Persaud (2004) investigate experiments from soil columns. Bear (1976) and Kinzelbach (1987) recommend the normal distribution for pollution problems in aquifers.

A dimensionless formulation can be derived easily if the transport (16.5) is modified to:

$$\frac{\partial c}{\partial \tau} = \frac{1}{Pe}\frac{\partial^2 c}{\partial \xi^2} - \frac{\partial c}{\partial \xi} \tag{16.6}$$

where $\xi = x/L$ denotes dimensionless space and $\tau = tv/L$ dimensionless time. $Pe = vL/D$ is the dimensionless Péclet number that was already introduced in

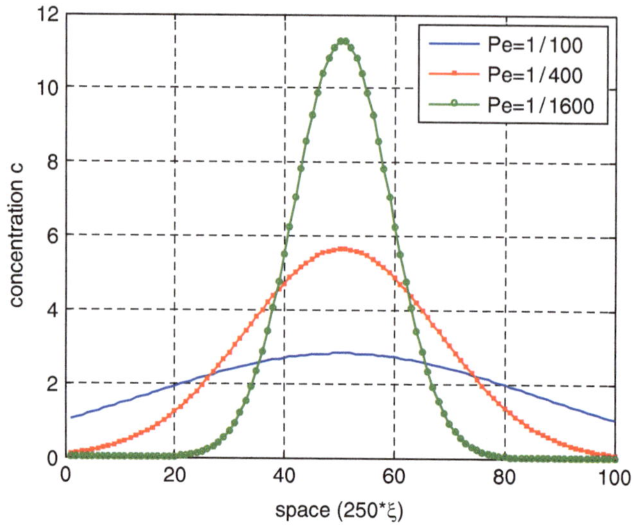

Fig. 16.2 Transport solution for instantaneous source in dependence of the Péclet number *Pe*

Chap. 5. The dependence of the transport solution for dimensionless time $\tau = 1$ on the Péclet number is visualized in Fig. 16.2. The figure was obtained using the M-file *'GaussianPuff.m'* with input data:

```
d = 1; v = 0; M = 1; xmin = -0.2; xmax = 0.2; t = 1; gplot = 1;
```

in three runs with different diffusivities:

```
Dx = 0.01; 0.0025; 0.000625;
```

It is instructive to view the development of the concentration profile in the space-time diagram. Such an illustration is given in Fig. 16.3, which was obtained using the M-file *'GaussianPuff.m'* with input data:

```
d = 1; Dx = 0.000625; v = 0.1; M = 1; xmin = -0.05; xmax = 0.2;
t = [.1:.025:1]; gcont = 2;
```

A more generalized formulation of the normal distribution is valid for substances that are subject to degradation or decay processes in addition. If λ refers to the decay coefficient, one obtains the formula:

$$c(x,t) = \frac{M}{\sqrt{4\pi t}\sqrt{D}} \exp\left(-\frac{(x-vt)^2}{4tD} - \lambda t\right) \tag{16.7}$$

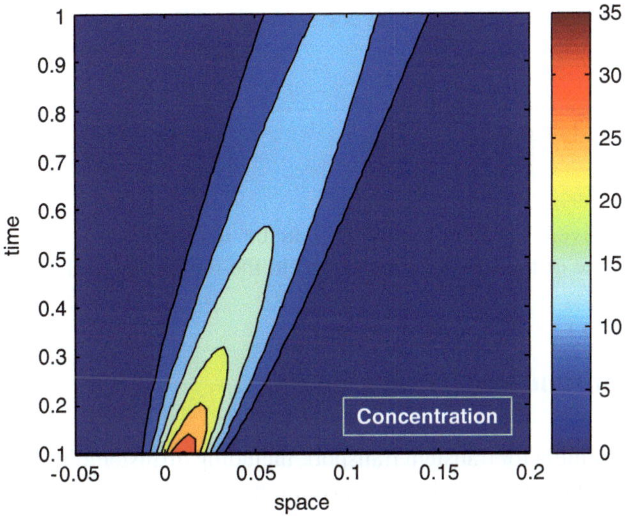

Fig. 16.3 Transport solution for an instantaneous source, represented in a time-space diagram

(see also: Hunt 1983; Kinzelbach 1987). Linear equilibrium sorption can be included following the derivations in Chap. 6. For a constant retardation coefficient R, the solution is given by:

$$c(x,t) = \frac{M}{\sqrt{4\pi t}\sqrt{D/R}} \exp\left(-\frac{(x - vt/R)^2}{4tD/R} - \lambda t\right) \tag{16.8}$$

In analogy to the 1D situation analytical solutions can be derived for the higher dimensional cases. The generalization of the 1D normal distribution (16.1) for 2D is:

$$f(x,y) = \frac{1}{2\pi\sigma_x\sigma_y} \exp\left(-\frac{1}{2}\left[\left(\frac{x-\mu_x}{\sigma_x}\right)^2 + \left(\frac{y-\mu_y}{\sigma_y}\right)^2\right]\right) \tag{16.9}$$

with standard deviations σ_x and σ_y and mean values μ_x and μ_y for x- and y-directions. Formula (16.9) gives the solution of the differential equation

$$\frac{\partial c}{\partial t} = \frac{\partial}{\partial x} D_x \frac{\partial c}{\partial x} + \frac{\partial}{\partial y} D_y \frac{\partial c}{\partial y} \tag{16.10}$$

with a δ-peak initial condition (formula (16.3)) at position (μ_x,μ_y) and zero boundary condition at infinity. Standard deviations and diffusivities are related by the equations $\sigma_x = \sqrt{2D_x t}$ and $\sigma_y = \sqrt{2D_y t}$.

Analogously, the solution for the corresponding 3D situation is given by:

$$f(x,y) = \frac{1}{\sqrt[3]{2\pi}\sigma_x\sigma_y\sigma_z} \exp\left(-\frac{1}{2}\left[\left(\frac{x-\mu_x}{\sigma_x}\right)^2 + \left(\frac{y-\mu_y}{\sigma_y}\right)^2 + \left(\frac{z-\mu_z}{\sigma_z}\right)^2\right]\right) \tag{16.11}$$

From the given formulae solutions can be derived for various 2D and 3D situations. Some of these will be presented in the remainder of this chapter

16.2 2D Instantaneous Line Source

The explicit formula for transient transport, including diffusion/dispersion, constant advection in x-direction and decay is given by:

$$c(x,y,t) = \frac{M}{4\pi t\sqrt{D_x D_y}} \exp\left(-\frac{1}{4t}\left(\frac{(x-vt)^2}{D_x} + \frac{y^2}{D_y}\right) - \lambda t\right) \tag{16.12}$$

(see also: Fried 1975; Kinzelbach 1987). M denotes the total mass per unit length in that situation. Figure 16.4 depicts the surface plot of an example 2D Gaussian puff.

The plot was produced using the M-file *'GaussianPuff.m'* with input data:

```
d = 2; Dx = 0.01; Dy = 0.000625; v = 0.1; M = 1;
xmin = -0.2; xmax = 0.5; ymin = -0.2; ymax = 0.2; t = 1; gsurf = 1;
```

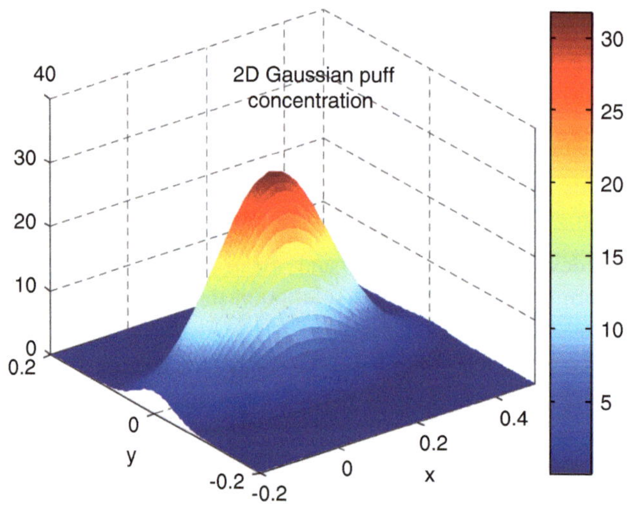

Fig. 16.4 Transport solution for an instantaneous source in 2D

The concentration distribution c, given by (16.12), is a solution of the differential equation:

$$\frac{\partial c}{\partial t}=\frac{\partial}{\partial x}D_x\frac{\partial c}{\partial x}+\frac{\partial}{\partial y}D_y\frac{\partial c}{\partial y}-v\frac{\partial c}{\partial x}-\lambda c \tag{16.13}$$

The formula (16.12) can be derived from the statistical formulation (16.9) with the help of the relationship between diffusivities as transport parameters on one side and standard deviations as statistical characteristics on the other side: $\sigma_x=\sqrt{2D_xt}$ and $\sigma_y=\sqrt{2D_yt}$, already encountered above. Various more complex models are based on more general relationships between these two types of parameters. Smith (1989) gives an overview and more details on this topic.

16.3 2D Constant Line Source

The 2D steady state analytical solution, describing the effect of a constant line source on an infinitely extended plane, is given by:

$$c(x,y)=\frac{c_0Q}{2\pi\sqrt{D_xD_y}}\exp\left(\frac{xv}{2D_x}\right)K_0\left(\frac{v^2x^2}{4D_x^2}+\frac{v^2y^2}{4D_xD_y}\right) \tag{16.14}$$

(see: Fried 1975; Bear 1976) with K_0 being the modified Bessel function of the second kind and zero order.

The formula can easily be implemented in MATLAB®. The Bessel function of second kind is reached by the command

```
bessely(nu,Z)
```

where `nu` is the order and `Z` the array of arguments.

16.4 3D Instanteneous Source

The effect of diffusion, advection in a constant unidirectional flow field, and decay is given by the analytical solution:

$$c(x,y,z,t)=\frac{M}{\left(\sqrt{4\pi t}\right)^3\sqrt{D_xD_yD_z}}\exp\left(-\frac{1}{4t}\left(\frac{(x-vt)^2}{D_x}+\frac{y^2}{D_y}+\frac{z^2}{D_z}\right)-\lambda t\right) \tag{16.15}$$

(see also: Kinzelbach 1987, Wexler 1992), which can be derived in analogy to (16.7) and (16.12). There are many applications of this equation, mainly for pollution spreading in the atmosphere. Richter and Seppelt (2004) apply the

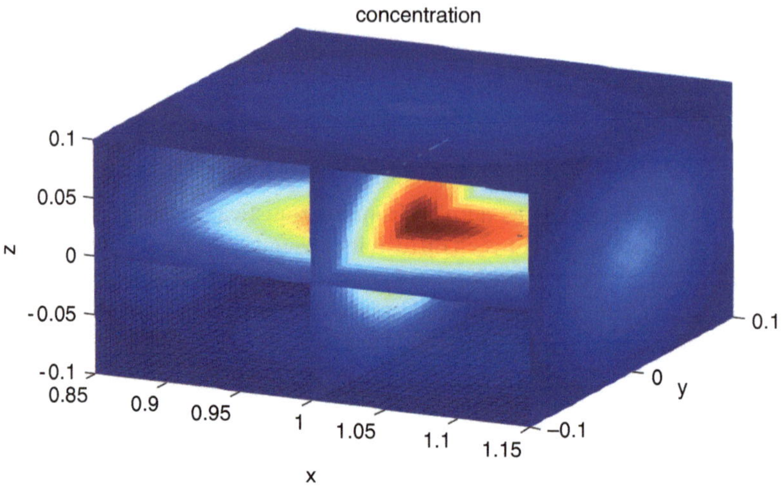

Fig. 16.5 Concentration distribution from a 3D Gaussian puff

Gaussian puff solution with no advection in order to evaluate pollen disposal from genetically modified crops in agricultural ecosystems. The method can be chosen as an option in the '*GaussianPuff.m*' file.

$$\frac{\partial c}{\partial t} = \frac{\partial}{\partial x} D_x \frac{\partial c}{\partial x} + \frac{\partial}{\partial y} D_y \frac{\partial c}{\partial y} + \frac{\partial}{\partial z} D_z \frac{\partial c}{\partial z} - v \frac{\partial c}{\partial x} - \lambda c \tag{16.16}$$

A 3D Gaussian puff is visualized in Fig. 16.5. The result was obtained using the '*GaussianPuff.m*' file with input data:

```
d = 3; Dx = 0.004; Dy = 0.001; Dz = 0.001; v = 1; M = 1;
xmin = 0.85; xmax = 1.15; ymin = -0.1; ymax = 0.1; zmin = -0.1; zmax =
0.1; t = 1; gcont = 2
```

For applications in the atmosphere, the formula is extended to take the ground surface into account. The ground surface is located at $z = 0$, while the release of the pollutant appears at height H. In applications H can correspond with the stack or chimney height, but often an increased value has to be adopted in order to account for the initial rise of an emission plume with an increased temperature compared to the ambient environment. H is also referred to as *effective stack height*.

At the ground surface, it is common to require a no-flow boundary condition. As demonstrated in Chap. 15 the no-flow condition can be fulfilled by adding another source in mirror position on the other side of the boundary, i.e. for the situation here at location $(x,y,z) = (0,0,-H)$. For a linear transport equation the solution can be constructed by using the principle of superposition (see Chaps. 13 and 14):

$$c(x,y,z,t) = \frac{Q}{\left(\sqrt{4\pi t}\right)^3 \sqrt{D_x D_y D_z}} \exp\left(-\frac{1}{4t}\left(\frac{(x-vt)^2}{D_x} + \frac{y^2}{D_y}\right) - \lambda t\right)$$
$$\times \left[\exp\left(-\frac{(z-H)^2}{4D_z t}\right) + \exp\left(-\frac{(z+H)^2}{4D_z t}\right)\right] \tag{16.17}$$

(see also: Kathirgamanathan et al. 2002; Mitsakou et al. 2003). Overcamp (1983) describes this method for a meteorological model.

16.5 3D Constant Source

Neglect of diffusion in horizontal direction leads to the differential equation:

$$\frac{\partial c}{\partial t} = \frac{\partial}{\partial y} D_y \frac{\partial c}{\partial y} + \frac{\partial}{\partial z} D_z \frac{\partial c}{\partial z} - v\frac{\partial c}{\partial x} - \lambda c \tag{16.18}$$

for which the steady state can be reformulated as:

$$\frac{\partial c}{\partial x} = \frac{\partial}{\partial y} \frac{D_y}{v} \frac{\partial c}{\partial y} + \frac{\partial}{\partial z} \frac{D_z}{v} \frac{\partial c}{\partial z} - \frac{\lambda}{v} c \tag{16.19}$$

From the analytical point of view the differential (16.19) is identical to (16.13). Only parameters and variables have different names. For the latter equation, the solution was already denoted in (16.12). Reformulation of the solution in terms of the new variables and parameters yields:

$$c(x,y,z) = \frac{Mv}{4\pi x\sqrt{D_x D_y}} \exp\left(-\frac{v}{4x}\left(\frac{y^2}{D_y} + \frac{z^2}{D_z}\right) - \frac{\lambda}{v}x\right) \tag{16.20}$$

In fact, (16.20) is the steady state solution for a constant source in 3D space. The product Q $= Mv$ in the nominator of the first term on the right side represents the emission rate in unit [mass/time].

Formula (16.20) can be modified to account for a source at height H and a no-flow surface boundary condition along the line $z = 0$. The procedure, using an image source, was already described in Sect. 16.4. In the same manner a steady state solution for a constant source in 3D is obtained:

$$c(x,y,z) = \frac{Q}{4\pi x\sqrt{D_y D_z}} \exp\left(-\frac{vy^2}{4xD_y}\right) \left[\begin{array}{c} \exp\left(-\dfrac{v(z-H)^2}{4xD_z}\right) \\ +\exp\left(-\dfrac{v(z+H)^2}{4xD_z}\right) \end{array}\right] \exp\left(\frac{\lambda}{v}x\right) \tag{16.21}$$

For the concentration distribution at ground level ($z = 0$) follows:

$$c(x,y) = \frac{Q}{2\pi x\sqrt{D_y D_z}} \exp\left(-\frac{v}{4x}\left(\frac{y^2}{D_y} + \frac{H^2}{D_z}\right)\right) \exp\left(\frac{\lambda}{v}x\right) \tag{16.22}$$

For a tracer component (i.e. without decay) in terms of standard deviations the corresponding formula reads:

$$c(x,y) = \frac{Q}{\pi v \sigma_y \sigma_z} \exp\left(-\frac{1}{2}\left(\frac{y^2}{\sigma_y{}^2} + \frac{H^2}{\sigma_z{}^2}\right)\right) \tag{16.23}$$

(see: Vaz and Ferreira 2004). Applying the principle of superposition (see Chaps. 14 and 15) the formula can be used to account for several stacks. The contributions from the different sources, calculated by formula (16.21), have to be added.

Such models are used extensively for estimations of the local development of a plume in the atmosphere. For the most common application of release from a stack, the parameters are visualized in Fig. 16.6. The Gaussian models take into account diffusive processes, advection with a mean air flow direction (wind), and first order decay. The term diffusion here is used as an umbrella term for various processes which have in common the tendency to lower concentration or temperature gradients. Diffusion at the molecular scale can surely be neglected in the atmosphere, while variations and fluctuations at various scales within the velocity field are the cause for the observation of diffusion at a larger scale. Moreover, turbulence adds as another origin of diffusion.

Idealized conditions are assumed for these processes if formula (16.21) has to be applied. Other processes, which often play an important role in the atmospheric

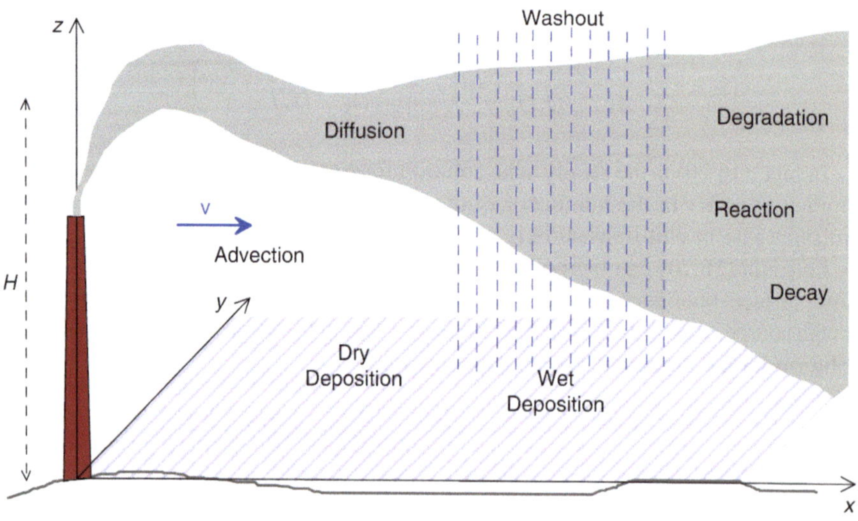

Fig. 16.6 Release from a stack; Schematic illustration of parameters and processes

environmental compartment, are not taken into account: dry and wet deposition on the ground, washout due to precipitation, and multi-species reactions. More specifically, all decay processes are neglected for which a first order description does not suffice. For example, photochemical degradation, that is relevant in air pollution problems, requires a more detailed approach.

The following commands deliver the result of a Gaussian plume according to formula (16.23).

```
Dy = 0.2; Dz = 1;                % diffusivities
v = 0.5;                         % velocity
lambda = 0;                      % decay constant
Q = 1;                           % emission rate(s)
xstack = 0; ystack = 50;         % stack location(s)
xmin = 10; xmax = 1000;          % x-axis interval
ymin = 0; ymax = 100;            % y-axis interval
H = 50;                          % effective stack height(s)
z = 0;                           % height of observation (=0 for ground
                                 surface)
gplot = 1;                       % plot option (=1 yes; =0 no)
gcont = 2;                       % contour plot option (=2 filled; =1 yes;
                                 =0 none)

%-----------------------------execution--------------------------
[x,y] = meshgrid (linspace(xmin,xmax,100),linspace(ymin,ymax,100));
c = zeros (size(x)); e = ones(size(x));
for i = size(Q,2)
   xx = x - xstack(i); yy = y - ystack(i);
   c =c+Q(i)*e./(4*pi*xx*sqrt(Dy*Dz)).*exp(-v*yy.*yy./(4*Dy*xx)).*...
   (exp(-v*(z-H(i))*(z-H(i))*e./(4*Dz*xx))+exp(-v*(z+H(i))*...
   (z+H(i))...*e./(4*Dz*xx)).*exp(-lambda*xx/v);
end

%---------------------------------output-------------------------
if gplot
    for i = 10:10:100
        plot (c(:,i)); hold on;
    end
end
if gcont
    figure;
    if gcont > 1
        contourf (x,y,c); colorbar;
    else
        contour (x,y,c);
    end
end
```

The command sequence is included in the accompanying software in file '*GaussianPlume.m*'.

The results of the commands, slightly modified in the MATLAB® figure editor, are shown in Figs. 16.7 and 16.8. The first figure depicts the concentration on the ground. Highest concentrations on the ground can be observed approximately 300 m downstream from the stack. In analogy to the 1D situation analytical

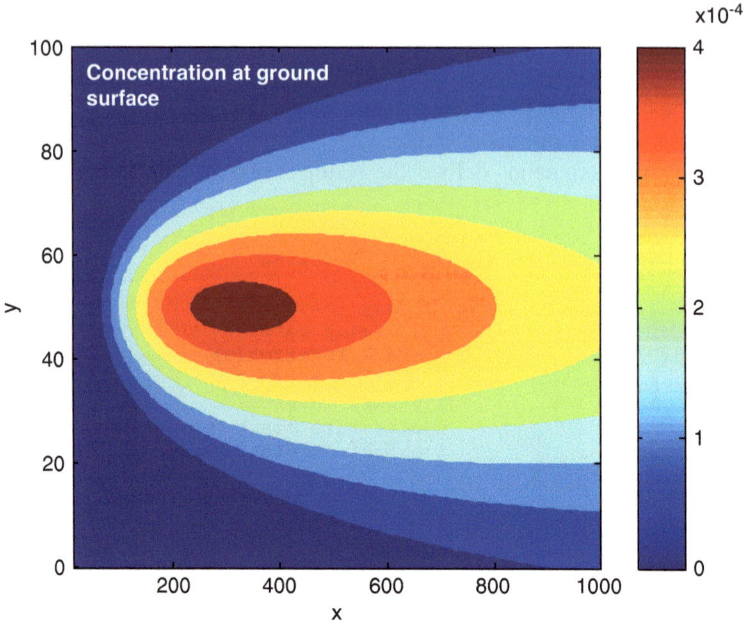

Fig. 16.7 Concentration distribution at ground surface according to a Gaussian plume

solutions can be derived for the 2D situation. They decrease slowly after the peak is reached at ≈350 m.

Figure 16.8 depicts ground concentrations along slices with constant x and varying y. Clearly, all distributions are of Gaussian type. Global peak concentration is found in the slice at $x = 300$. For lower $x < 300$, the bell shaped concentrations have small standard deviations, and local peak concentrations increase with x. Beyond the slice with peak concentration, i.e. for $x > 300$, the standard deviations increase, and the local maximum decreases with x.

Finally let us list the assumptions for the application of the Gaussian plume model for the estimation of stack release:

- The smokestack emission is continuous and constant
- The terrain is relatively flat
- The wind speed is constant in time and in elevation
- In main wind direction, advection dominates over diffusion and dispersion
- The diffusivities in horizontal and vertical direction are constant, i.e. they do not change spatially and temporally
- No settling velocity for the species
- The pollutant is not involved in reactions, i.e. the species is neither produced nor consumed; in particular there is no degradation
- When the pollutants hit the ground, they are reflected and not absorbed

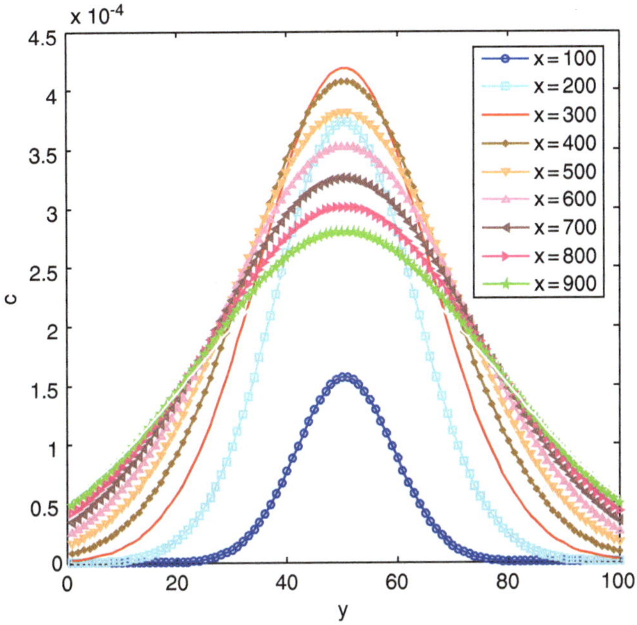

Fig. 16.8 Concentration distribution at ground surface according to a Gaussian plume; Concentrations at cross-sections, parallel to *y*-axis at various distances from the source

It is not only the potential hazards from stacks or chimneys that can be estimated by Gaussian' plumes. As mentioned above the normal distribution has applications in all environmental compartments. Methods, similar to the ones described in this subchapter, were applied in other fields as well. Karol et al. (1997) simulate the exhaust composition of a subsonic aircraft and combine Gaussian plumes with chemical reaction modeling. Raupach et al. (2001) deal with the insecticide endosulfan that, applied by spraying on agriculture land, finally enters and pollutes rivers and lakes. Dayan and Koch (2002) deal with the dispersion of PCB following a release caused by fire.

References

Bear J (1976) Hydraulics of groundwater. Mc Graw Hill, New York, p 569

Dayan U, Koch J (2002) Dispersion of PCB in the environment following an atmospheric release caused by fire. Sci Total Environ 285:147–153

Fried JJ (1975) Groundwater pollution. Elsevier, Amsterdam, p 330

Hunt B (1983) Mathematical analysis of groundwater resources. Butterworths, London, p 271

Karol IL, Ozolin YE, Rozanov EV (1997) Box and Gaussian plume models of the exhaust composition evolution of subsonic transport aircraft in- and out of the flight corridor. Ann Geophysicae 15:88–96

Kathirgamanathan P, McKibbin R, McLachlan RI (2002) Source term estimation of pollution from an instantaneous point source. Res Lett Inf Math Sci 3:59–67
Kinzelbach W (1987) Numerische Methoden zur Modellierung des Transports von Schadstoffen im Grundwasser. Oldenbourg, München, p 317, in German
Maloszewski P, Maloszewski P, Benischke R, Harum T, Zojer H (1994) Estimation of solute transport parameters in heterogen groundwater system of a karstic aquifer using artificial tracer experiments, XXV congress IAH 'Management to sustain shallow groundwater Systems'. National Conference Publication, Adelaide, pp 105–111
Mitsakou C, Eleftheriadis K, Housiadas C, Lazaridis M (2003) Modeling of the dispersion of depleted uranium aerosol. Health Phys 84(4):538–544
Overcamp TJ (1983) A surface-corrected gausian model for elevated sources. J Applied Meteorology 22(6):1111–1115
Raupach MR, Briggs PR, Ahmad N, Edge VE (2001) Endosulfan transport II: modeling airborne dispersal and deposition by spray and vapor. J Environ Qual 30:729–740
Richter O, Seppelt R (2004) Flow of genetic information through agricultural ecosystems: a generic modeling framework with application to pesticide-resistance weeds and genetically modified crops. Ecol Model 174:55–66
Smith RW (1989) Review of recent development in mixing and dispersion. In: Falconer RA, Goodwin P, Matthew RGS (eds) Hydraulic and environmental modeling of coastal, estuarine and river waters. Gover Techn, Aldershot, pp 277–290
Sukhodolov AN, Nikora VI, Rowinski PM, Czernuszenko W (1997) A case study of longitudinal dispersion in small lowland rivers. Water Environment Res 69(7):1246–1253
Vaz AIF and Ferreira EC (2004) Air pollution control with semi-infinite programing, XXVIII Congreso Nat. de Estadística e Investigatión Operativa, Cádiz
Wang H, Persaud N (2004) Miscible displacement of initial distributions in laboratory columns. J Soil Sci Soc of Am 68(5):1471–1478
Wexler EJ (1992) Analytical solutions for one-, two-, and three-dimensional solute transport in groundwater systems with uniform flow, Techniques of Water-Resources Investigations of the United States Geological Survey, Book 3, Chapter B7, 190p

Chapter 17
Image Processing and Geo-Referencing

17.1 Introduction

In many applications of environmental modeling it makes sense to process images, to read them into a MATLAB® figure for certain types of processing. In some applications it is necessary to process the images themselves. Satellite images for example contain information about some environmental variables. The distribution of botanical and sometimes also of zoological species within a geographical area can be determined by a computational algorithm based on images. Infrared images are used as they contain information about the temperature distribution.

MATLAB® allows various types of image processing. Only few commands are treated here. Trauth (2010) tackles the topic in much more details. For those who need an extended functionality, Math Works offers an Image Processing Toolbox, containing tools to analyze and visualize images, develop algorithms, and share results; see: http://www.mathworks.com/products/image/. Interested readers are referred to the related textbook of Gonzales et al. (2004).

In core MATLAB®, the *'hdftool.m'* is available to explore, extract, and display satellite remote sensing data sets, distributed by the National Aeronautical and Space Administration (NASA) in Hierarchical Data Format (HDF). The reader can view the details in the MATLAB® help index under the keywords 'hdf' or 'hdftool'. In this chapter these options will not be presented further.

In the following, examples are restricted to few other purposes. Imagine a map or a cross-section plot through some environmental system. The figure contains some valuable information about a relevant variable or parameter distribution. Using a map the location of positions, of lines and of areas can be determined. Lakes, rivers, shorelines, and land-use patterns are examples of such distributions which can be located with the help of a map. A geological cross-section, in which rock- or soil-type layers are visualized, is another example of a 2D illustration of information. The thickness of these layers, as well as their location, represented in the cross-sectional view, are data that the user may need to extract for modeling soil, aquifer or geological processes.

E. Holzbecher, *Environmental Modeling*,
DOI 10.1007/978-3-642-22042-5_17, © Springer-Verlag Berlin Heidelberg 2012

It is the aim of the modeller to transfer some information from the map into a computer model, which may itself be implemented in MATLAB® or at least connected to MATLAB®. Different kinds of connections between MATLAB® and a Geo Information System (GIS) have already been discussed in literature. Marsili-Libelli et al. (2001) couple a river quality model with a GIS software, whereas Marsili-Libelli et al. (2002) present a pure MATLAB® approach for a task of similar kind. Raterman et al. (2001) describe an integrated approach, using MATLAB® with GIS for groundwater modeling.

Moreover, it is often convenient to view model output results in front of a map or a cross-section plot. Calculated concentration distributions of environmental species, temperature, hydraulic head or pressure are often visualized in front of an area background. In the following it is outlined how the task can be achieved with MATLAB®.

In the first subchapter it is shown how the user can include a bitmap image in a MATLAB® figure. There are two steps: the image must be read (1) and displayed on the screen (2). The second part is a guide to the correct coordinate frame. In the GIS literature such a task is referred to as *Geo-referencing*. In the third subchapter it is demonstrated how information from the map is transferred to the computer model. We summarize this work under the header *Digitizing*.

17.2 Reading and Display

The main new features within an M-file for geo-referencing are explained using an example. As an example image, we choose a map that is read from MATLAB®. It is shown how the map is depicted in the MATLAB® figure editor. Thereafter, the image is geo-referenced in order to be able to extract basic features from the map image figure. The following command sequence performs the first task.

The complete code, outlined in this chapter, is included in the accompanying software under the name *'georef.m'*.

```
[infile,path] = uigetfile('*.jpg','Select graphics file...');
infilepath = strcat(path,infile);
lx = 1000; ly = 1000;
figure;
[X,map] = imread(infilepath);
imagesc(X);
axis off; hold on;
ax1 = gca;
ax1 = axes ('Position',get(ax1,'Position'),'Color',...
    'none','XLim',[0 lx],'Ylim',[0 ly],'XTick',[],'YTick',[]);
```

The first line opens a file select box using the MATLAB® `uigetfile` command. Parameters in the command are the type of the file (here: *.jpg), and the header 'Select graphics file...'. The file-select box looks as follows:

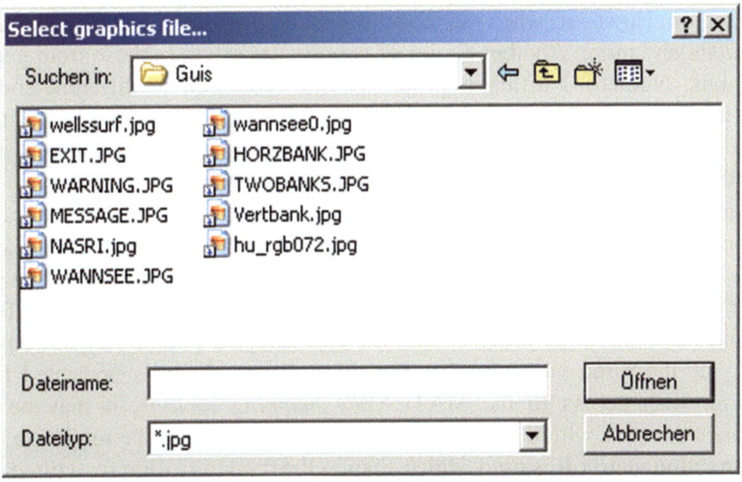

Fig. 17.1 The MATLAB® file-select box for graphic files

The output of the `uigetfile` command is the filename, which is stored in the variable `infile`. The second output is the directory path of the graphics file, stored in the variable `path`. Both these text strings are concatenated in the second command by using `strcat`. The initial length `lx` and width `ly` scale of the image are set to 1,000. The `imread` demands the chosen file to be read.

After reading into the workspace, the image file is represented by two variables, `X` and `map`. `X` contains the color information for each pixel and `map` the colormap information. The dimension of `X` is 3; the content of the first and second entries is equal to the corresponding pixel-numbers of the image. For RGB color images, the third dimension is 3 because the color information is internally represented by three color values. Altogether there is thus one R, one G and one B value for every pixel.

For those modelers wishing to work extensively with graphic files, it is recommended to study the information in the MATLAB® help, the various types of image representation and the various corresponding commands for input, output, display and conversion. Using the `imread` command all standard type bitmap images can be read by MATLAB®, such as 'bmp', 'cur', 'gif', 'ico', 'jpg', 'jpeg', 'pbm', 'pcx', 'pgm', 'png', 'pnm', 'ppm', 'ras', 'tif', 'tiff', and 'xwd'. The complete list of formats is found in the MATLAB® help.

With the `imagesc` command the map is displayed in the figure that was opened before. The following final commands of the presented sequence manipulate the outlook of the coordinate axes.

17.3 Geo-Referencing

When the map appears on the display, the user may like to have the real world coordinates displayed on the axes. The real world coordinates are of course not unique. Very often the origin is chosen in the lower left or upper left corner of the model region. However, when the model region is irregular, the choice may not be appropriate any more. Another choice of coordinates refers to those from geodetic projections. Numerous different projections are common in different countries of the world, even within countries, and this is not the place to go into details. The procedure which enables the user to have any real world coordinates available on the display, showing the map, is known as geo-referencing.

There are various ways of geo-referencing. One way is to specify the coordinates of two opposite corners. Often the positions for which exact coordinates are known are not identical with the corners of the map. Then two positions within the image can be chosen to which all others are related. There is one condition for the procedure to work well, which is that the two selected points may not have the same x- or the same y-coordinates. For this task an M-file is presented below. For those with access to the MATLAB@ *mapping toolbox*, it may be more convenient to work with the more general approach there. In the toolbox a 3×2 transformation matrix **R** is used, which enables the transformation for a tilted image and for a geoid (see: http://www.mathworks.com/products/mapping/).

In the following commands, which continue the command listing from the previous subchapter, the user chooses the image positions by mouse-click and enters the real world coordinate values in an input box.

```
%------------------------ Geo-reference --------------------------
h = gca; hold on;

h0 = text (0,-ly*0.05,'Set referencepoint 1','BackgroundColor'...
    ,'y','EdgeColor','red','LineWidth',2);
[x0,y0,but] = ginput(1);
h1 = plot (x0,y0,'k+');
coords = inputdlg({'horizontal','vertical'},'Field
  position',1,{'0','0'});
xx0 = str2double(coords(1)); yy0 = str2double(coords(2));
delete (h0);

h0 = text (0,-ly*0.05,'Set referencepoint 2','BackgroundColor',...
    'y','EdgeColor','red','LineWidth',2);
[x1,y1,but] = ginput(1);
h2 = plot (x1,y1,'k+');
coords = inputdlg({'horizontal','vertical'},'Field
  position',1,{'4000','3000'});
xx1 = str2double(coords(1)); yy1 = str2double(coords(2));
delete (h0,h1,h2);

mx = (xx1-xx0)/(x1-x0); xmin = xx0-mx*x0; xmax = xx1+mx*(lx-x1);
my = (yy1-yy0)/(y1-y0); ymin = yy0-my*y0; ymax = yy1+my*(ly-y1);
ax1 = axes ('Position',get(ax1,'Position'),...
    'Color','none','XLim',[xmin xmax],'Ylim',[ymin ymax]);
```

In the second command line, the text 'Set reference point 1' is displayed below the bottom axis of the figure. The user is informed that a location has to be set on the image by mouse-click. `ginput` is the MATLAB® command that delivers the location of the mouse during a mouse click. The third command demonstrates its use. `ginput` stores the position in the current coordinate system in the variables `x0` and `y0`. The input parameter `1` in the command tells that a single point is read. Note that the command waits for user input, i.e. the execution of the M-file does not continue before the user has not finished the task.

The following `plot` command places a black ' + ' at the location on the display where the mouse click occurred (see Fig. 17.2). `inputdlg` is the MATLAB® command for an input dialog, which here asks the user to input the coordinates of the marked location in the real coordinate system. The dialog box is shown in Fig. 17.2, with the values '900' and '1,000', which have been entered by the user. The input parameters of the command are easy to relate: 'horizontal' and 'vertical' are text strings associated with the input fields; 'Field position' is the header text. Both input fields are located in a single line, and default values for the two variables are '0'.

After the input dialog is closed, the coordinate values, chosen by the user, are stored in the `coords` variable, containing two strings. As for further processing the values (not the strings) are required, the strings have to be transformed into numbers. The two `str2double` commands transform the strings to double precision values. After the reference point 1 is read, the following `delete` command deletes

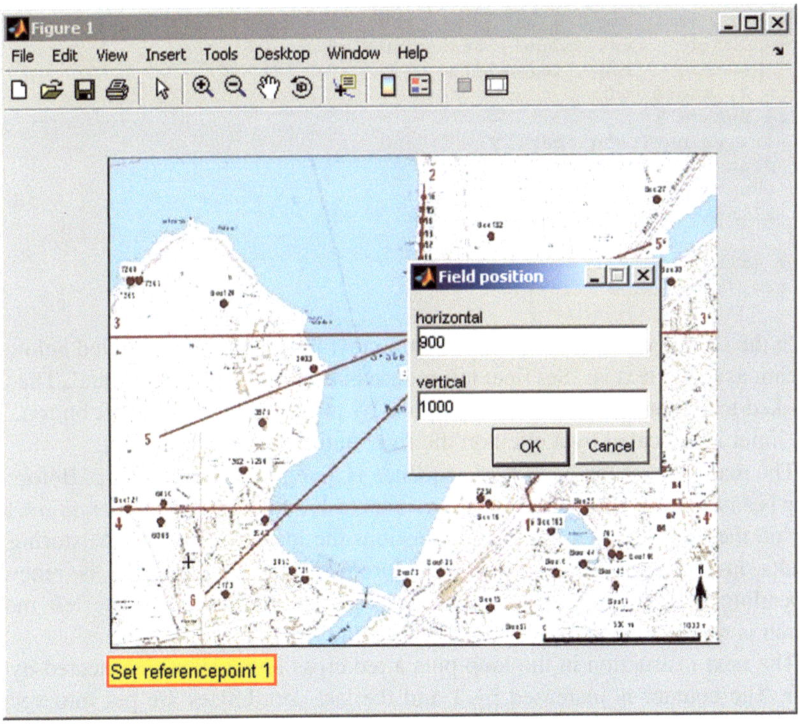

Fig. 17.2 Illustration of work with the geo-referencing example

the text string h0, which is used in the sequel for other purposes. The following commands (until the next delete) repeat the previous commands for reference point 2. As these correspond exactly to the commands, outlined for reference point 1, they need not to be commented again.

In the final four lines the new coordinate system is computed and displayed on the figure axes. The user may check that the linear transformation of coordinates, gathered in the first two lines, is correct. The final command adds the real world coordinates under XLim and Ylim axes properties. The specification of the 'Color' property as 'none' ensures that the image remains visible.

17.4 Digitizing

In the example it is shown how some characteristic structures, which are visualized on the map, can be made known to MATLAB® in real-world coordinates. That is done once for a line (structure) and once for locations. The command sequence within the M-file is basically the same for both tasks.

```
% --------------------------- Set Line ----------------------------
h = gca; hold on;
h0 = text (xmin,ymin-0.05*(ymax-ymin),'Set line: left mouse button:...
   set; right: last value','BackgroundColor','y',...
   'EdgeColor','red','LineWidth',2);
but = 1; count = 0;
while but == 1
    [xi,yi,but] = ginput(1);
    plot (xi,yi,'rx');
    count = count+1;
    xline(count) = xi; yline(count) = yi;
end
line (xline,yline,'Color','r');
delete (h0);
```

In the second command, the information text for the user is displayed below the bottom axis; the text is: 'Set line: left mouse button: set; right: last value'. The user is asked to set the locations of the polyline by mouse clicks on the left button. The last input is indicated by a click on the right button instead of the left.

The main part of the command sequence is given in the while loop. Before the loop is entered, the button indicator but is set to 1, and the point counter count to 0. Within the loop, the first command questions the location of the click, storing the results in the variables xi and yi. Moreover, the but variable is renewed, depending on the mouse button that was used by the modeller. The left mouse button is represented by a 1 in variable but, the right button by a 3.

The next instruction in the loop puts a red cross at the location selected by the user. The counter is increased by 1 and the last coordinates are put into vectors xline and yline, which represent the current polyline.

The loop is ended when the right mouse button is used and the button variable `but` contains a 3. Finally, after the loop, the polyline is plotted on top of the map (in red color) and the info-text is deleted.

```
% --------------------------- Set Locations------------------------
h0 = text (xmin,ymin-0.05*(ymax-ymin),'Set locations: left mouse...
   button: set; right: last value','BackgroundColor','y',...
   'EdgeColor','red','LineWidth',2);
h = gca; hold on;
but = 1; n = 0;
while but == 1
     [xj,yj,but] = ginput(1);
     plot (xj,yj,'bo');
     count = count+1;
     xloc(count) = xj; yloc(count) = yj;
end
delete (h0)
```

The procedure, described for a polyline, is repeated for the location set. There are minor differences: instead of a red cross, a blue circle becomes the indicator on the map, and there is no connecting line drawn finally.

An example result of digitizing work with the presented M-file is given in Fig. 17.3.

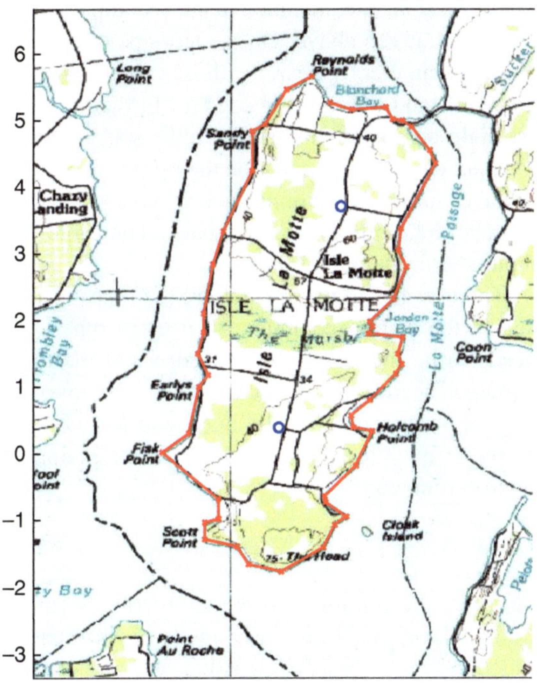

Fig. 17.3 Effect of digitizing work on the map of an island (shoreline in red, locations in blue)

17.5 MATLAB® Functions

In the accompanying software the *'georef.m'* M-file has a different header than most of the other M-files developed in the book. The header line

```
function [xx0,yy0,xx1,yy1,xline,yline,xloc,yloc] = georef ()
```

defines a MATLAB® function with name `georef`. The name appears on the right side of the equals sign. As a rule, the file has the same name, with extension '*.m*'. A function may be called from the MATLAB® command line or another M-file. That is not the main point, as all M-files can be called that way. The difference lies in the connection between calling module and function via both input- and output parameters. When an M-file, which is not a function, is called, parameters and variables in the called file are taken from the calling program. If there is a variable in the M-file, it is taken from the pool of *global variables*, defined at an upper level by the calling program. If it does not exist there, an error message results. In MATLAB® such files are called *scripts*.

Functions are distinguished from scripts, because functions are working within their own local data environment with variables that are locally defined only: *local variables*. Of course, data can be transferred from and to the calling M-file. This is done via a list of reference arguments. This point is best discussed by an example. The *'georef.m'* file, included in the accompanying software, starts with the function command, given above. On the left side of the equals sign, in square brackets, the output parameters appear, i.e. those values that are computed or manipulated within the M-file and which are needed for further processing in the calling module. There are eight parameters in this example, starting with `xx0` and ending with `yloc`.

Input parameters appear on the right side of the function name in round brackets. In the example there are no such variables: the brackets are empty. Empty brackets can also be omitted in m-language. The same parameter may appear both in the input and in the output list.

The user has to make sure that the types of the variables in the function M-file and in the calling command fit to each other. Not only the number of variables needs to be the same, the sequence and the types of each of the corresponding variables need to be identical as well. Exceptions from that general rule exist, but are not discussed here. The interested MATLAB® user may have a look into the help system under `nargin`, `nargout`. The example `georef` function could be called by the command:

```
[A,B,C,D,xbound,ybound,xwell,ywell] = georef;
```

by which the variable `xx0` of the function becomes `A` in the calling routine. In this example command, all variables have different names within and outside of the function. `A` must be a double value, as `xx0` is one, and `xbound` must be a 1D array, as the corresponding `xline` is a 1D array. The function `georef` can be called from the MATLAB® command window, or from any M-file in which geo-referencing is needed. It may be useful in any M-file in which data from scanned graphs are processed. It is no problem, if the variables in both files have different names: `georef` can be called from both, each using its own names for the variables, as explained. That's the advantage of using functions in comparison to using scripts.

The majority of M-files accompanying this book are written as functions, i.e. the first command is the function command without input and output parameters. In most modules the first line is not necessary: the reader of this book will already have recognized that the `function` line is omitted in the printed listings. The major reason for using the first command in the files is to ensure the user that the right program is called. Because function-name and file-name are identical, the module can be recognized by the file name.

The `function` commands, printed in the book, are always necessary. Sometimes the reason is that input and/or output parameters have to be specified. It is also necessary to use the `function` command if there are *subfunctions* within the M-file. Otherwise MATLAB® gives an error message, as can be demonstrated by the following lines:

```
A = 1;
function demo (A)
A = 2;
Error: A function declaration cannot appear within a script M-file.
```

Subfunctions are functions that are called within an M-file. There have been several examples already. Within the '*pdepetrans.m*' file (see Chapter 4), there are calls to three subfunctions:

```
function [c,f,s] = transfun(x,t,u,DuDx,D,v,lambda,...)
function u0 = ictransfun(x,D,v,lambda,sorption,k1,k2,c0,cin)
function [pl,ql,pr,qr] = bctransfun(xl,ul,...)
```

The calling rules, explained in this chapter, have already been applied there.

References

Gonzales RG, Woods RE, Eddins SL (2004) Digital image processing using MATLAB. Prentice Hall, Upper Saddle River, p 624

Marsili-Libelli S, Caporali E, Arrighi S, Becattelli C (2001) A georeferenced quality model. Water Sci Tech 43(7):223–230

Marsili-Libelli S, Pacini G, Barresi C, Petti E, Sinacori F (2002) An interactive georeferenced water quality model. In: Falconer RA, Lin B, Harris EL, Wilson CAME, Cluckie ID, Han D, Davis JP, Heslop S (eds) Proc. Hydroinformatics 2002. Cardiff

Raterman B, Schaars FW, Griffioen M (2001) GIS and MATLAB integrated for groundwater modeling (2001) ESRI User Conference, Proc, San Diego

Trauth MH (2010) MATLAB recipes for earth sciences, 3rd edn. Springer, Berlin, Heidelberg, p 336

Chapter 18
Compartment Graphs and Linear Systems

18.1 Compartments and Graphs

Seen from the process perspective, compartment models are the simplest type of environmental models. This type of model is based on quite rigorous conditions. There is probably no environmental system at all, where the conditions are fulfilled exactly. Nevertheless, as a first guess and in order to give a rough idea about the interactions between compartments, the simplicity justifies the application.

Compartment models consist of a network of compartments. An example for such a network is given in Fig. 18.1, representing the terrestrial part of the hydrological cycle.

There are four compartments to be modelled in the system of Fig. 18.1: interception, soil moisture, groundwater and surface water, visualized by rectangular boxes. Systems of compartments are the simplest concept in environmental modeling. A compartment is part of an environmental system which is spatially not further resolved. The analogue in chemical engineering is the continuously stirred chemical reactor. A compartment model can thus only be an approximation of a real system if there are no steep gradients within the real environment.

Several processes induce fluxes between compartments, visualized in Fig. 18.1 by arrows. Groundwater recharge is a flux from the soil to the groundwater, overland flow from interception to the surface water compartment. There is interflow from unsaturated soil moisture to the surface water compartment.

Rounded boxes illustrate processes connecting to the outer world, which is not explicitly taken into account in the model. Ocean and atmosphere are compartments within the hydrological cycle not being treated within the conceptual model demonstrated by Fig. 18.1.

A system of compartments, shown in a flux diagram, can be represented by a matrix. The *adjacency matrix* represents each compartment in one line and one column. It has a one-entry at the corresponding position, if there is a flux from the row compartment to the column compartment; otherwise there is a 0-entry.

The adjacency matrix for the system of Fig. 18.1 is:

E. Holzbecher, *Environmental Modeling*,
DOI 10.1007/978-3-642-22042-5_18, © Springer-Verlag Berlin Heidelberg 2012

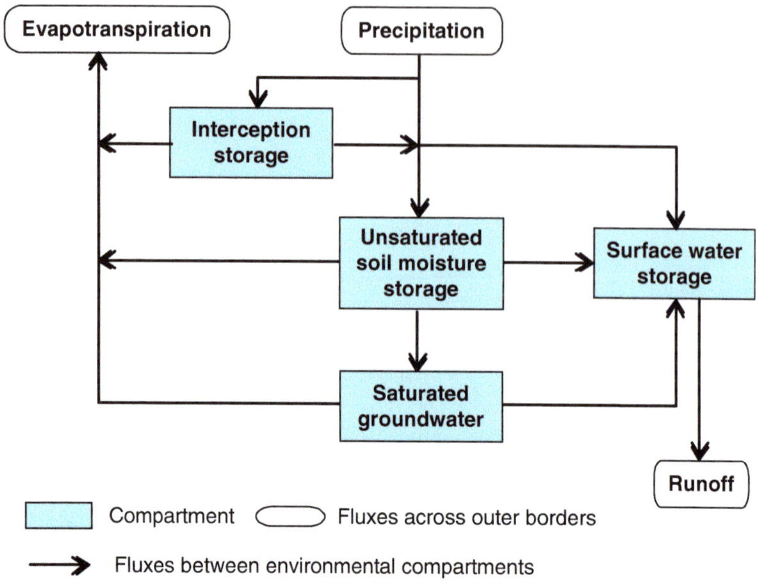

Fig. 18.1 Network of compartments for modeling the terrestrial part of the hydrological cycle according to Freeze and Cherry (1979), modified by E.H.

Table 18.1 Tabular representation of connections within the compartment system of Fig. 18.1

Storage	Interception	Soil moisture	Ground-water	Surface water
Interception				x
Soil moisture			x	x
Groundwater				x
Surface water				

$$\mathbf{A} = \begin{pmatrix} 0 & 0 & 0 & 1 \\ 0 & 0 & 1 & 1 \\ 0 & 0 & 0 & 1 \\ 0 & 0 & 0 & 0 \end{pmatrix} \tag{18.1}$$

which is directly related to the tabular representation of Table 18.1.

A more complex representation of the terrestrial part of the hydrological cycle is given in Fig. 18.2. Vegetation and surface are included as new compartments, interception is omitted. Several processes, which were neglected for simplicity in Fig. 18.1, are included. For example, due to capillary rise there may be fluxes from groundwater to the soil and from soil to the surface. Instead of surface water the less general term 'channel storage' is preferred in this graph.

The adjacency matrix for the system of Fig. 18.2 is:

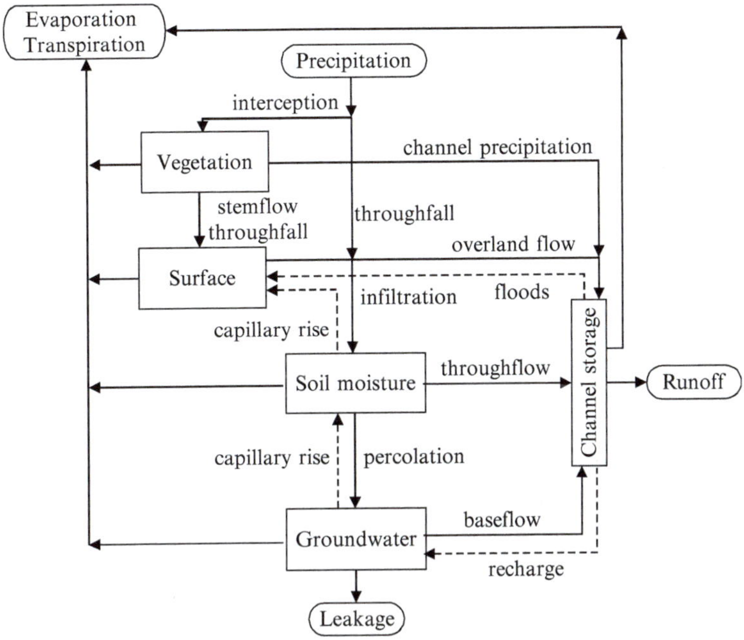

Fig. 18.2 Network model of a part of the hydrological cycle containing secondary fluxes (Ward & Robinson, 1990, modified by E.H.)

Fig. 18.3 Graph, representing a flux network between compartments, visualized using MATLAB®

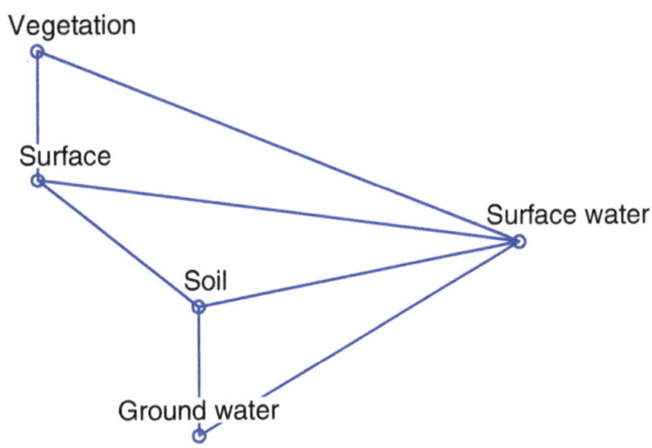

$$\mathbf{A} = \begin{pmatrix} 0 & 1 & 0 & 0 & 1 \\ 0 & 0 & 1 & 0 & 1 \\ 0 & 1 & 0 & 1 & 1 \\ 0 & 0 & 1 & 0 & 1 \\ 0 & 1 & 0 & 1 & 0 \end{pmatrix} \quad (18.2)$$

In MATLAB®, the graph corresponding to an adjacency matrix can be plotted directly using the `gplot` command. The following command sequence demonstrates how that is done. Figure 18.3 shows the graph, as given by MATLAB®.

```
A = [0 1 0 0 1; 0 0 1 0 1; 0 1 0 1 1; 0 0 1 0 1; 0 1 0 1 0];
xy = [1 4; 1 3; 2 2; 2 1; 4 2.5];
gplot (A,xy,'-o');
axis ([0.5 4.5 0.5 4.5]);
text (0.8,4.2,'Vegetation'); text (0.9,3.2,'Surface');
text (1.9,2.2,'Soil'); text (1.5,1.2,'Groundwater');
text (3.7,2.7,'Surface water');
axis off;
```

In `A` the adjacency matrix is specified. The vector `xy` contains the positions representing the compartments in the plot. The `'-o'` string in the `gplot`-command specifies how lines and nodes in the plot (*vertices* and *edges* in graph theory – see for example: Chartrand 1977) are represented in the figure. The `axis`-command defines the limits of the coordinate system. The final `text` calls set the text-strings near to the corresponding edges.

In order to complete a compartment model, fluxes need a mathematical quantification. Formulae for such fluxes are more or less complex, which determines the degree complexity at this point. In the following sub-chapters fluxes of different mathematical complexity will be treated, starting with the simplest linear case.

Holzbecher et al. (2005) describe some other properties of the adjacency matrix, especially in relation to the feedback property within compartment systems. In the following we examine linear systems as most simple mathematical examples of compartment models. Finally, there is to mention that Avila et al. (2003) present the ECOLEGO toolbox for radiological risk assessment, which is a MATLAB® implementation based on the compartment idea. The model was used for modeling the accumulation of radionuclides in the arctic Barents Sea (Dommasnes et al. 2001). A simplified version of the food chain as a graph is shown in Fig. 18.4

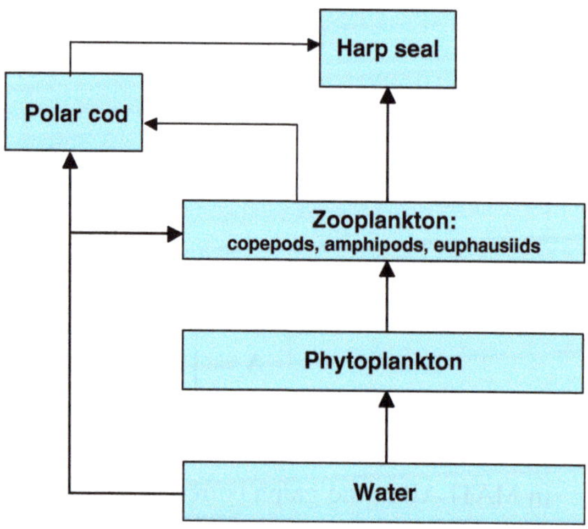

Fig. 18.4 Graph, representing the foodchain in the Barents Sea; simplified by E.H. from Dommasnes et al. (2001)

18.2 Linear Systems

A first, most simple approach comprises component mass fluxes as linear functions of some state variable of the involved compartments. The procedure is best demonstrated on a sequence of lakes, connected by streams, as depicted in Fig. 18.5. If some substance is introduced into the lake most upstream, it is subject of mixing processes first. It is assumed that mixing is fast in comparison to the *residence time* of the lake[1]. In the argumentation we assume all fluxes Q_i and volumes V_i to be known, and the concentrations c_i being the unknowns to be determined.

The concentration in the outflow of the lake is equal to the concentration c_1. The total mass leaving the lake per time is thus $Q_1c_1(t)$, where Q_1 denotes the mean water flux. Neglecting further losses, that mass enters the next lake downstream.

Again the same argumentation can be used to set $Q_2c_2(t)$ as the flux out of the second lake into the third. The procedure can be extended to the entire system of lakes. Following the principle of mass conservation (Chap. 2), one obtains a differential equation for each lake. For the ith lake holds:

$$V_i\frac{\partial c_i}{\partial t} = Q_{i-1}c_{i-1} - Q_ic_i \tag{18.3}$$

where V_i denotes the mean volume of the lake. In systems theory such a set-up is called *donor controlled*, as the input for the following compartment (lake) is determined by the state variable (concentration) of the previous compartment. The contrasting term is *recipient controlled*, i.e. the flux is determined by the concentration of the receiving compartment (for the lake sequence that approach does not make sense). Each equation can be divided by V_i. Using matrix notation the resulting equations can be written in one system, representing all lakes:

$$\frac{\partial}{\partial t}\mathbf{c} = \mathbf{Bc} \tag{18.4}$$

with elements $B_{i-1,i} = Q_{i-1}/V_i$ and $B_{i,i} = -Q_i/V_i$. The matrix $\mathbf{B}$ is a generalization of the adjacency matrix. The off-diagonal locations of the zeroes in both matrices

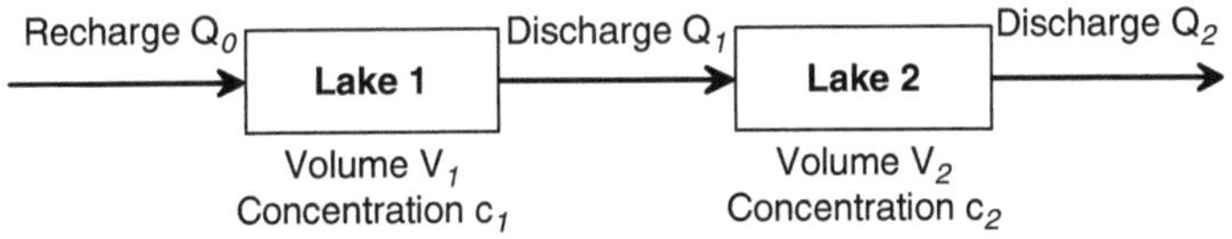

Fig. 18.5 Scheme for a sequence of lakes

[1] Under steady state conditions (inflow Q_{i-1} = outflow Q_i) the residence time is given by V_i/Q_i; the notation is given in the text.

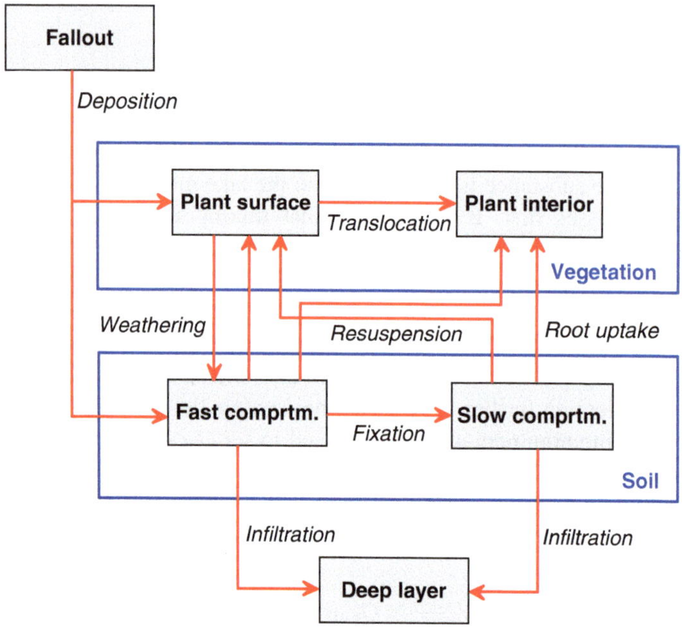

Fig. 18.6 Graph for the migration of nuclides in the plant-soil environment after a fallout, as used by Amano et al. (2003); modified by E.H.

coincide. Where the transpose of the adjacency matrix $\mathbf{A}^{\mathrm{T}}$ has unit entry, there is a positive entry in $\mathbf{B}$ of (18.4), too. Opposite to the adjacency matrix there are nonzero entries in the diagonal of the compartment matrix. In the donor-controlled cases these entries are negative. The sign of the matrix entries corresponds to the sign in the formulae given above, because the fluxes Q_i denote absolute values and are thus positive.With such compartment matrices general networks of connected compartments can be described. Each upstream compartment corresponds with a positive off-diagonal entry in the matrix. Each downstream compartment leads to a negative contribution in the diagonal. Sidebar 18.1 outlines the idea how a linear model can be used for indoor air quality modeling. Figure 18.6 depicts a graph of the compartment concept used for the migration of radionuclides after the fallout in the plant-soil environment. The concept, which was presented by Amano et al. (2003), leads to a linear system of equations.

We note the compartment matrix for the system, shown in Fig. 18.2, as an example. It looks as follows:

$$\mathbf{B}=\begin{pmatrix} -(Q_{VS}+Q_{VC})/V_V & 0 & 0 & 0 & 0 \\ Q_{VS}/V_V & -(Q_{SSo}+Q_{SC})/V_S & 0 & 0 & 0 \\ 0 & Q_{SSo}/V_S & -(Q_{SoG}+Q_{SoC})/V_{So} & 0 & 0 \\ 0 & 0 & Q_{SoG}/V_{So} & -Q_{GC}/V_G & 0 \\ Q_{VC}/V_V & Q_{SC}/V_S & Q_{SoC}/V_{So} & Q_{GC}/V_G & 0 \end{pmatrix} \tag{18.5}$$

For a more general description it is assumed that there is an additional vector of fluxes $\mathbf{f}_i$ into the lakes (i for influx). Additional fluxes out of the system can be taken into account.

There are connections to compartments, which are not included in the model, as the atmosphere and the ocean in the examples represented by Figs. 18.1 and 18.2. These fluxes are also donor controlled. For that reason they are best represented by a matrix-vector product $\mathbf{E}_o\mathbf{c}$, with a diagonal matrix $\mathbf{E}_o$(o for outflow). The mass conservation equations within the network can then be expressed in the more general form

Sidebar 18.1: Indoor Air Quality Modeling

Various compounds from different sources affect Indoor air quality. There are random short-term on/off sources, like cigarettes for example, long-term on-off sources like heaters, long-term steady-state sources, like moth crystals. Sources may have a high initial emission rate, which is decreasing in time with quite different rates. Wax or painted surface emissions decline within hours, while others show modest decay. Sources for volatile organic compounds (VOCs) may be located outdoors (air quality in vicinity of industrial emissions, landfills or contaminated sites) or within the building (combustion, human activities, surface emissions).

Problems of indoor air quality, due to VOCs, can be treated by linear compartment models, as demonstrated by Bouhamra and Elkilani (1999). The approach is presented here briefly. In analogy to (18.3) for each room the concentration of VOC is described by two differential equations:

$$V\frac{\partial c}{\partial t} = Qc_{in} - Qc - k_aAc_i + k_dAc_s + q$$
$$\frac{\partial c_s}{\partial t} = k_ac - k_dc_s$$

where c denotes the concentration within the room, c_{in} the inflow concentration, V the volume of the room, Q air inflow and outflow, and q the rate of the sources within the room, if there is any. Two terms and the second differential equation are introduced to account for sorption effects. Especially furniture and soft tissues may act as temporary sinks of sources for the VOC due to ad- and desorption processes. The corresponding ad- and desorption coefficients are denoted by k_a and k_d, while c_s denotes the sorbed concentration and A the area of the reacting surface.

When the source is described by $q = k_c(c_{source} - c)$ a set of two linear equations results. The approach can be extended to a complete apartment, floor or building, when the set of equations, given above, is formulated for a network of room compartments:

(continued)

$$V_i \frac{\partial c_i}{\partial t} = \sum_{j_{\text{inflow}}} Q_j c_j - \sum_{k_{\text{outflow}}} Q_k c_k - k_{a,i} A_i c_i + k_{d,i} A_i c_{s,i} + q_i$$

$$\frac{\partial c_{s,i}}{\partial t} = k_{a,i} c_i - k_{d,i} c_{s,i}$$

where the index i is used to indicate the *ith* room. Inflow and outflow terms are extended in order to consider that several other rooms may contribute to the total inflow and outflow. Sorption coefficients are assumed to be room-specific, as the involved surfaces may be of different kind. Of course it would also be possible to consider different types of surfaces with respect to sorption in each room. With all these extensions the presented approach still leads to a linear system connecting compartments, which can be solved by the methods described in the second sub-chapter.

Some caution concerning the applicability of the approach should be mentioned. Bouhamra and Elkilani (1999) use the model to determine sorption coefficients within an experimental test chamber with controlled in- and outflow and an installed toluol source. For real apartments or buildings, the number of uncontrolled parameters increases quite fast and surely limits the model's applicability for predictive purposes. However, the presented approach can be useful in hypothetical studies exploring the relevance and interaction of processes. In any case, the compartment approach is more justified for gaseous environments, where mixing occurs fast in comparison to an aquatic environment.

$$\frac{\partial}{\partial t}\mathbf{c} = \mathbf{Bc} + \mathbf{f}_i - \mathbf{E}_o\mathbf{c} \tag{18.6}$$

where $\mathbf{c}$ denotes the vector of the state variables. The number of elements of $\mathbf{c}$ corresponds thus with the number of compartments. In the matrix $\mathbf{B}$ the exchange coefficients are gathered. $\mathbf{E}_0$ is a diagonal matrix, representing outflow into the exterior, which is also proportional to the state variable. $\mathbf{f}_\mathbf{i}$ is a source/sink vector, defining a constant sink or source for each compartment which is independent of any state variable. For a compartment source the sign of the corresponding element is positive, for a sink it is negative.

For the following we define:

$$\mathbf{C} = \mathbf{B} - \mathbf{E}_o \tag{18.7}$$

For non-constant input vector $\mathbf{f}_i$ the general solution is given as:

$$\mathbf{c}(t) = \exp(\mathbf{C}t)\tilde{\mathbf{c}} + \exp(\mathbf{C}t)\int_0^t \exp(-\mathbf{C}s)\mathbf{f}_i(s)ds \tag{18.8}$$

(Jordan and Smith 1977; Walter and Contreras 1999). The expressions with the exponential function need further explanation, as the arguments $\mathbf{C}t$ and -$\mathbf{C}s$ are matrices. The exponential value of a square matrix is defined by the infinite series:

$$\exp(-\mathbf{C}t) = I - \mathbf{C}t + \frac{\mathbf{C}^2t^2}{2!} + \ldots + (-1)^k \frac{\mathbf{C}^k t^k}{k!} + \ldots \quad (18.9)$$

which is the analogue to the infinite series of the exponential function for single numbers. The powers of **C** in the higher order terms are results of matrix multiplication. The result of the operation is also a matrix. The so computed matrix is different from the element wise evaluation of the infinite series or the exponential function. In MATLAB® the expression $\exp(\mathbf{C}t)$ can be programmed easily, as the exponential function of a matrix is available. In m-code it is written using the matrix exponential call `expm` (compare Sidebar 18.2):

```
c = expm(C*t)
```

The `exp` command in MATLAB® is reserved for the elementwise evaluation of the exponential function. The vector $\tilde{\mathbf{c}}$ in formula (18.8) contains unknown parameters, which have to be determined from boundary or initial conditions.

The computation of the general solution (18.8) is quite complex. It includes the evaluation of an integral, which usually needs special analysis. We keep it simple by assuming constant $\mathbf{f}_i$. For constant $\mathbf{f}_i$ the solution of (18.6) can be written as:

$$\mathbf{c}(t) = \exp(\mathbf{C}t)\tilde{\mathbf{c}} - \mathbf{C}^{-1}\mathbf{f}_i \quad (18.10)$$

The result can be derived easiest by the application of the integration formula for the exponential:

$$\int_0^t \exp(\mathbf{C}s)ds = \mathbf{C}^{-1}\exp(\mathbf{C}t)$$

which is common knowledge for single values, but also holds for matrices. $\mathbf{C}^{-1}$ is the inverse matrix of **C**. Thus, the formula (18.10) is applicable only for regular matrices (for which an inverse exists). The first term in (18.10) is the general solution of the problem $\partial\mathbf{c}/\partial t = \mathbf{C}\mathbf{c}$, which mathematicians call *homogeneous*. The second term in (18.10) is a particular solution of the differential (18.6), representing the *equilibrium* solution with $\partial\mathbf{c}/\partial t = 0$, which can be verified easily.

For an initial value problem with the condition $\mathbf{c}(t=0)=\mathbf{c}_0$, it can be verified that the solution is:

$$\mathbf{c}(t)=\exp(\mathbf{C}t)\mathbf{c}_0-(\exp(\mathbf{C}t)-\mathbf{I})\mathbf{C}^{-1}\mathbf{f}_i \tag{18.11}$$

For a zero source vector $\mathbf{f}_i$ the special solution is:

$$\mathbf{c}(t)=\exp(\mathbf{C}t)\mathbf{c}_0 \tag{18.12}$$

Sidebar 18.2: MATLAB® Matrix Functions

Matrix functions are common functions that allow matrices as arguments. There are several matrix functions directly implemented in MATLAB®:

```
expm ()
logm ()
sqrtm ()
```

Other matrix functions can be defined by using the MATLAB® *'fun.m'* M-file or `funm` command. How it works is best explained by an example. Instead of using `expm (A)` as command with matrix `A` one may write:

```
funm (A,@exp)
```

In the same manner `sin, cos, sinh` or `cosh` can be called with a matrix argument. Unfortunately, the same procedure does not work for functions with more than one argument. Thus the call

```
funm (0,A,@besselk)
```

does not work. Maas and Olsthoorn (1997) propose the following nice trick, which works by introducing two new M-files:

```
function f = besselk0 (x)
f = besselk (0,x)

function f = K0(A)
f = funm(A,@besselk0)
```

An example for the application of (18.11) is given by the following command sequence:

```
T = 10;                    % maximum time
C = [-1 1; 1 -3];          % matrix
f = [1; 0];                % input vector
c0 = [1; 1];               % initial concentrations
N = 60;                    % discretization of time
t = linspace (0,T,N);
c = c0;
for i = 2:N
    E = expm(C*t(i));
    c = [c E*c0-(eye(size(C,1))-E)*inv(C)*f];
end
plot (t,c');
legend ('1','2');
text (T/2,1.2,'Eigenvalues:'); text (T/2,1.1,num2str(eig(C)'));
text (T/2,0.8,'Steady state:');
text (T/2,0.7,num2str(-(inv(C)*f)'));
xlabel ('time');
```

Sidebar 18.3: Chains of Radionuclides

The safety analysis for a repository of radioactive waste includes the modeling of the fate of radionuclides. In such a work it is necessary to look at several radionuclides simultaneously, as these are connected in chains of radionuclides. Let's explain the general behavior here without going too much into detail.

With a characteristic half-life the concentration of a mother nuclide declines by radioactive decay. Instead of half-life it is convenient to work with a decay coefficient, as introduced in Chap. 5. The differential equation (5.3) with $n = 1$ is the basis for mathematical modeling of the mother nuclide. As a result of the radioactive decay a daughter nuclide is produced. The daughter nuclide usually is unstable itself, i.e. it also decays with a characteristic constant, which is expressed by the differential equation:

$$\frac{\partial c_{daughther}}{\partial t} = \lambda_{mother} c_{mother} - \lambda_{daughter} c_{daughter}$$

The daughter nuclide itself is a mother nuclide for a next daughter nuclide. Thus a chain of radionuclides can be identified. Let's denote the concentrations within that chain by c_i, where the index i indicates the ith member in the chain. The entire system of species may thus be described by a system of linear differential equations:

$$\frac{\partial}{\partial t}\begin{pmatrix} c_1 \\ c_2 \\ c_3 \\ \cdots \end{pmatrix} = \begin{pmatrix} -\lambda_1 & 0 & 0 & \cdots \\ \lambda_1 & -\lambda_2 & 0 & \cdots \\ 0 & \lambda_2 & -\lambda_3 & \cdots \\ \cdots & \cdots & \cdots & \cdots \end{pmatrix}\begin{pmatrix} c_1 \\ c_2 \\ c_3 \\ \cdots \end{pmatrix}$$

(continued)

briefly written as $\frac{\partial}{\partial t}\mathbf{c} = \mathbf{Bc}$, which is a special case of (18.6). The solution of the system is given by (18.12) with $\mathbf{C} = \mathbf{B}$. The eigenvalues of the matrix are identical to the decay rates with negative sign. If source rates for the nuclides are gathered in the vector **f**, the resulting system

$$\frac{\partial}{\partial t}\mathbf{c} = \mathbf{Bc} + \mathbf{q}$$

is still a special case of (18.6). The general solution is given by (18.11) with $\mathbf{C} = \mathbf{B}$ and $\mathbf{f}_{\mathrm{i}} = \mathbf{q}$.

An M-file for the simulation of the nuclide chain can be constructed in close relation to the compartment simulation, described and listed in the text. In the specification part of the M-file the decay rates, the initial concentration and the source rates have to be defined. The matrix **B** is constructed by the following commands:

```
B = -diag(lambda);
for i = 2:size(lambda,1)
    B(i,i-1) = lambda(i-1);
end
```

The solution is evaluated using the exponential matrix function `expm`.

The complete code is included in the accompanying software under the name *"nuclides.m"*

In the code the matrix exponential is evaluated only once, and stored as matrix `E`. This is good programming practice, as for the computation of `c` in the next command the term is needed twice. Such programming saves time, which is surely not relevant in the simple example presented here, but it can be crucial in an elaborated code, where in the interior of nested loops the same command sequence is executed again and again. The term `eye(size(C,1))` in the expression for `c` denotes the unit matrix, represented in the mathematical formulae by **I**.

In the final lines the eigenvalues and steady state values of the given system are calculated and displayed in the figure. The plot resulting from the M-file is shown in Fig. 18.7 (markers were added afterwards using the figure editor). The reader may notice that the `text` command adds textstrings to the figure. The first `text` places the 'Eigenvalues' text into the figure. Within the second `text` command eigenvalues (see the following subchapter) are calculated, values

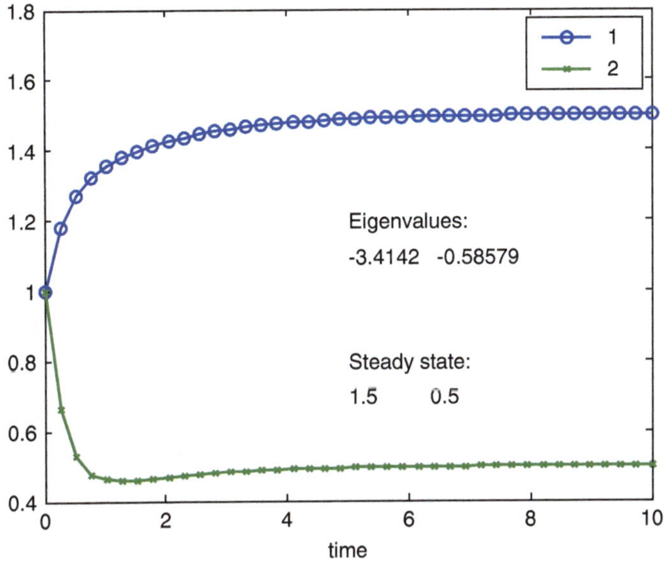

Fig. 18.7 Transient and steady state solution of a two-compartment model using MATLAB®

(numbers) are converted to strings (command `num2str`), and the result is added to the figure.

It was already mentioned above that the formula $\mathbf{C}^{-1}\mathbf{f}_i$ represents the steady equilibrium. That term is evaluated within the last `text` statement. The transient development, depicted in the figure, shows that the steady state, obtained by the evaluation of the formula, is approached. We are in the lucky position that the steady state can be obtained in two ways: by evaluating an analytical formula and by regarding the temporal development at long times. In many other models only the second alternative is available.

The development of the transient simulation against the steady state (here $\mathbf{c} = (1.5, 0.5)^T$) does not become as obvious as in Fig. 18.7, if the time interval is not sufficiently long. Note that there are systems which do not approach the equilibrium, independent of the length of the time interval. In that case we speak of an *unstable equilibrium*. It can also be checked by the sign of the biggest eigenvalue (see below), whether the system converges towards a steady state, i.e. if the equilibrium is stable or unstable. For the next sub-chapter keep in mind that in the example the maximum eigenvalue is ≈ -0.59, and thus negative.

The complete code is included in the accompanying software under the name *"comparts.m"*

Sidebar 18.4: Systems of Aquifers

Systems of aquifers or layers of permeable porous media in the subsurface are often connected. If the flow within these layers is to be studied, one has to take into account the connections. Using matrix notation for the entire setting all layers can be represented in one model. If the porous matrix within each aquifer is homogeneous one can note the following differential equation for the *i*th layer:

$$\nabla^2 h_i = \frac{h_i - h_{i-1}}{K_i H_i c_{i-}} + \frac{h_i - h_{i+1}}{K_i H_i c_{i+}}$$

where h_i denotes the piezometric head, K_i the hydraulic conductivity and H_i the thickness, all for in the *i*th layer, whereas $c_{i-}$and c_{i+} denote the conductances of the overlying and underlying aquitards. In this notation we have: $c_{i-} = c_{(i-1)+}$ and $c_{i+} = c_{(i+1)-}$. The entire system can be noted in vector notation as:

$$\nabla^2 \mathbf{h} = \mathbf{A}\mathbf{h}$$

with matrix **A**, which for a three-layer system is given by:

$$\mathbf{A} = \begin{pmatrix} \frac{1}{K_1 H_1}\left(\frac{1}{c_{1-}} + \frac{1}{c_{1+}}\right) & -\frac{1}{K_1 H_1 c_{1+}} & 0 \\ -\frac{1}{K_2 H_2 c_{2-}} & \frac{1}{K_2 H_2}\left(\frac{1}{c_{2-}} + \frac{1}{c_{2+}}\right) & -\frac{1}{K_2 H_2 c_{2+}} \\ 0 & -\frac{1}{K_3 H_3 c_{3-}} & \frac{1}{K_3 H_3 c_{3+}} \end{pmatrix}$$

Using the exponential and square-root matrix functions, the solution for the system is given by:

$$\mathbf{h}(x) = \exp(-x\sqrt{\mathbf{A}})\mathbf{h}_0$$

for all positions x (Maas 1986). In the vector $\mathbf{h}_0$ for each aquifer the reference heights at position $x = 0$ are stored. The explicit formula can be programmed easily:

```
h = expm (-x*sqrtm(A))*h0
```

In the same manner the formula of de Glee for semi-confined aquifers, which was introduced for a single aquifer in Chap. 12, can be extended for a system of aquifers. One may write

$$\mathbf{h}(x) = \frac{1}{2\pi} K_0(r\sqrt{\mathbf{A}})\mathbf{h}_0$$

(continued)

(Maas and Olsthoorn 1997) where K_0 is the Bessel function, already discussed in Chap. 12. $\mathbf{h}_0$ here is the element wise product of pumping rate with the reciprocal of conductivity and thickness.

The MATLAB® command is simply:

```
h = 1/(2*pi)*K0 (-r*sqrtm(A))*(Q./K/H)
```

More details are given by Maas (1986) and Maas and Olsthoorn (1997). For an extension of the method for unsteady flow towards wells see Hemker (1985) and Hemker and Maas (1987).

18.3 Eigenvalues and Phase Space

Eigenvalues of a matrix allow a much deeper insight in the behavior of a system of differential equations. For every square matrix $\mathbf{C}$, the eigenvalues λ are values for which the system of linear equations:

$$\mathbf{Cx} = \lambda\mathbf{x} \tag{18.13}$$

has a non-zero solution vector $\mathbf{x}$. The vector $\mathbf{x}$ is called eigenvector for the eigenvalue λ.

Eigenvalues and eigenvectors as basic characteristics of matrices are discussed in every textbook on linear algebra or matrix algebra (for example: Robbin 1995). It is not the place here to recall properties of eigenvalues and eigenvectors; the reader who is not yet familiar with these terms should refer to a textbook on linear algebra. As MATLAB®'s origin is numerical linear algebra, the determination of eigenvalues is one of the most basic tasks for which this software can be used.

In MATLAB® eigenvalues are calculated using the `eig` command, for example:

```
eig ([1 2; 3 4])

ans =
   -0.3723
    5.3723
```

i.e. the matrix $\begin{pmatrix} 1 & 2 \\ 3 & 4 \end{pmatrix}$ has the eigenvalues $\lambda_1 = -0.3723$ and $\lambda_2 = 5.3723$. The eigenvalues of a diagonal matrix are the elements in the diagonal:

```
eig ([1 0; 0 4])

ans =
     1
     4
```

For a square matrix with N rows and columns there are N eigenvalues, of which some have to be counted several times. How many times an eigenvalue has to be counted is not arbitrary, of course: we call it the order of the eigenvalue. In the following example

```
eig ([1 0 0; 0 2 0; 0 0 1])
ans =
     1
     1
     2
```

there are the eigenvalues '1' and '2'. '1' has the order 2 and appears twice in the MATLAB® output list. The unit matrix has only one eigenvalue ('1'), but with the order of N. Try the command:

```
eig (eye(5))
```

The nice property that the number of eigenvalues, if counted according to their order, is equal to N is only true, if complex valued eigenvalues are allowed. In general, eigenvalues are complex numbers, which is no problem for MATLAB®, of course:

```
eig ([1 2; -2 1])
ans =
   1.0000 + 2.0000i
   1.0000 - 2.0000i
```

The eigenvalues in this example are $\lambda_{1,2} = 1 \pm 2i$, where i denotes the imaginary unit $i = \sqrt{-1}$ (for complex numbers see also Chap. 15). Also, the eigenvectors can be obtained from the `eig` command; see the MATLAB® help system for details. The reader interested in the numerics of the calculation of eigenvectors should consult a textbook on linear algebra.

The connection with linear systems of differential equations is that the eigenvalues tell something about the behavior of the solutions. Negative real parts tell that the unsteady solution is converging towards the equilibrium (which is given by:$\mathbf{c}_{eq} = \mathbf{C}^{-1}\mathbf{f}_i$, see the preceding subchapter). The equilibrium in that case becomes a steady state, as demonstrated in Fig. 18.7. As both eigenvectors are negative in that example, the solutions with increasing time tend towards the steady state.

The importance of eigenvalues for the analysis of the development of linear systems is most apparent in *phase diagrams*. Phase diagrams, such as plots of the *phase space*, are a tool for the visualization of the behavior of linear and nonlinear systems. In a phase diagram two dynamic variables or derivations are plotted against each other. The representation is unique if there are only two unknown variables. For the matrix **C**, treated in a listing above, the procedure is performed and demonstrated by the following commands:

```
T = 10;                    % maximum time
C = [-1 1; 1 -3];          % matrix
f = [1; 0];                % input vector
c0 = [1; 1];               % initial concentrations
N = 60;                    % discretization of time

t = linspace (0,T,N);
c = c0;
for i = 2:N
    E = expm(C*t(i));
    c = [c E*c0-(eye(size(C,1))-E)*inv(C)*f];
end
plot (c(1,:)',c(2,:)');
```

The result is a 'flowpath' in the phase space towards the steady state, a so-called *trajectory*. In the phase space such lines are called trajectories. In order to obtain a more illustrative figure, we chose several starting positions around the equilibrium solution. One obtains:

```
T = 10;                    % maximum time
C = [-1 1; 1 -3];          % matrix
f = [1; 0];                % input vector
cc = 1;                    % initial concentrations (absolute value of the
                             vector)
N = 60;                    % discretization of time
M = 16;                    % no. of trajectories

%----------------------execution & output-------------------------
equilibrium = -(inv(C)*f);
t = linspace (0,T,N);
for angle = linspace (0,pi+pi,M)
    c0 = equilibrium + cc*[sin(angle); cos(angle)]; c = c0;
    for i = 2:N
        E = expm(C*t(i));
        c = [c E*c0-(eye(size(C,1))-E)*inv(C)*f];
    end
    plot (c(1,:)',c(2,:)'); hold on;
end

plot (equilibrium(1),equilibrium(2),'s');
xlabel ('variable 1'); ylabel ('variable 2')
title ('phase diagram')
```

The complete code is included in the accompanying software under the name '*phasediag.m*'.

The graphical output of the M-file is depicted in Fig. 18.8. The graphic was extended manually by arrows in order to indicate the temporal direction of the trajectories. A textbox was added close to the position of the stable equilibrium. Figures similar to the one depicted arise, whenever two variables are plotted in the vicinity of a stable equilibrium. Obviously, the equilibrium is approached differently from different angles, which is a result of the different eigenvalues. For the more seldom case of equal eigenvalues, the resulting figure looks differently, although the qualitative behavior remains the same. In case of an unstable

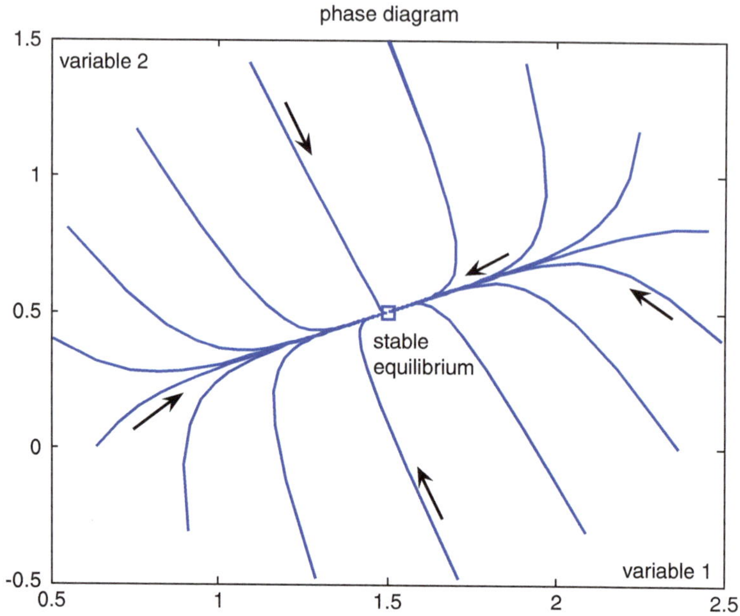

Fig. 18.8 Phase diagram for a simple compartment model using MATLAB®

equilibrium similar trajectories result, but the flow direction is opposite. It is left to the reader to explore these different cases by modifying the *'phasediag.m'* M-file.

Table 18.2 provides a classification of equilibria for a two variable system, based on the real parts of the eigenvalues (see also: Hale and Koçak (1991)

Figure 18.9 provides a view of the stable oscillations that are obtained if both eigenvalues are purely imaginary, i.e. if both eigenvalues have vanishing real parts. The trajectories are circles around the origin, describing a cycling of the corresponding variables: if variable 1 increases, variable 2 decreases and vice versa. Turning points between these two situations are reached, when one of the variables has a zero value and changes its sign.

For higher values of N the characterization of stable and unstable situations can easily be extrapolated from the simple $N = 2$ case. If there is at least one eigenvalue with a positive real part, the equilibrium is unstable. If real parts of all eigenvalues are negative, there is convergence towards a stable solution. Degenerate situations, in which there is at least one purely imaginary eigenvalue, can be interpreted analogously to the five lower rows in Table 18.2.

For 2D phase space calculation and visualization, the MATLAB® M-file *'pplane.m'* by Polking is available on the web (http://math.rice.edu/~dfield/). It sets up a graphical user interface (GUI) and has several other convenient features. The manual for the program is available as a book (Polking 2004). We demonstrate an application example in Chap. 19.

Table 18.2 Classification of equilibria stability according to eigenvectors, for $N = 2$

Eigenvalues λ_1, λ_2	Matrix of equivalent reference system	stable (s) unstable (u)	Comment
$\mathrm{Re}(\lambda_1)<0, \mathrm{Re}(\lambda_2)<0$	$\begin{pmatrix} -1 & 0 \\ 0 & -1 \end{pmatrix}$	s	Hyperbolic sink
$\mathrm{Re}(\lambda_1)>0, \mathrm{Re}(\lambda_2)>0$	$\begin{pmatrix} 1 & 0 \\ 0 & 1 \end{pmatrix}$	u	Hyperbolic source
$\mathrm{Re}(\lambda_1)>0, \mathrm{Re}(\lambda_2)<0$	$\begin{pmatrix} 1 & 0 \\ 0 & -1 \end{pmatrix}$	u	Hyperbolic saddle
$\mathrm{Re}(\lambda_1)<0, \lambda_2 = 0$	$\begin{pmatrix} -1 & 0 \\ 0 & 0 \end{pmatrix}$	s	Equilibria for all c_2, $c_1 = 0$
$\mathrm{Re}(\lambda_1)>0, \lambda_2 = 0$	$\begin{pmatrix} 1 & 0 \\ 0 & 0 \end{pmatrix}$	u	Equilibria for all c_2, $c_1 = 0$
$\lambda_1 = \lambda_2 = 0$ 2 eigenvectors	$\begin{pmatrix} 0 & 0 \\ 0 & 0 \end{pmatrix}$	s	All locations equilibria
$\lambda_1 = \lambda_2 = 0$ 1 eigenvector	$\begin{pmatrix} 0 & 1 \\ 0 & 0 \end{pmatrix}$	u	Equilibria for all c_1, $c_2 = 0$
$\mathrm{Re}(\lambda_1) = \mathrm{Re}(\lambda_2) = 0$	$\begin{pmatrix} 0 & 1 \\ -1 & 0 \end{pmatrix}$	s	Oscillations

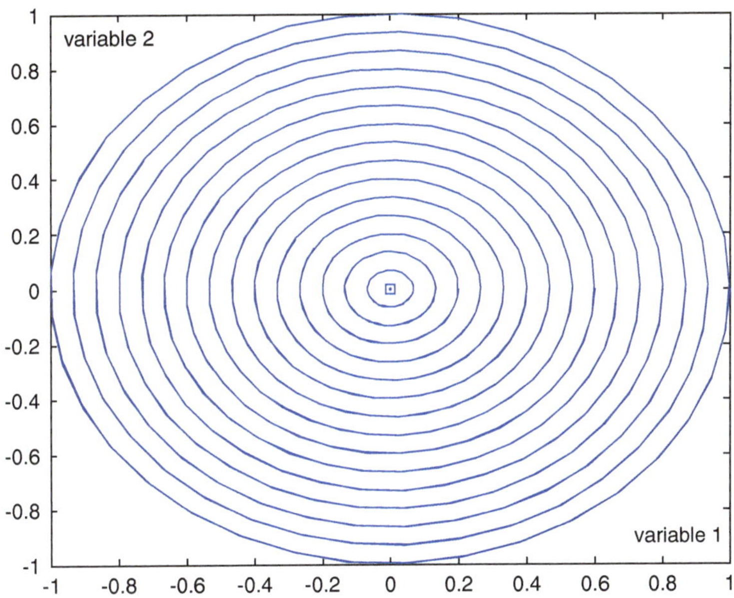

Fig. 18.9 Phase diagram for the reference matrix of two purely imaginary eigenvalues, representing oscillations around the equilibrium; obtained using *phasediag.m* file with reference matrix from Table 18.2, zero right hand side and equidistant start positions on the c_2-axis

References

Amano H, Takahashi T, Uchida S, Matsuoka S, Ikeda H, Hayashi H, Kurosawa N (2003) Development of a code MOGRA for predicting the migration of ground additions and its application to various land utilization areas. J Nucl Sci Eng 40(11):975–979

Avila R, Broed R, Pereira A (2003) ECOLEGO – toolbox for radiological risk assessment. In: IAEA, Proceedings international conference on protection of the environment from Ionizing radiation, Stockholm, pp 229–232. http://www.facilia.se/ecolego/

Bouhamra W, Elkilani A (1999) Development of a model for the estimation of indoor valitile organic compounds concentration based on experimental sorption parameters. Environ Sci Technol 33:2100–2105

Chartrand G (1977) Introductory graph theory. Dover Publ, New York, p 294

Dommasnes A, Christensen V, Ellertsen B, Kvamme C, Melle C, Nøttestad L, Pedersen T, Tjelmeland S, Zeller D (2001) An ECOPATH model for the Norwegian Sea and Barents Sea. In: Guénette S, Christensen V, Pauly D (eds) Fisheries impacts on North Atlantic ecosystems: models and analyses. Univ. of British Columbia, Canada, pp 213–240

Freeze RA, Cherry JA (1979) Groundwater. Prentice-Hall, Englewood Cliffs, p 604

Hale J, Koçak H (1991) Dynamics and bifurcations. Springer, New York, p 568

Hemker CJ (1985) Transient well flow in leaky multiple-aquifer systems. J Hydrology 81:111–126

Hemker CJ, Maas C (1987) Unsteady flow to wells in layered and fissured aquifer systems. J Hydrology 90:231–249

Holzbecher E, Bonell M, Bronstert A, Vasiliev O (2005) Fluxes, compartments, and ordering of feedbacks. In: Bronstert A, Carrera J, Kabat P, Lütkemeier S (eds) Coupled models for the hydrological cycle – integrating atmosphere, biosphere, and pedosphere. Springer, Berlin, pp 76–97, Chapter 2.1

Jordan DW, Smith P (1977) Nonlinear ordinary differential equations. Clarendon, Oxford, p 360

Maas C (1986) The use of matrix differential calculus in problems of multiple aquifers flow. J Hydrology 88:43–67

Maas C, Olsthoorn T (1997) Snelle oudjes gaan MATLAB. Stromingen 3(4):21–42, in Dutch

Polking J (2004) Ordinary differential equations using MATLAB. Prentice Hall, Upper Saddle River, p 264

Robbin JW (1995) Matrix algebra. A K Peters, Wellesley, p 544

Walter GG, Contreras M (1999) Compartmental modeling with networks. Birkhäuser, Boston, p 250

Chapter 19
Nonlinear Systems

Linear systems, as examined in the previous chapter, represent the simplest type of models. But linear models are often too simplistic from the process aspect. The set-up of a linear model is often motivated by the fact that few characteristics, parameters or variables, of the system have been observed and that few data are available to check the model approach, whatever that may be. This chapter describes models slightly more complex than the linear ones. It is demonstrated that even simple nonlinear terms in the differential equation open the door to a much greater variety of phenomena than experienced by the work with linear systems.

There are several mathematical textbooks on nonlinear systems of ordinary differential equations. Jordan and Smith (1977) provide a wide range of examples, not only for environmental systems. Hale and Koçak (1991) focus on bifurcations in nonlinear systems, but with hardly a connection to environmental sciences. For MATLAB® users the book of Polking (2004) is highly recommended, because the accompanying software is extremely user-friendly and can be obtained via internet.

19.1 Logistic Growth

The linear differential equation $\partial c/\partial t = rc$ for a single species describes exponential growth, for $r>0$. With reference to discussions on earth's population this is sometimes referred to as *Malthus*[1]*ian growth*. However, there is no environmental system in which any species can grow infinitely. The model, described by the simple linear equation above, can thus be valid for a limited range of parameter or variable values only. If the model is to be valid for an extended parameter range,

[1] Thomas Robert Malthus (1766–1834), English demographer and political economist.

E. Holzbecher, *Environmental Modeling*,
DOI 10.1007/978-3-642-22042-5_19, © Springer-Verlag Berlin Heidelberg 2012

the equation needs to be extended itself. The basic formulation for the development of a biological species is the *logistic growth* equation:

$$\frac{\partial c}{\partial t} = rc\left(1 - \frac{c}{\kappa}\right) =: f(c) \tag{19.1}$$

with growth rate r and carrying capacity κ. For populations we maintain to use the symbol c, which was introduced in previous chapters (with an eye on concentrations to be described). The term 'logistic growth' was introduced by Verhulst[2], who studied the equation already in the first half of the nineteenth century. For small populations the first (linear) rc term is dominant, describing first order growth $r>0$. For high concentrations the population approaches the carrying capacity κ, while the temporary growth rate $r(1 - c/\kappa)$ approaches zero. Equation (19.1) is a nonlinear differential equation. It has an analytical solution:

$$c(t) = \frac{c_0 \kappa \exp(rt)}{\kappa + c_0(\exp(rt) - 1)} \tag{19.2}$$

which can easily be implemented using MATLAB®. Here an example command sequence for parameter input, execution and graphical output:

```
%---------------------- input --------------------------------
T = 10;                  % maximum time
r = 1;                   % rate
kappa = 1;               % capacity
c0 = 0.01;               % initial value

%---------------------- execution --------------------------------

t = linspace (0,T,100);
e = exp(r*t);
c = c0*kappa*e./(kappa+c0*(e-1));

%---------------------- graphical output ------------------------

plot (t,c); grid;
xlabel ('time'); legend ('population');
title ('logistic growth');
```

The complete code is included in the accompanying software under the name *"logistic.m"*.

The graphical output of the M-file is given in Fig. 19.1, showing the increase of the population from an initial value towards a maximum value, at which the carrying capacity of the system is reached.

Also for nonlinear equations it makes sense to examine equilibria, as introduced in Chap. 18. The logistic equation has two equilibria that are obtained by finding

[2] Pierre François Verhulst (1804–1849), Belgian mathematician.

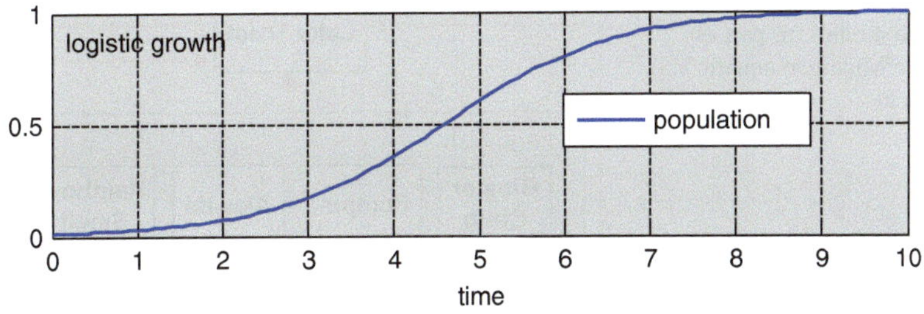

Fig. 19.1 Logistic growth; Computed form analytical solution using MATLAB®

values c, for which the left hand side of (19.1) vanishes: $f(c) = 0$. The population $c = 0$ is an unstable equilibrium, while $c = \kappa$ is stable. Small deviations from the unstable equilibrium, in the example $c_0 = 0.01$, lead to increased deviations at later times. The user may check easily for the given parameters that an initial value near to the stable equilibrium of $c = 1$ produces an almost constant system development.

The stability of equilibria can be examined by the use of the derivative. For a negative value of the derivative the system is stable, while it is unstable for a positive derivative. For the logistic (19.1) there is $\partial f/\partial c = r(1 - 2c/\kappa)$, which is positive at the origin but negative for $c = \kappa$. In the following we treat systems of equations instead of single equations. For systems the eigenvalues of the Jacobi-matrix take the just described role of the derivative and have to be examined in order to check the equilibria for stability.

There is almost no branch of environmental modeling in which nonlinear systems do not appear. In ecological sciences ecosystems are in the focus with interactions between species populations. The structure of a foodweb model is often visualized in a compartment graph, in the way hydrological systems were represented in Figs. 18.1 and 18.2. An example of a foodweb graph is depicted in Fig. 19.1. Species or groups of species are represented by compartments. Arrows in foodweb graphs indicate the direction of the food-chain; in the example lake trouts consume forage fish, which themselves live on zooplankton.

In the example graph, representing a part of the Lake Michigan aquatic eco-system, there are four trophic levels. Detritus and phytoplankton are the lowest level and lake trout alone represents the highest level in this model. Most modeling efforts of lake eco-systems end up with less than five or six trophic levels. Foodweb structures can be represented by an adjacency matrix and visualized using the MATLAB® `gplot` command, as shown in Chap. 18.

In the sequel some simple foodweb models are examined as examples for the treatment of nonlinear systems by MATLAB®. Analytical solutions can be obtained for simple networks and interactions only, as for example for the logistic growth (19.1). Therefore, numerical methods will be used for the solution for more complex set-ups. First we study species of the same trophic level, like in the forage fish or the zooplankton compartments of Fig. 19.2.

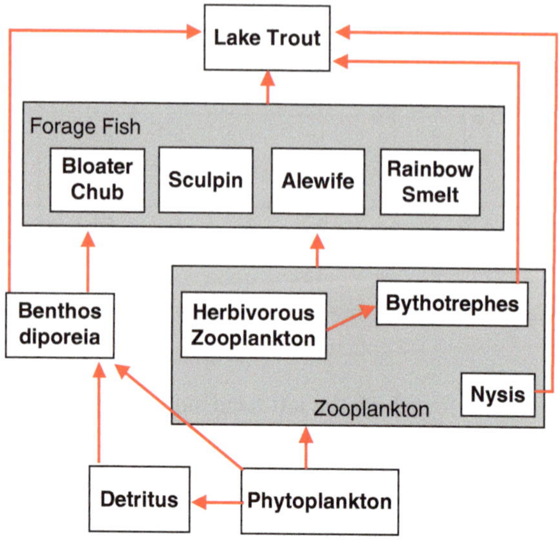

Fig. 19.2 Example of a foodweb model; for part of the Lake Michigan aquatic ecosystem

19.2 Competing Species

More specifically we are interested in the development of two species that are competing for exactly the same resources. For very low populations the growth rates are given by the specified values r_1 and r_2. For increased populations the growth rate is reduced by a term which takes into account the reduced foodstock of both species. If the foodstock is reduced by Δh, the following system of two differential equations can be used to describe the temporal development:

$$\left.\begin{aligned}\frac{\partial c_1}{\partial t} &= c_1(r_1 - \alpha_1 \Delta h)\\ \frac{\partial c_2}{\partial t} &= c_2(r_2 - \alpha_2 \Delta h)\end{aligned}\right\} \text{ with } \Delta h = h_1 c_1 + h_2 c_2 \tag{19.3}$$

The system of (19.3) can be re-written in analogy to formulation (19.1):

$$\frac{\partial}{\partial t}\begin{pmatrix} c_1 \\ c_2 \end{pmatrix} = \begin{cases} r_1 c_1\left(1 - \frac{c_1}{\kappa_1}\right) \text{ with } \kappa_1 = \dfrac{\kappa_{10}}{1 + \lambda c_2/c_1} \\ r_2 c_2\left(1 - \frac{c_2}{\kappa_2}\right) \text{ with } \kappa_2 = \dfrac{\kappa_{20}}{1 + \lambda^{-1} c_1/c_2} \end{cases} \tag{19.4}$$

with capacities $\kappa_{i0} = r_i/\alpha_i h_i$ for single species cases and the dimensionless system parameter $\lambda = \frac{h_2}{h_1} = \frac{r_2}{r_1}\frac{\alpha_1}{\alpha_2}\frac{\kappa_{10}}{\kappa_{20}}$ (compare: Richter 1985). When the ratios r_i/α_i are equal, the capacities are related by the formula: $\kappa_{10} = \lambda\kappa_{20}$. If species 1 is more efficient than species 2 concerning resource consumption, it holds: $\lambda < \kappa_{10}/\kappa_{20}$; while the opposite inequality holds if species 2 is more efficient.

The following M-file explores the situation in a phase diagram. As example a situation is studied in which the parameters are non-dimensionalized (Murray 2002), i.e. in which rates and equilibria are set to unity.

```
T = 1000;                        % maximum time
r = [1; 1];                      % rates
e = [1; 1];                      % equilibria
lambda = 0.2;                    % lambda parameter
c0 = [0.1; 0.1];                 % initial concentrations

%---------------------execution-----------------------------------

options = odeset('AbsTol',1e-20);
[t,c] = ode15s(@CS,[0 T],c0,options,r,e,lambda);

%--------------------- graphical output --------------------------

plot (c(:,1)',c(:,2)'); hold on;
plot (e(1),0,'s'); plot (0,e(2),'s');
legend ('trajectory');
xlabel ('species 1'); ylabel ('species 2');
title ('competing species');

%--------------------- function ----------------------------------

function dydt = CS(t,y,r,e,lambda)
dydt = zeros(2,1);
k = [e(1)/(1+lambda*y(2)/y(1)); e(2)/(1+y(1)/y(2)/lambda)];
dydt = r.*y.*(1-y./k);
```

The complete code is included in the accompanying software under the name *"compspec.m"*

Various trajectories (see Chap. 18.3) for the same starting populations, but with varying parameter λ, are shown in Fig. 19.3. Obviously, for almost all λ-values the solutions at coordinates (0,1) or (1,0) are approached.

In both equilibria cases one of the species becomes extinct. As both equilibria for single species are identical, the marginal parameter value is $\lambda = 1$. For $\lambda > 1$ species two uses resources more efficiently and species 1 becomes extinct. For $\lambda < 1$ the fate of the species is reversed. In Fig. 19.3 the two positions in the phase space, which represent these two situations, are marked by 'species 1' and 'species 2'.

In fact the two mentioned states are *equilibria*, because they fulfil the system (19.4) for vanishing left hand side, i.e. for zero time derivatives. A refined examination of the two-equations- system (19.4) reveals that there are three equilibria in the competing species model, which are given by:

$$\begin{pmatrix} 0 \\ 0 \end{pmatrix}, \begin{pmatrix} 0 \\ \dfrac{r_2}{\alpha_2 h_2} \end{pmatrix}, \begin{pmatrix} \dfrac{r_1}{\alpha_1 h_1} \\ 0 \end{pmatrix} \tag{19.5}$$

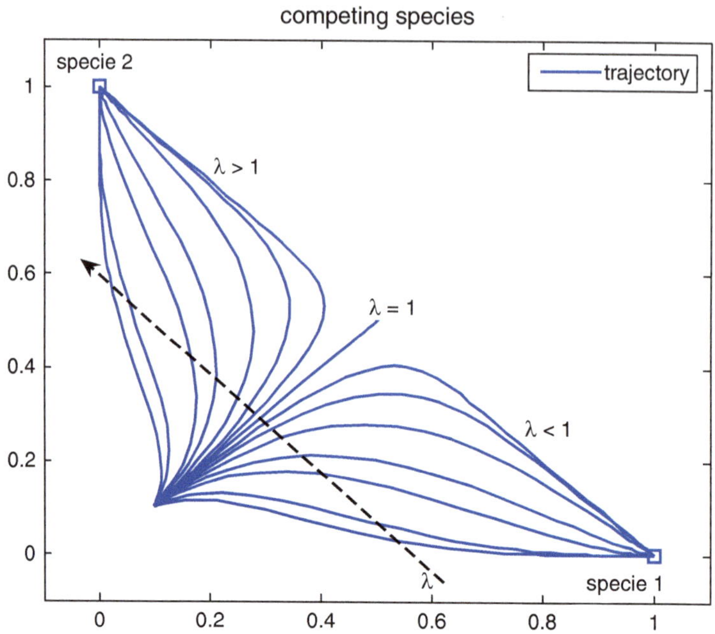

Fig. 19.3 Trajectories in phase space for the competing species model; $\lambda \in \{0.2, 0.25, 0.4, 0.5, 0.667, 0.8, 0.9, 1, 1.11, 1.25, 1.5, 2, 2.5, 4, 5\}$; use the corresponding M-file for other input values

Even more equilibria are obtained for the degenerate case of parameter values, in which both brackets on the right side of (19.3) vanish. That condition is characterized by the equality $r_1/\alpha_1 = r_2/\alpha_2$, and all positions on the line, given by $h_1c_1 + h_2c_2 = r_1/\alpha_1$, become equilibria. The second and third equilibria represent those states in which one of the species dominates over the other, and the latter becoming extinct.

The mathematical analysis offers more than just the number of equilibria and their position in the phase space. Also the behavior of the system concerning the equilibria can be determined by analytical means. The crucial condition includes the eigenvalues, as outlined in more detail in the sequel.

The behavior of the system at the equilibria can be analyzed by the examination of the *Jacobi*[3]*-matrix* at the equilibrium position. For a (multicomponent) vector function **f** with components f_i, depending on the variables c_j, the Jacobi matrix is given by

$$\mathbf{Df} := \left(\frac{\partial f_i}{\partial c_j}\right)_{i,j=1\ldots N} \tag{19.6}$$

[3] Carl Gustav Jacob Jacobi (1804–1851), German mathematician.

where i denotes the row index, and j the column index. The Jacobi matrix changes with the variables c_i and thus with time t, as the c's are functions of t. The Jacobi matrix depends on the position in the phase space where it is evaluated. **Df** is a generalization of the derivative for multi-valued functions, depending on several variables.

For the competing species system (19.3), the function **f** is given by:

$$\mathbf{f}(\mathbf{c}) = \mathbf{f}(c_1, c_2) = \begin{pmatrix} f_1(c_1, c_2) \\ f_2(c_1, c_2) \end{pmatrix} = \begin{pmatrix} c_1(r_1 - \alpha_1 \Delta h) \\ c_2(r_2 - \alpha_2 \Delta h) \end{pmatrix} \tag{19.7}$$

All four expressions in (19.7) are different writings of the same thing. MATLAB® users, familiar with vector notation, probably prefer the most compact writing on the left. As the reader may easily verify, the Jacobi matrix at location (c_1,c_2) is given by:

$$\mathbf{Df}(c_1, c_2) = \begin{pmatrix} r_1 - \alpha_1(2h_1c_1 + h_2c_2) & -\alpha_1 h_2 c_1 \\ -\alpha_2 h_1 c_2 & r_2 - \alpha_2(2h_2c_2 + h_1c_1) \end{pmatrix} \tag{19.8}$$

For $c_1 = c_2 = 0$ the Jacobi matrix is diagonal with r_1 in the upper-left position and r_2 in the lower right. The eigenvalues of the diagonal matrix are given by two reaction rates r_1 and r_2 (compare Chap. 18). These are always positive; the eigenvalues are positive, indicating that the equilibrium at zero concentrations is unstable.

At the other two equilibrium positions the Jacobi matrices look as follows:

$$\mathbf{Df}(r_1/\alpha_1 h_1, 0) = \begin{pmatrix} -r_1 & -h_2 r_1/h_1 \\ 0 & r_2 - \alpha_2 r_1/h_1 \end{pmatrix} \mathbf{Df}(0, r_2/\alpha_2 h_2)$$
$$= \begin{pmatrix} r_1 - \alpha_1 r_2/\alpha_2 & 0 \\ -h_1 r_2/h_1 & -r_2 \end{pmatrix} \tag{19.9}$$

The eigenvalues for these two triangular matrices can be read directly from the diagonal:

$$\lambda_1 = \begin{cases} -r_1 \text{ at } (r_1/\alpha_1 h_1, 0) \\ r_1 - \alpha_1 r_2/\alpha_2 \text{ at } (0, r_2/\alpha_2 h_2) \end{cases} \quad \lambda_2 = \begin{cases} r_2 - \alpha_2 r_1/\alpha_1 \text{ at } (r_1/\alpha_1 h_1, 0) \\ -r_2 \text{ at } (0, r_2/\alpha_2 h_2) \end{cases} \tag{19.10}$$

The eigenvalues are real in all cases, because all parameters are real numbers. The stability depends on the sign of both eigenvalues. The equilibria are stable if both eigenvalues are negative. As the rates r_1 and r_2 are positive, there remains only one condition. Stability is equivalent to the conditions:

$$\frac{r_1}{\alpha_1} < \frac{r_2}{\alpha_2} \text{ at } (r_1/\alpha_1 h_1, 0) \quad \text{and} \quad \frac{r_1}{\alpha_1} > \frac{r_2}{\alpha_2} \text{ at } (0, r_2/\alpha_2 h_2) \tag{19.11}$$

The Jacobi-matrix at $(r_1/\alpha_1 h_1, 0)$ has negative eigenvalues for $r_1/\alpha_1 > r_2/\alpha_2$ and positive eigenvalues if the inequality holds in opposite direction. For the Jacobi-matrix at $(0, r_2/\alpha_2 h_2)$ the conditions are exactly reversed. The analysis shows that the result, obtained above for the example values from a phase space analysis, is generally valid.

For the degenerate case with $r_1/\alpha_1 = r_2/\alpha_2$ one eigenvalue becomes zero; as the other one is negative the equilibrium is still stable. However, it needs to be noted that all positions on a line are also equilibria in that case. The eigenvalues for those points are $-\alpha_1 h_1 c_1$ and $-\alpha_2 h_2 c_2$ and thus also negative. All these equilibria are stable.

In order to visualize this in the phase space, various trajectories have been plotted as described above. Input parameter within the '*compspec.m*' were specified as follows:

```
T = 1000;                 % maximum time
r = [1; 1];               % rates
e = [1; 1];               % equilibria
lambda = 1.;              % lambda parameter
```

Several starting positions were chosen and the M-file was run several times with the `hold on` option being active. As shown in Fig. 19.4 all trajectories are straight lines, which end at the diagonal that connects the upper left and the lower right corners of the plotted region. Using the mathematical analysis from above it is easy to verify that this line represents all equilibria (except zero). The illustration allows the obvious conclusion that the equilibria on the line are stable, which we already know from the eigenvalues.

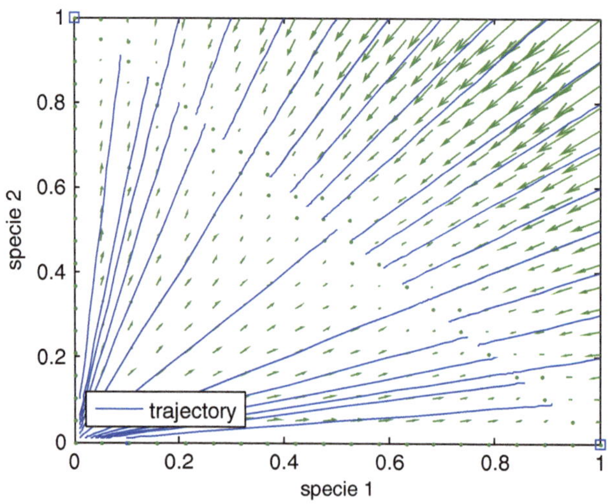

Fig. 19.4 Trajectories in the phase space for the degenerate case λ=0 of the competing species model; all equilibria on the diagonal line (from *upper left* to *lower right*) are stable

In order to improve the graphical representation of the phase space, an arrow field is added, illustrating the direction and the velocity along the trajectories. The option for such an extension is implemented in the '*compspec.m*' file. In the input part several option parameters are added:

```
gquiv = 20;                % arrow field plot;
                            value for no. of arrows in 1D
xmin = 0; xmax = 1;        % x- interval for arrow field plot
ymin = 0; ymax = 1;        % y-     "     "    "     "    "
scale = 2;                 % scaling factor for arrows
```

In the output part of the file find the following instruction block:

```
if (gquiv)
    [x,y] = meshgrid
    (linspace(xmin,xmax,gquiv),linspace(ymin,ymax,gquiv));
    dy = zeros(gquiv,gquiv,2);
    for i = 1:gquiv
        for j = 1:gquiv
            dy(i,j,:) = CS(0,[x(i,j);y(i,j)],r,e,lambda);
        end
    end
    quiver (x,y,dy(:,:,1),dy(:,:,2),scale);
end
```

With the `meshgrid` command a regular mesh is constructed, which has already been demonstrated in previous chapters. The array `dy` is initialized with two values for each mesh node. In the interior of the `for` loops these two values, which represent the time derivatives of both variables, are calculated. The final `quiver` request initiates the plot of the arrow field. The `scale` parameter is related to the length of the arrows. The outcome of the M-file with the given values can be recognized in Fig. 19.4.

The described situation of non-coexistence in an ecological system can be observed in reality, especially in aquatic eco-systems, where disfavored species have no chance to emigrate to places with more favourable conditions. One of the such catastrophes happened in Lake Victoria, when the nile perch was introduced into Africa's biggest lake in 1960 (Goldschmidt 1998). At the top of the food-chain the single predator gradually replaced more than one hundred species of predators. As a result, the complete foodweb changed. The prawn, which was rare before the appearance of the Nile perch, replaced several detritus-eating species, the sardine replaced more than 20 species of zooplankton-eating fish. Algae-grazing fish disappeared without replacement. This radical change of the species population in the lake may have contributed to increased blue-green algae blooms and eutrophication, which has been observed hitherto. While the catch of Nile perch strongly increased after its introduction, the overall productivity of the lake was reduced by 80% in 1984, compared to pre-1960 levels. The consequences not only concerned the aquatic system. Problems arose in communities around the lake, which were strongly dependent on the catch. Ecological problems worsened the situation in addition to economical problems. Large perch is oily and cannot be dried in the sun.

Fish preservation by smoking over wood led to a decline in the stock of trees. The spread of the infectious and often lethal bilharzia disease can also be related to the introduction of the Nile perch (Murray 2002).

19.3 Predator–Prey Models

Another type of model describes the development of populations of a predator and its prey. The model covers two trophic levels. The simplest approach goes back to publications in the 1920s of the last century. In 1926, Volterra[4] tried to explain the observation of oscillatory fish catches in the Adriatic Sea with the set of two ordinary differential equations; the first for prey population c_1, the second for the predator concentration c_2:

$$\begin{aligned}\frac{\partial c_1}{\partial t} &= c_1(r_1 - \alpha_1 c_2)\\ \frac{\partial c_2}{\partial t} &= c_2(\alpha_2 c_1 - r_2)\end{aligned} \qquad (19.12)$$

with positive parameters r_1, r_2, α_1 and α_2.

While Lotka[5] developed the same approach for a system of chemical species (Lotka 1956) Volterra was the first who applied the system to an ecological problem (Murray 2002). Often the term *Lotka-Volterra*-models is found. The assumptions that lead to the system (19.12) are rather simplistic. Without the predator-prey interaction the prey population would increase exponentially with rate r_1, accompanied by the exponential decrease of the predator population with rate r_2. However, it is assumed that for the overall behavior the interaction is crucially relevant. Prey consumption and predator population growth, both increase or decrease with c_1 and c_2, which is expressed by the two terms including the product c_1c_2. In the corresponding M-file the MATLAB® `ode15s` solver is again utilized for the calculation of the ordinary differential equations. The M-file in large parts coincides with the previous examples, so that few comments are necessary only.

---

[4] Vito Volterra (1860–1940), Italian mathematician and physicist.

[5] Alfred James Lotka (1880–1949), Austrian-US-American mathematician, chemist and ecologist.

```
T = 100;                    % maximum time
r = [.5; .5];               % single species rates
a = [1; 1];                 % alpha parameter
c0 = [0.1; 0.1];            % initial population density

%---------------------- execution ---------------------------------

options = odeset('AbsTol',1e-20);
[t,c] = ode15s(@PP,[0 T],c0,options,r,a);

%---------------------- graphical output -------------------------

subplot (2,1,1);
plot (c(:,1)',c(:,2)'); hold on;
legend ('trajectory');
xlabel ('predator'); ylabel ('prey');
subplot (2,1,2);
plot (t,c(:,1)','-',t,c(:,2)','--');
legend ('predator','prey');
xlabel ('time');

%---------------------- function ----------------------------------

function dydt = PP(t,y,r,a)
dydt = zeros(2,1);
dydt(1) = y(1)*(r(1)-y(2)*a(1));
dydt(2) = y(2)*(-r(2)+y(1)*a(2));
```

The complete code is included in the accompanying software under the name *"predprey.m"*

In the output section the `subplot` command is used to place two plots with different coordinate systems in the same figure. `subplot` can be used to place sub-plots in one or more rows or columns (like a matrix) within the same MATLAB® figure. The command has three integer parameters. The first integer specifies the number of rows, the second the number of columns, the third gives the 'serial number' if sub-plots are counted along the rows first, starting with the uppermost row. See the MATLAB® help for some instructive examples. Here two plots are drawn, one above the other; the upper containing the phase space plot, the other a visualization of the population time series. The output of the M-file with the data set printed above is shown in Fig. 19.5.

Both sub-plots show the characteristic oscillations of predator and prey populations. When prey populations increase, the predator finds favourable conditions, which leads to an increase of the predator population, too. With increasing predatory stress the situation changes gradually for the prey, leading to a decline of prey population, which finally results in unfavourable conditions for the predator and a corresponding decline of the population. When the situation becomes favourable for the prey, the entire loop starts again. The oscillations can be observed in the second subplot, where the oscillating behavior of the prey population is followed closely by the predator population curve.

The corresponding figure in the phase space is a closed curve, as depicted in the upper subplot. The populations follow that curve in anti-clockwise direction. Along

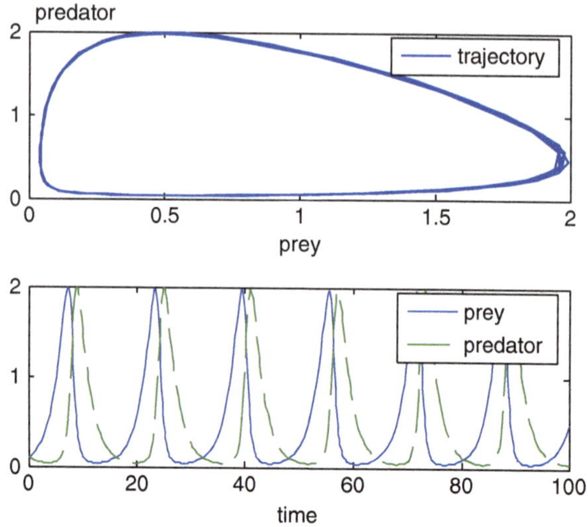

Fig. 19.5 Example output of a predator-prey model

the lower boundary the predator population remains small, while the prey has favourable conditions. Then, the number of predators starts to increase, which at the rightmost positions leads to a decline of the prey. With declining prey the predator population can increase only for a limited time, until, at the uppermost position of the closed curve, these decline, too. Finally, near the origin the low abundance of predators makes the prey population rise again.

The result, obtained by the numerical methods included in MATLAB®, coincides with findings from analytical methods for the simple and much examined classical Lotka-Volterra system. In fact, it is easy to see that there are two equilibrium solutions:

$$\mathbf{c}_1 = \begin{pmatrix} 0 \\ 0 \end{pmatrix} \text{ and } \mathbf{c}_2 = \begin{pmatrix} r_2/\alpha_2 \\ r_1/\alpha_1 \end{pmatrix} \tag{19.13}$$

The Jacobi-matrix can be used to examine the stability of the equilibria. Simple calculations yield that the eigenvalues of the Jacobi-matrix at $\mathbf{c}_1$ are r_1 and $-r_2$. As the second eigenvalue is real and positive, the equilibrium is unstable. At $\mathbf{c}_2$ the eigenvalues are pure imaginary $\pm i\sqrt{r_1 r_2}$, which indicates a stable equilibrium with oscillations around the equilibrium in its vicinity. In the phase space there are circular motions around the equilibrium. The reader may explore this using the MATLAB® M-file by changing the initial value `c0`.

The solutions in the phase space can also be obtained by integration (Richter 1985, Murray 2002). In non-dimensional form they are given implicitly by the equation:

$$\ln(c_1) - c_1 = \frac{r_1}{r_2}(\ln(c_2) - c_2) + c_0 \tag{19.14}$$

The calculation and visualization of such curves requires also numerical means, which are in most cases more sophisticated than the presented solution method using numerical MATLAB® solvers for ordinary differential equations.

In practical applications it has been observed that the classical Lotka-Volterra-system has several drawbacks. The simplistic assumptions have already been mentioned. Nevertheless, it is a jumping-off place for more realistic models, which are obtained by extensions of the original system (19.12). Such extensions have been proposed in the literature. Murray (2002) provides an overview. The most nearby idea is to use a logistic growth term, as in (19.1), instead of the linear 1.order term.

In the presented approach is quite easy to implement extensions of the Lotka-Volterra equations. All that needs to be done is to extend the formulae in the `function` of the '*predprey.m*' M-file. If additional parameters are required, these should be added in the input specifications part of the M-file and considered as formal parameters in the function call. Using various starting positions in following runs, it is possible to examine, whether an equilibrium is stable or unstable or if a limit cycle exists.

For 2D problems, i.e. settings with two variables, the '*pplane.m*' model can be a very useful tool for the MATLAB® user. Briefly, we present an example application for the phase space M-file '*pplane.m*', which was already mentioned in Chap. 18. The M-file is available from the web (http://math.rice.edu/~dfield/).

Figure 19.6 depicts the set-up and the output for a predator pray problem, for which we used the MATLAB® version 7 file: '*pplane7.m*'. Four windows are depicted. In the 'Setup' window the differential equations are specified. There are edit-boxes for the input and change of parameters as well as for the basic settings concerning outlook and axes of the display window. For the example case we did not edit the system of differential equations, but chose the 'predator prey' entry from the 'Gallery' menu.

The display window shows the phase space. A field of grey arrows as a first visualization depicts the trajectories. Blue lines for trajectories appear by mouse click within the displayed phase space where the cursor location is taken as starting value for a trajectory. The red dot, indicating an equilibrium location, appears when the corresponding sub-menu entry 'Find an equilibrium point' under 'Solutions' is selected. The user has to click into the displayed phase space and to select a starting point for the search.

When an equilibrium position has been found, some of its characteristics are shown in a separate small window with text output, depicted on the right side of Fig. 19.6. The exact position, the corresponding Jacobi matrix, the eigenvalues and eigenvectors can all be read from the window. There is an additional button reading 'Display the linearization'. After pressing that button, the linearized system is shown in another window named 'Linearization', into which trajectories are plotted after a mouse click (Fig. 19.6). In the example we see circles around the origin. The pattern of the trajectories of the linearized system corresponds with the last row of Table 18.2, with non-zero entries only in the off-diagonal of the matrix and two purely imaginary eigenvalues.

The user may further explore the '*pplane7.m*' program by her/himself. There are numerous new entries in the menu, which are added to the MATLAB® figure

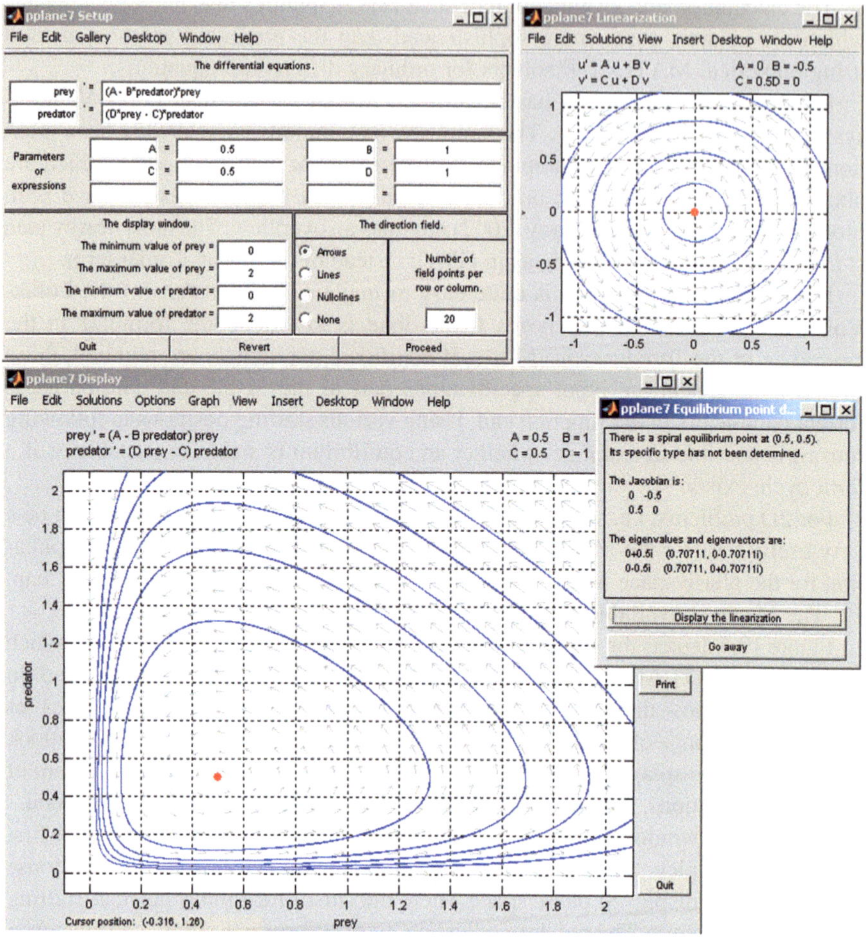

Fig. 19.6 Example windows of '*pplane.m*' (Polking 2004) for the predator-prey model (see text)

editor. In the 'Solutions' there is the option to 'Show nullclines'. Nullclines are curves along which only one of the two functions becomes zero. Under the 'Edit' menu several options allow to erase unwanted graphical objects or to include text. The 'Solver' options, under menu 'Options', may be relevant to obtain better trajectories, especially closed orbits.

19.4 Chaos (Lorenz Attractor)

Finally, in this chapter we want to give a short impression of a more complex behavior than shown in the previous sub-chapters. Even quite simple nonlinear systems may show such behavior which we nowadays call *chaotic* in the scientific literature.

In the first publication on the topic, Lorenz (1963) dealt with the following relatively simple system:

$$\begin{pmatrix} u_1 \\ u_2 \\ u_3 \end{pmatrix} = \begin{pmatrix} \sigma(u_2 - u_1) \\ \rho u_1 - u_2 - u_1 u_3 \\ u_1 u_2 - \beta u_3 \end{pmatrix} \tag{19.15}$$

Lorenz was concerned with convective motions in the atmosphere. Facing the problem of the limited capacity of computers at that time, he had set up a very simplified model with which he hoped to explain real motions of air masses. The system (19.15) has three unknown variables u_1, u_2 and u_3 and three parameters: the Prandtl number σ, the Rayleigh number ρ and β.

What Lorenz discovered was a completely new behavior of a nonlinear system, which had not been known before that time. The trajectories did not converge to a stable equilibrium position or an oscillating orbit neither did they run to infinity. Within a limited region the trajectories showed a chaotic behavior. The shown phase space picture is now well-known and the system which it represents is called the *Lorenz attractor*.

The Lorenz attractor, depicted in Fig. 19.7, is produced by the following command sequence:

```
sigma = 16; rho = 45.92; beta = 4;  % parameters
N = 1000;               % no. of time steps
span = 0.05;            % inner iteration time span
AbsTol = 1.e-5;         % absolute tolerance for ODE solver
RelTol = 1.e-5;         % relative tolerance for ODE solver

H = figure; set(H,'DefaultLineLineWidth',1.0);
options = odeset('RelTol',RelTol,'AbsTol',ones(1,3)*AbsTol);
u0 = [1;1;1];
for i = 1:N
    [t,u] = ode45(@lornz,[0 span],u0,odeset,beta,rho,sigma);
    hold on;
    plot(u(:,1),u(:,2),'r');
    u0 = u(end,:);
end

title('Attractor of Lorenz System');
xlabel('Component 1'); ylabel('Component 2');
axis off; hold off;

function dydt = lornz(t,y,beta,rho,sigma)
dydt = [sigma*(y(2)-y(1)); rho*y(1)-y(2)-y(1)*y(3); y(1)*y(2)-
beta*y(3)];
```

The code is included in the accompanying software under the name *"lorenza.m"*

The code does not need lengthy explanations. The first five lines represent the specification part of the file. The main numerical computation is performed within the `for` loop. First the Lorenz system (19.15) is solved, using the `ode45` solver of MATLAB®. The system itself is specified as a subfunction in the very last lines of

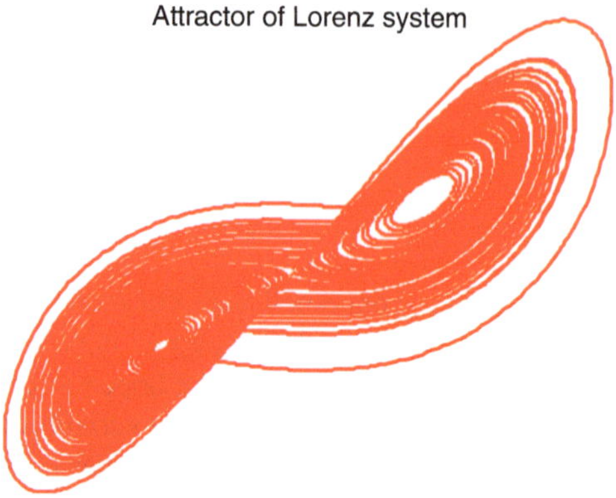

Fig. 19.7 The Lorenz attractor, produced using MATLAB®

the listing. Then the trajectory in the phase space, which is spanned by the first two variables, is plotted (computation and output are not strictly separated in this M-file). The advantage of the chosen procedure is that the storage space for the solutions remains small. If the `span` is small, only a limited number of storage is required: the old solution values are overwritten after the corresponding part of the trajectory has been plotted.

The example shows that even simple nonlinear systems may exhibit chaotic behavior. Also systems of chemical or biological species may lead to observations similar to the Lorenz system (see for example: Fussmann and Heber 2002).

If a model system is chaotic, it can not be used for predictions. In fact, chaos in the mentioned sense can also be characterized by the property that tiny deviations in an initial state lead to significant deviations at later times (Devaney 1987). The modeler, working with dynamical systems, should be aware that models may show such a 'strange' behavior.

References

Devaney RL (1987) An introduction to chaotic dynamical systems. Addison-Wesley, Redwood City, p 320

Fussmann G, Heber (2002) Food web complexity and chaotic population dynamics. Ecol Lett 5:394–401

Goldschmidt T (1998) Darwin's Dreampond. MIT Press, Boston, p 274

Hale J, Koçak H (1991) Dynamics and bifurcations. Springer, New York, p 568

Jordan DW, Smith P (1977) Nonlinear ordinary differential equations. Clarendon, Oxford, p 360

Lorenz EN (1963) Deterministic nonperiodic flow. J Atmos Sci 20:130–141

Lotka AJ (1956) Elements of mathematical biology. Dover Publ, New York, p 465

Murray JD (2002) Mathematical biology I: introduction. Springer, New York, p 551

Polking J (2004) Ordinary differential equations using MATLAB. Prentice Hall, Upper Saddle River, p 264

Richter O (1985) Simulation des Verhaltens ökologischer Systeme, VCH Verlagsgesellschaft, Weinheim, 219 p (in German)

Chapter 20
Graphical User Interfaces

In the previous chapter the reader found a graphical user interface ('*pplane7.m*'), which allows easy input of parameters as well as direct output of results, numerically or graphically. The technical term for these types of 'man-machine-interfaces' is *Graphical User Interface*, and the abbreviation is *GUI*. Nowadays, computer users are already used to GUIs, as there is hardly any software that does not utilize its capacities, including the operating system.

Looking back to Fig. 19.6, the user may ask how difficult it may be to construct such an interface. In this final chapter it is shown how a GUI can be implemented using MATLAB®. It will be demonstrated that GUIs can be set up using core MATLAB®, and that it is easier than the novice may have imagined beforehand, although the functionality of the examples does not reach the one of the '*pplane7.m*' file of the previous chapter.

20.1 The MATLAB® Guide

There is a special MATLAB® tool for the set-up of GUIs. It is called from the MATLAB® command window by:

```
guide
```

What appears is a graphical user interface to construct GUIs. First the user is asked whether she/he wants to open an existing GUI, or whether to create a new one (Fig. 20.1).

We proceed with an example, and therefore choose the 'Blank GUI' default. The window that appears on the display next is reproduced here in Fig. 20.2.

E. Holzbecher, *Environmental Modeling*,
DOI 10.1007/978-3-642-22042-5_20,

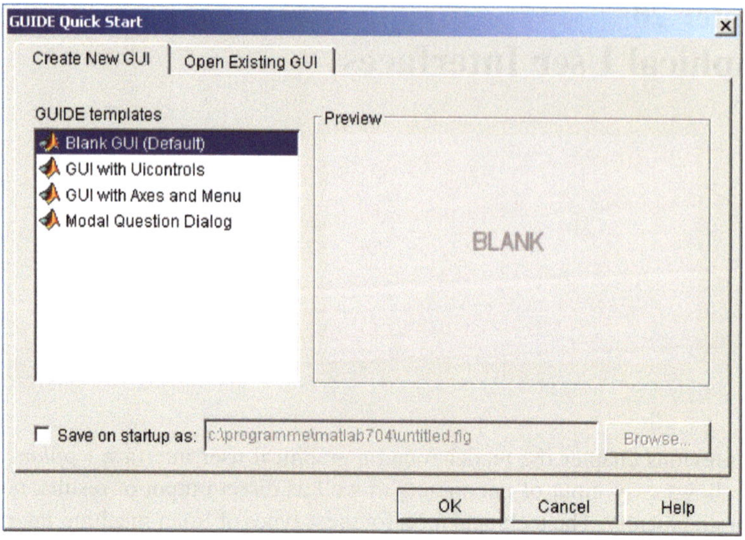

Fig. 20.1 Start window for the MATLAB® graphical user interface for GUIs

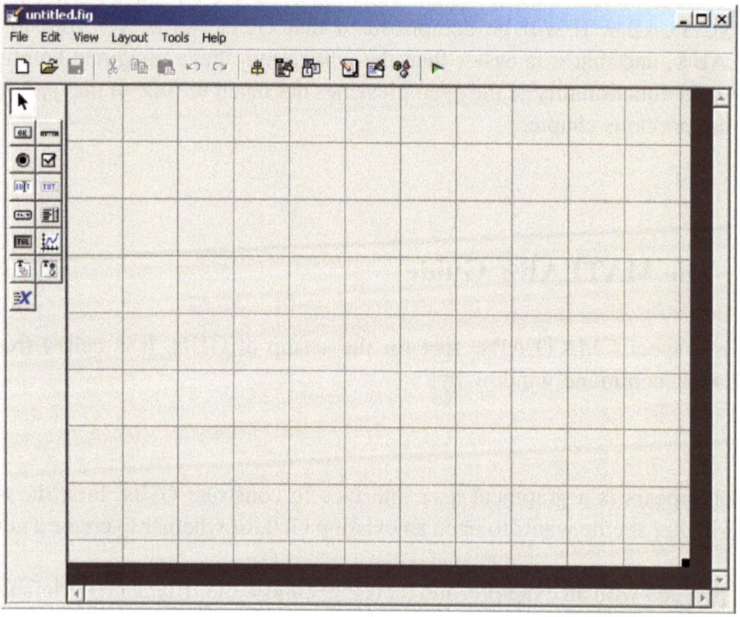

Fig. 20.2 The MATLAB® 'guide' GUI

On the left side bar of the window we find various elements which constitute a GUI. Major elements that are used in the following demonstration are:

 Push button

 Edit text

 Static text

 Axes

 Pop-up menu

 Radio button

 Panel

There are further elements we do not use in this example: slider, radio button, check-box, list box, toggle button, panel, button group and ActiveX control. Click at the 'static text' variable and move the cursor into the big GUI workbench on the right. The workbench represents the GUI window, which finally will become the interface to the user. The 'static text' appears and is fixed at a certain location. In order to change and format the text element, double click on it (see right side of Fig. 20.3). As a result another window appears in which all properties of the selected element are shown and can be edited. This so-called 'Property Inspector' is shown on the left side of Fig. 20.3.

We want to change the text to 'Diffusivity'. The new text string is entered as 'String' property, where before the 'Static Text' was found (see Fig. 20.3). As further exercise with the new tool the reader may change the 'Font Size' property to 12 and the 'FontWeight' property to 'bold'. Select a color for the 'Foreground Color' property. Clicking  the color select box appears that is depicted in Fig. 20.4.

Having entered the changes in the Property Inspector these become active in the workbench.

Now let's examine the previous procedure for a different GUI element. Choose the 'edit text' element and place it aside the static text element, which was already entered. Double-click on the element to reach the Property Inspector with the corresponding properties. Change the 'String' property to '1'; we want to use this element for the input of the diffusivity and change the 'Tag' property to 'D_edit'.

In the very same way we add some further text and edit elements, as illustrated in Fig. 20.5. The reader may recognize that the elements are related to the major input parameters of a transport model, as described in Chaps. 4, 5 and 6. In fact, when finished the example GUI will allow 1D transport simulation and visualization.

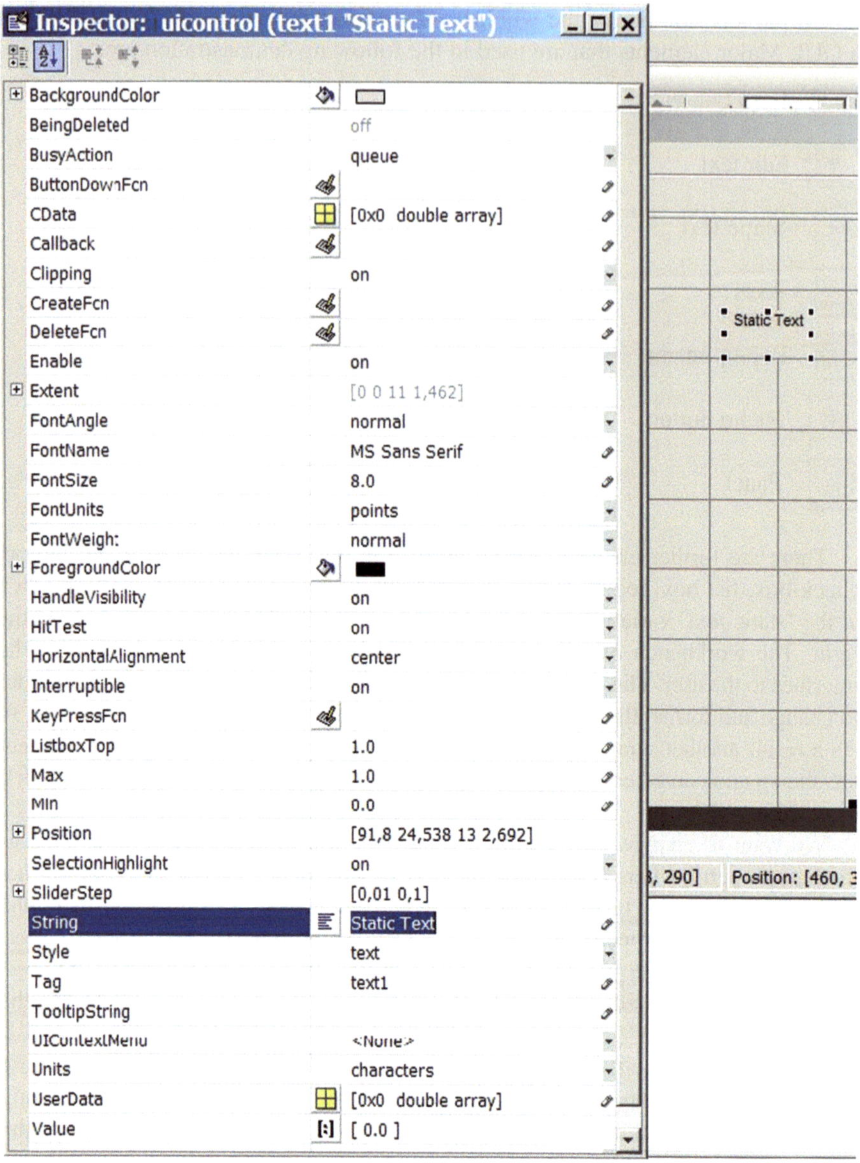

Fig. 20.3 The Property Inspector window within the MATLAB® guide

In addition an ‘axes’ element is introduced, covering nearly the entire space of the GUI. It is reserved for the graphical output and has the ‘Tag’ property ‘xaxes’.

When the user clicks on the ‘Save as…’ button, MATLAB® saves the work in two files. The user interface window is stored under a file with ‘*.fig*’

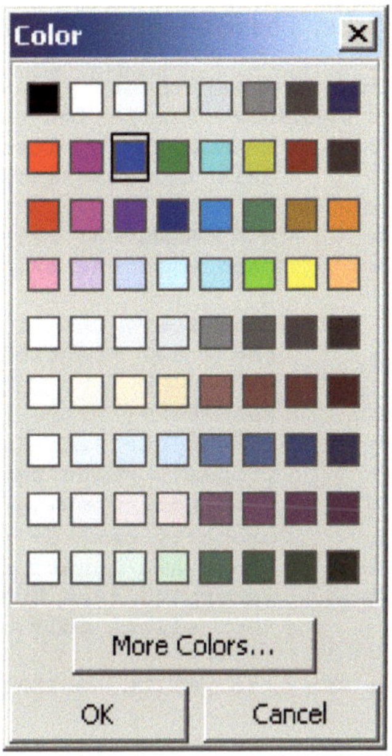

Fig. 20.4 The color select box of the MATLAB® guide

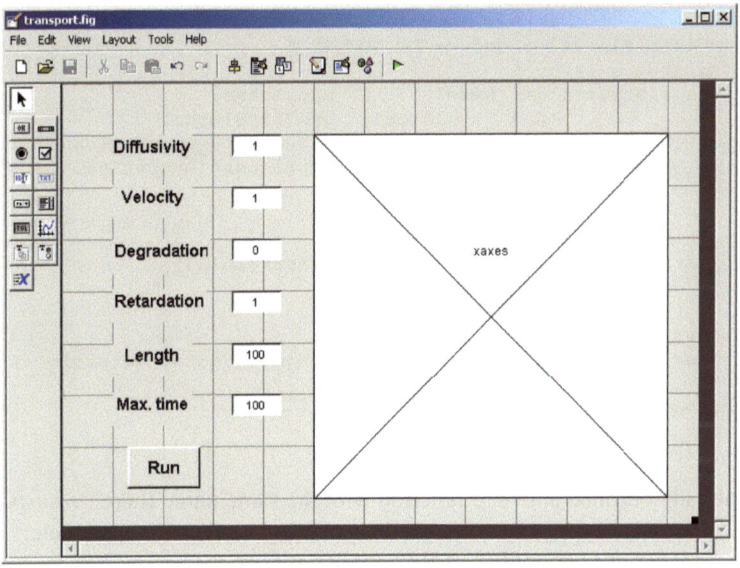

Fig. 20.5 The example GUI, in construction using MATLAB® guide

extension, a corresponding M-file with the same name is saved in addition. Within the exercise the user may choose the name '*transport.fig*' and a file '*transport.m*' is created too.

Let's have a look on the M-file. It starts with the following commands:

```
function varargout = transport(varargin)
% TRANSPORT M-file for transport.fig
%       TRANSPORT, by itself, creates a new TRANSPORT or raises the
%       existing singleton*.
%
%       H = TRANSPORT returns the handle to a new TRANSPORT or the
%       handle to the existing singleton*.
%
%       TRANSPORT('CALLBACK',hObject,eventData,handles,...) calls
%       the local function named CALLBACK in TRANSPORT.M with the
                       given input arguments.
%       TRANSPORT('Property','Value',...) creates a new TRANSPORT or
%       raises the existing singleton*.  Starting from the left,
%       property value pairs are applied to the GUI before
%       transport_OpeningFunction gets called. Anunrecognized
%       property name or invalid value makes property application
%       stop. All inputs are passed to transport_OpeningFcn via
%       varargin. *See GUI Options on GUIDE's Tools menu. Choose
%       "GUI allows only one instance to run (singleton)".
%
% See also: GUIDE, GUIDATA, GUIHANDLES

% Edit the above text to modify the response to help transport

% Last Modified by GUIDE v2.5 22-Sep-2006 16:50:52

% Begin initialization code - DO NOT EDIT
gui_Singleton = 1;
gui_State = struct('gui_Name',       mfilename, ...
                   'gui_Singleton',  gui_Singleton, ...
                   'gui_OpeningFcn', @transport_OpeningFcn, ...
                   'gui_OutputFcn',  @transport_OutputFcn, ...
                   'gui_LayoutFcn',  [] , ...
                   'gui_Callback',   []);
if nargin && ischar(varargin{1})
    gui_State.gui_Callback = str2func(varargin{1});
end

if nargout
    [varargout{1:nargout}] = gui_mainfcn(gui_State, varargin{:});
else
    gui_mainfcn(gui_State, varargin{:});
end
% End initialization code - DO NOT EDIT
```

As already outlined above, a function with the same name (here: '*transport*') is created. `varargin` and `varargout` are lists of input- and output parameters. What follows is a list of comments and the initialization of the entire structure. As the details are relevant for the specialist only, we skip any explanation but pinpoint to the very last comment of this part of the program. It tells the user that all statements above concern initialization and may under no circumstances be edited.

After the initialization part several subfunctions follow, starting with the opening function

```
function transport_OpeningFcn(hObject, eventdata, handles, varargin)
```

Some explanations concerning the functions are given in the comments. Most sub-functions are of the following types:

```
function D_edit_Callback(hObject, eventdata, handles)
% hObject    handle to D_edit (see GCBO)
% eventdata  reserved - to be defined in a future version of MATLAB
% handles    structure with handles and user data (see GUIDATA)

% Hints: get(hObject,'String') returns contents of D_edit as text
%        str2double(get(hObject,'String')) returns contents of
%        D_edit as a double

% -- Executes during object creation, after setting all properties.
function D_edit_CreateFcn(hObject, eventdata, handles)
% hObject    handle to D_edit (see GCBO)
% eventdata  reserved - to be defined in a future version of MATLAB
% handles    empty - handles not created until after all CreateFcns
% called

% Hint: edit controls usually have a white background on Windows.
%       See ISPC and COMPUTER.
if ispc && isequal(get(hObject,'BackgroundColor'),
get(0,'defaultUicontrolBackgroundColor'))
    set(hObject,'BackgroundColor','white');
end
```

There is one 'callback' function and one 'create' function. The reader may verify that there are these two sub-functions for each element of the GUI window. With the functions the functionality of the GUI is created. The 'callback' function is called every time when the element is selected during the execution of the program. In the example command sequence the function listings are empty, aside from an adjustment concerning the background color (`get` and `set` commands). I.e. the standard functionality of the GUI element is performed, with which the user is not really concerned here.

In the current example there is only one part of the program where the user has to add some functionality. That concerns the *'Run'*-button. When that button is pressed, several tasks have to be performed in order to draw a picture of the concentration profiles on the screen. The corresponding commands have to be added to the `runbutton_Callback` function, which is executed after pressing the run-button. Here we add the following list:

```
% --- Executes on button press in runbutton.
function runbutton_Callback(hObject, eventdata, handles)
% hObject    handle to runbutton (see GCBO)
% eventdata  reserved - to be defined in a future version of MATLAB
% handles    structure with handles and user data (see GUIDATA)

% Get user input from GUI
D = str2double(get(handles.D_edit,'String'));
v = str2double(get(handles.v_edit,'String'));
R = str2double(get(handles.R_edit,'String'));
lambda = str2double(get(handles.lambda_edit,'String'));
T = str2double(get(handles.T_edit,'String'));
L = str2double(get(handles.L_edit,'String'));
```

The data, given by the user in the GUI window, have to be allocated to corresponding variables. The reader recognizes the relevant variables on the left side of the assignments above. In the sequel we give some clues how to understand the remainder of the lines. `handles` is a structure which contains the entire GUI implementation. `handles.D_edit` is a call of the `D_edit` function of the given structure. The name of the function corresponds with the *'Tag'*, specified in the Property Inspector. The value, entered by the GUI user during program execution, can be found under *'String'* and is reached by the `get` command. The content of the *'String'* is of data type string, i.e. a listing of characters. As for the further processing in the program the numerical value is relevant (not the text-string), the string has to be converted to a numerical double type. In MATLAB® the command that performs such conversion is `str2double`.

Using the debugging mode, the user may check some parameter functions. For example the variable `D` contains the value found in the edit box for diffusivity. The given command sequence delivers the values of the parameters for further processing. In Chaps. 4–6 the same task was performed in the specification part of the M-files in a more obvious way.

Now that the parameters are available, the execution part can be adopted almost identically from former transport models. The remainder of the `runbutton_Callback` function is taken from the *'analtrans.m'*, outlined in Chap. 6:

```
c0 = 0;     % initial concentration
cin = 1;    % inflow concentration

%Calculate data
y ='rgbcmyk';
e = ones (1,100);
u = sqrt(v*v+4*lambda*R*D);

% Create space plot
t = linspace (T/10,T,10);
axes(handles.xaxes);
x = linspace(0,L,100);
for i = 1:size(t,2)
    h = 1./(2.*sqrt(D*R*t(i)));
    hh = plot (x,c0*exp(-lambda*t(i))*(e-0.5*erfc(h*(R*x-...
       e*v*t(i)))-0.5*exp((v/D)*x).*erfc(h*(R*x+e*v*t(i)))) +...
       (cin-c0)*0.5*(exp((v-u)/(D+D)*x).*erfc(h*(R*x-e*u*t(i)))+...
       exp((v+u)/(D+D)*x).*erfc(h*(R*x+e*u*t(i)))),y(mod(i,7)+1));
    set (hh,'LineWidth',2)
    hold on;
end
grid on
hold off
```

As the mathematical aspects of the command sequence are explained in Chap. 6, we can restrict our description to the graphics commands. After having set initialization values for `c0`, `cin`, `y`, `e` and `u`, the `t` vector is computed. `t` contains those time instants at which a concentration profile is plotted. Here the entire time period of length `T` is equidistantly divided into 10 parts (the starting time $t = 0$ is not included). The following `axes` command tells that the following graphic

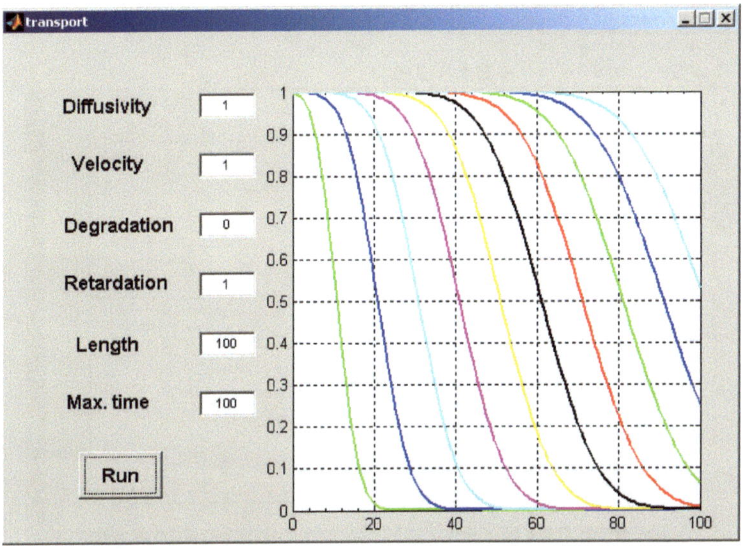

Fig. 20.6 The example GUI, first step

statements are executed within the coordinate system, named `xaxes`, which is part of the `handles` structure. Later we introduce another coordinate system, making the necessity to specify the axes more obvious.

The `x` vector is a discretization of the entire model length. Within the following `for` loop, for each time instant of the `t` vector, the temporary 'help' parameter `h` is calculated. `h` is used in the lengthy analytical formula that follows within the `plot` command. Note that the concentration values are not stored in a big matrix, unlike in the M-files of Chaps. 4–6. Here they are lost after plotting is executed, because they are not needed further.

The last formal parameter in the `plot` command (`y(mod(i,7) + 1)`) effects that the line color changes from one profile to the next, according to the characters given above in the `y` string. The `set` command specifies an increased line width (default is 0.5). The `hold on` statement is necessary in order to obtain all profiles in the same figure. The final graphics command, `grid on`, plots the grid. The resulting outlook of the GUI is depicted in Fig. 20.6.

20.2 The Transport GUI

Here we extend the M-file from the previous part. In the MATLAB® guide we add three graphical elements, all of which are shown in Fig. 20.7. A second 'Axes' element is introduced. In order to make both graphics fit on the same panel figure, the size of the coordinate systems is reduced. Using the property inspector, the new

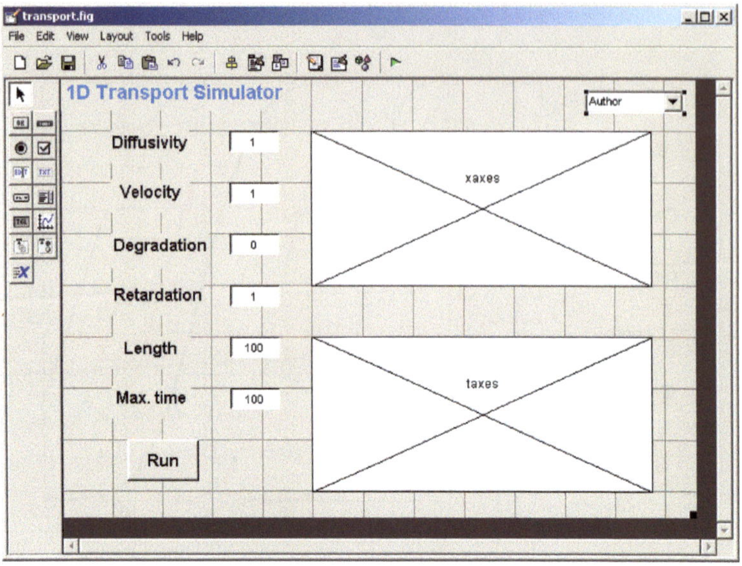

Fig. 20.7 The extended example GUI, in production using MATLAB® guide

axes gets the tag 'taxes' (for time axis). As header we introduce another static text element, for which font size, font weight and color are changed from default. As third adjustment we enter a pop-up menu. Using again the property inspector we specify 'Author', 'Book', 'Publisher' and 'Software' as four entries in the 'String' property.

After these adjustments of the last sub-chapter GUI, use 'Save' in the MATLAB® guide menu to store the new figure. Note that the *'transport.m'* is also changed as a result of the 'Save' click. The interested user may have a look into the M-file listing to see some new 'callback' and 'create' command blocks, corresponding with the new elements.

As a next step, the new M-file has to be extended to include a new functionality. Within the command block, related to the run-button, the following list should be added at the end, i.e. behind the commands concerning the space plot:

```
% Create time plot
x = linspace (L/10,L,10);
axes(handles.taxes);
t = linspace (T/100,T,100);
h = 1./(2.*sqrt(D*R*t));
for i = 1:size(x,2)
    hh = plot(t,c0*exp(-lambda*t).*(e-0.5*erfc(h.*(e*R*x(i)...
        -v*t)))-0.5*exp((v/D)*x(i))*erfc(h.*(e*R*x(i)+v*t))+...
        (cin-c0)*0.5*(exp((v-u)/(D+D)*x(i))*erfc(h.*(e*R*x(i)...
        -u*t))+exp((v+u)/(D+D)*x(i))*erfc(h.*(e*R*x(i)+u*t)))...
        ,y(mod(i,7)+1));
    set (hh,'LineWidth',2)
    hold on;
end
grid on
hold off
```

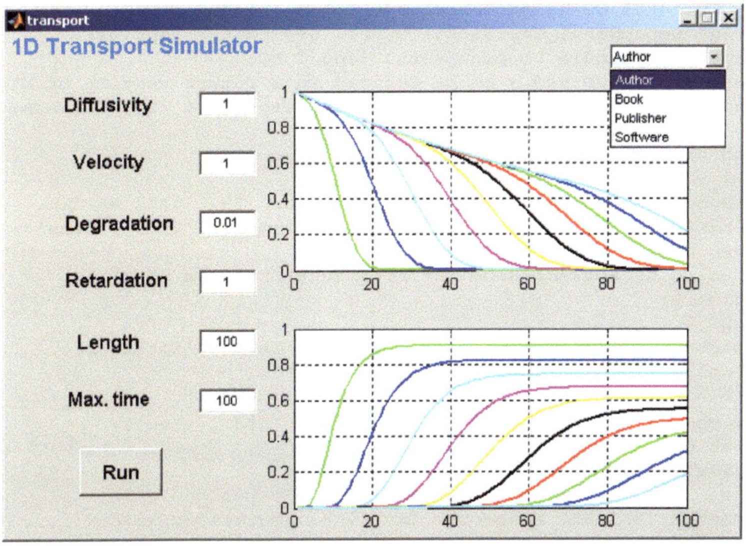

Fig. 20.8 The transport GUI, with output for a selected parameter set

In the sequel we follow the procedure outlined in the preceding sub-chapter. The concentration values are calculated from the analytical solution given in Chap. 6. Here the vector of concentration values is computed for each of ten equidistant locations along the model axis. After the calculation the corresponding curve is plotted. The graphs are depicted in the second coordinate system, where all graphics commands are directed after the `axes(handles.taxes)` command has become effective.

An example view of the GUI with the computed result for an example data set, including dispersion/diffusion, advection and degradation, is depicted by Fig. 20.8.

The algorithm is surely not time-optimized, as all values of the concentration matrix are calculated twice. On the other hand the computation algorithm is quite effective with respect to computer storage. Only two 100-element vectors are used aside from all other variables. The set-up of the GUI itself surely is the most time consuming part in the M-file. However, for problems in one space dimension performance questions are irrelevant, taking into account the performance of today's computers. For problems in two space dimensions performance questions may become relevant, in three dimensions they surely are.

At last we introduce some functionality to the pop-up menu. This is done within the corresponding 'callback' function. Function header and added commands are listed below:

```
% --- Executes on selection change in popupmenu1.
function popupmenu1_Callback(hObject, eventdata, handles)
% hObject    handle to popupmenu1 (see GCBO)
% eventdata  reserved - to be defined in a future version of MATLAB
% handles    structure with handles and user data (see GUIDATA)
val = get(hObject,'Value');
switch val
case 1
    msgbox('E. Holzbecher, WIAS, Mohrenstr. 39, 10117 Berlin,...
    GERMANY, E-mail: holzbecher@wias-berlin.de','Info','none');
    web http://www.igb-...
    berlin.de/abt1/mitarbeiter/holzbecher/index_e.shtml;
    uiwait;
case 2
    msgbox('Environmental Modeling using MATLAB, Springer...
    Publ.','Info','none');
case 3
    msgbox('Springer Publ., Heidelberg','Info','none');
    web www.springer.com;
    uiwait;
case 4
    msgbox('MATLAB, MathWorks Inc, see: www.mathworks.com',...
    'Info','none');
    web www.mathworks.com;
    uiwait;
end
```

The ‘callback’ function is reached when the user has clicked on a menu entry of the pop-up menu. In the example, the pop-up menu has four entries: ‘Author’, ‘Book’, ‘Publisher’ and ‘Software’. During the execution of the program, the menu entry, which is selected by the user, is obtained by the command `get (hObject,'Value')`. The returned variable is an integer between 1 and 4, by which the selection is uniquely determined. In the listing the current value is stored in variable `val`.

It is convenient to use the `switch` keyword for the `val` variable in order to jump to different command blocks. If the value is 1, a message box opens with information about the author. The corresponding MATLAB® command is: `msgbox`. Additionally, using the `web` command, a browser opens with the personal site of the author. The `uiwait` command makes the process wait until the user clicks at the closure button.

The other cases are handled in exactly the same way; there are other message texts and other web-sites.

The code is included in the accompanying software under the name *‘transport.m’*

As an example for a more elaborated graphical user interface, the *‘pplane7.m’* file for processing ordinary differential equations was already mentioned in Chap. 19; see also: Polking (2004). Another example implementation for environmental science and technology is the 'Menyanthes' software, which is used to manage, analyze, model, and present groundwater level data in the Netherlands (van Asmuth 2006); see also: (http://www.mathworks.com/res/menyanthes).

References

Polking J (2004) Ordinary differential equations using MATLAB. Prentice Hall, Upper Saddle River, p 264

van Asmuth J (2006) Kiwa water research builds a tool for monitoring groundwater levels. The Mathworks, News&Notes

Chapter 21
Numerical Methods: Finite Differences

It was demonstrated in previous chapters that the combination of fundamental physical laws, as the conservation of mass or conservation of energy in combination with empirical laws, such as Darcy's Law, Fick's Law or Fourier's Law, leads to a mathematical equalities expressed as partial differential equations (pdes) – or the ordinary differential equations, which can be considered here simple as a specific simpler type of pde. With the addition of initial and/or boundary conditions a partial differential equation usually has a specific solution (which not be always unique, but this should not be a question here).

In the described way many application cases are transformed to the problem of finding the solution of a differential equation. So far we have given several examples of such solutions. In Chap. 1 a very simple differential equation was solved by programming the analytical solution, describing exponential growth. In Chap. 4 we gave the analytical solution for the 1D transport equation.

In the first edition of the book we presented all types of solutions, as just listed, which are given by explicit formulae of mathematical analysis. This type of solutions is gathered under the term *analytical solutions*. However, analytical solutions exist only for a limited set of differential equations. The solution differs with pde and with each variation of boundary or initial conditions. They are often valid only under special conditions concerning the coefficients of the differential equations. If the coefficient varies in dependence of space, time or the function itself, a formula derived for the constant coefficient, is not valid any more.

Another example: if the additional conditions change in space of time, a new formula has to be sought. This may be very tedious, as the derivation of the formula for the constant condition case is not valid for varying conditions anymore. Modifying a simple pde with analytical solution one comes very easily to the situation in which an analytical solution cannot be found, neither by analytical derivation, nor by literature search. In that sense the method of analytical solutions is rather limited.

Nevertheless, here are other solution methods, which are incomparable more powerful and flexible. These are gathered under the term *numerical solutions*, and there are finite differences, finite elements, finite volumes, boundary elements, just

E. Holzbecher, *Environmental Modeling*,
DOI 10.1007/978-3-642-22042-5_21,

to name the most important ones, and cellular automata, just to name a more nowadays still exotic example. The specific branch of mathematics that deals with this type of solutions is *numerics* – which by the way concerns more than the solution of odes or pdes.

A first example of a numerical method was already given in Chap. 4.2. However that worked for 1D only. In this chapter we show, how numerical methods can be used to solve problems in higher dimensions. The pdepe solver was described in Chap. 4.4. In fact there the MathWorks programmers have included sophisticated knowledge from numerics, which cannot be outlined at this place. However, also pdepe works for 1D spatial problems only. The MATLAB® finite element toolbox may be used for higher dimensional spaces. However, in this book it was intended to use core MATLAB® only, and thus we show how here how to do it.

Starting point for a numerical solution is the differential equation itself. The equality, given in the equation is in some way mimicked by a numerical algorithm. The method of that mimicing makes the difference for most of the different numerical methods mentioned above. In order to make the method powerful the approximate mimicking of the differential equation is performed in small parts of the model region and the boundary. I.e. the model region and/or the boundary have to be sub-divided in many small parts, which may be called blocks, cells, elements or volumes, depending on the method. This process of discretization is outlined in a first introductory example.

21.1 Introductory Example

The differential equation

$$\frac{\partial c}{\partial t} = -\lambda c \tag{21.1}$$

is the simplest one to describe the decay or degradation of a chemical specie due to whatever processes. c is the concentration and λ the decay constant. This type of problem was already introduced in Chap. 1 (with a positive coefficient it describes exponential growth). There is a well-known analytical solution for that differential equation for the case when the concentration is known at an initial time, i.e. for the condition

$$c(t = 0) = c_0 \tag{21.2}$$

with a concentration c_0. The solution is:

$$c(t) = c_0 \exp(-\lambda t) \tag{21.3}$$

Imagine that we don't know the analytical solution, or would not have the possibility to evaluate it; maybe we have a supermarket calculator only without access to the exponential function. Are there other means to obtain a solution value for a given time t?

Luckily there are numerical methods. For the current problem we will outline the simplest numerical method. The idea is to approximate the differential term on the left side of the (21.1) by a finite difference term. The numerical method is thus termed '*finite differences*' (FD). Here the finite difference looks like this:

$$\frac{\partial c}{\partial t} \approx \frac{c(t + \Delta t) - c(t)}{\Delta t} \tag{21.4}$$

Δt is the so called timestep, and we can use the approximation for any t that we like. Geometrically the approximation can be visualized by taking the secant through the two points (t,c(t)) and ($t + \Delta t$,$c(t + \Delta t)$) instead of the tangent at the point (t,c(t)). From (21.1) we obtain:

$$\frac{c(t + \Delta t) - c(t)}{\Delta t} = -\lambda c(t) \tag{21.5}$$

Assume that $c(t)$ is known. The we have to resolve equation for the unknown $c(t + \Delta t)$, and obtain:

$$c(t + \Delta t) = (1 - \lambda \Delta t)c(t) \tag{21.6}$$

Thus we see that we have to multiply the concentration at time t with the factor $1 - \lambda \Delta t$, to obtain an approximation of the concentration at time $t + \Delta t$. The factor depends, aside from the given material property λ, only on the numerical parameter Δt.

We utilize the formula (21.6) in the following 'timestepping' approach. We start at initial time $t = 0$, at which the concentration is known (formula (21.2)) and apply formula (21.6). In that way we obtain an approximate value for the concentration at time Δt:

$$c(\Delta t) = (1 - \lambda \Delta t)c_0 \tag{21.7}$$

Now that we have a value for c at $t = \Delta t$, we can apply the formula (21.6) once more to obtain a value for c at time $t = 2\Delta t$:

$$c(2\Delta t) = c(\Delta t + \Delta t) = (1 - \lambda \Delta t)c(\Delta t) \tag{21.8}$$

In that manner we use the formula (21.6) over and over again until we cover the entire time interval in we are interested and reach the final time instant that we want.

The entire procedure is implemented in the following M-file. Starting time is zero and final time is given by the variable `Tmax`. After the initialization of all variables, we plot the analytical solution, as given by (21.3). Then the numerical solutions are computed as described. In the outer loop (with counter variable `i`), in each iteration we halfen the timestep. In the inner loop (with counter variable `j`) the numerical solution is computed. The numerical solutions are plotted after calculation, in the outer loop. Here we use only four different timesteps, i.e. we refine three times, in order to see the differences from one numerical solution to the next in a single plot.

```
% numerics demonstration (for decay equation)
Tmax = 1;
lambda = 2;
c0 = 1;
marker='sod.';
% plot analytical solution
plot (Tmax*[0:.01:1],c0*exp(-lambda*(Tmax*[0:.01:1])),'-r');
hold on

% compute and plot numerical solutions
deltat = .5*Tmax;
for i = 1:4
    f = 1-lambda*deltat;
    c(1) = c0;
    for j = 2:2^i+1
        c(j)=c(j-1)*f;
    end
    plot (linspace(0,Tmax,2^i+1),c,['-' marker(i)]);
    deltat=deltat/2;
end
legend ('analytical',['\Delta' 't=.5'],['\Delta' 't=.25'],['\Delta'
't=.125'],['\Delta' 't=.0625'])
```

The complete code is included in the accompanying software under the name '*numdemo.m*'

In the following figure we show the results of the numerical method for $c_0 = 1$, $\lambda = 2$ and the time interval $t \in [0, 1]$. The solution at $t = 1$ is 0.1353. The unmarked graph depicts the analytical solution. If we use the timestep $\Delta t = 1/2$ we obtain $c(1) = 0$, which is a very bad approximation. But we can easily see that we obtain much better results if we use smaller values for Δt (Fig. 21.1).

The demonstration shows clearly that the solution is better approximated for smaller timesteps. In general we may use the term *discretization*. Here the time period is discretized into smaller peaces, the timesteps: the smaller the timestep, the finer the discretization; and the better results can be expected.

Mathematicians say that the numerical solutions *converge* towards the wanted solution with refinement of the discretization. All numerical solutions are approximations for the 'real' solution. For practical purposes it is always sufficient to obtain an approximate solution. It depends very much on the problem which degree of approximation is wanted, and on the programmer as well.

There is a simple procedure, which is usually applied in order to check the accuracy of the solution. If the modeller uses different discretizations, she/he can check the change from the finest to the previous discretization. If the change is lower than the wanted accuracy, the programmer is usually satisfied. It is not strictly guaranteed that the required accuracy is reached, but the described criterion to end the discretization refinement usually delivers an approximation that deviates from the real solution by

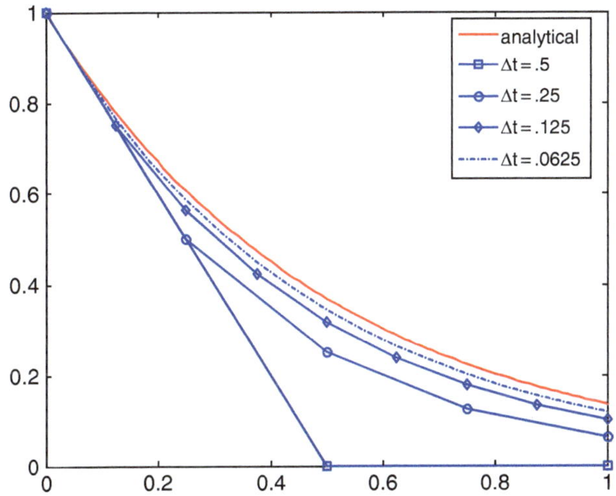

Fig. 21.1 Analytical solution (*red*) and numerical solutions for different timesteps

the wanted accuracy. In the example above, if the modeller is interested in $c(t = 1)$, the following approximations are obtained in 12 refinement steps:

0 0.0625 0.1001 0.1181 0.1268 0.1311 0.1332 0.1343 0.1348 0.1351 0.1352 0.1353 0.1353

If she/he is interested to obtain the first 4 digits of $c(1)$ accurate, she/he can stop the algorithm after the 12th refinement, as the result shows no change in the first four digits any more.

21.2 Finite Differences

In the previous chapter the numerical method of finite differences has been used for the approximate solution of the decay equation. The method can in general be used for the solution of ordinary or partial differential equations. The recipe to replace differentials by finite differences can be applied for time derivatives and spatial derivatives in all space directions. In case of spatial differentials we speak of a *grid spacing* instead of a timestep, and we use the symbols $\Delta x, \Delta y$ and/or Δzinstead of Δt.

For first order derivatives there are various alternatives; forward, backward and central FD:

$$\begin{aligned} \frac{\partial u}{\partial x} &\approx \frac{u(x+\Delta x)-u(x)}{\Delta x} \qquad \text{forward} \\ \frac{\partial u}{\partial x} &\approx \frac{u(x)-u(x-\Delta x)}{\Delta x} \qquad \text{backward} \\ \frac{\partial u}{\partial x} &\approx \frac{u(x+\Delta x)-u(x-\Delta x)}{2\Delta x} \qquad \text{central} \end{aligned} \tag{21.9}$$

here formulated for a function u depending on the independent variable x.

For second order derivatives the central FD scheme is usually without alternative. It is obtained by using the finite difference approximation of the first order FD approximations:

$$\frac{\partial^2 u}{\partial x^2} \approx \frac{1}{\Delta x}\left(\frac{u(x+\Delta x)-u(x)}{\Delta x}-\frac{u(x)-u(x-\Delta x)}{\Delta x}\right) \tag{21.10}$$

or (compare Sidebar 4.2)

$$\frac{\partial^2 u}{\partial x^2} \approx \frac{u(x+\Delta x)-2u(x)+u(x-\Delta x)}{\Delta x^2} \tag{21.11}$$

A simple example may illustrate the entire procedure. FD transforms the differential equation for 1D steady state transport

$$D\frac{\partial^2 c}{\partial x^2} - v\frac{\partial c}{\partial x} - \lambda c = 0 \tag{21.12}$$

into

$$D\frac{c(x+\Delta x)-c(x)+c(x-\Delta x)}{\Delta x^2} - v\frac{c(x)-c(x-\Delta x)}{\Delta x} - \lambda c(x) = 0 \tag{21.13}$$

Here the backward FD for the first order derivative has been chosen. In fact the backward FD delivers a numerical algorithm that usually converges much better method the version, in which forward or central schemes are used. For details about this consult an introductory textbook on numerics (for example: Thomas 1995; Moler 2004).

Equation (21.13) serves as basis for a numerical solution of (21.12) if we divide the model region of interest, that is an interval $x_{\min} \le x \le x_{\max}$on the x-axis, into peaces of length Δx. Using the numerical method we can obtain approximate values for $c(x)$ at the mesh positions, called *nodes* $\mathbf{x} = x_{\min} + \Delta x, x_{\min} + 2\Delta x, \ldots .x_{\max} - \Delta x$ (lets take $\mathbf{x}$ as a vector here). For each position x (21.13) states a relation between the (unknown) function value $c(x)$ and the (also unknown) function values at the neighboring two nodes.

Boundary conditions have to be considered separately, but without problem. A Dirichlet boundary condition at the left side of the interval for example, $c(x_{\min}) = c_0$, according to (21.13) leads to:

$$D\frac{c(x_{\min}+2\Delta x)-2c(x_{\min}+\Delta x)+c_0}{\Delta x^2} - v\frac{c(x_{\min}+\Delta x)-c_0}{\Delta x} - \lambda c(x_{\min}+\Delta x) = 0 \tag{21.14}$$

A Neumann condition at the right hand side, $\partial c/\partial x = 0$, transferred to the first order FD leads to:

$$D\frac{c(x_{\max}-2\Delta x)-c(x_{\max}-\Delta x)}{\Delta x^2}-v\frac{c(x_{\max}-\Delta x)-c(x_{\max}-2\Delta x)}{\Delta x}-\lambda c(x_{\max}-\Delta x)=0 \tag{21.15}$$

In that way we obtain as many equations as there are unknowns in the vector **x**.

Before we proceed with the description of the numerical procedure lets take a partial differential equation as another example. Using the approximations (21.11) one obtains for the 2D Laplace equation in *x*,*y*-coordinates:

$$\begin{aligned}\frac{\partial^2 u}{\partial x^2}+\frac{\partial^2 u}{\partial y^2} \approx & \frac{u(x+\Delta x, y)-2u(x,y)+u(x-\Delta x,y)}{\Delta x^2} \\ & +\frac{u(x,y+\Delta y)-2u(x,y)+u(x,y-\Delta y)}{\Delta y^2}=0\end{aligned} \tag{21.16}$$

which can be applied In 2D space domains, which are discretized into a rectangular *mesh* or *grid*, as shown in the following figures.

At nodes at the boundary the discretization (21.16) changes according to the boundary condition. Let's give one example for a Dirichlet boundary condition. If at the right boundary we have a value u_{bnd} specified, then we obtain instead:

$$\frac{u_{bnd}-2u(x,y)+u(x-\Delta x,y)}{\Delta x^2}+\frac{u(x,y+\Delta y)-2u(x,y)+u(x,y-\Delta y)}{\Delta y^2}=0 \tag{21.17}$$

Other node values have to be replaced, if the Dirichlet condition is for a node on the other sides. For a Neumann condition at the right side, we have to make the following modification:

$$\frac{-u(x,y)+u(x-\Delta x,y)}{\Delta x^2}+\frac{u(x,y+\Delta y)-2u(x,y)+u(x,y-\Delta y)}{\Delta y^2}=0 \tag{21.18}$$

This results from the first order approximation $0=\frac{\partial u}{\partial x}\approx\frac{u(x+\Delta x,y)-u(x,y)}{\Delta x}$ (see (21.9)), which is equivalent to $u(x+\Delta x,y)-u(x,y)=0$(Fig. 21.2).

In case of an equidistant mesh with equal grid spacing $h=\Delta x=\Delta y$the approximation of (21.16) simplifies to the well-known five-point stencil:

$$\frac{1}{h^2}\left(u(x+\Delta x,y)+u(x-\Delta x,y)+u(x,y+\Delta y)+u(x,y-\Delta y)-4u(x,y)\right)=0 \tag{21.19}$$

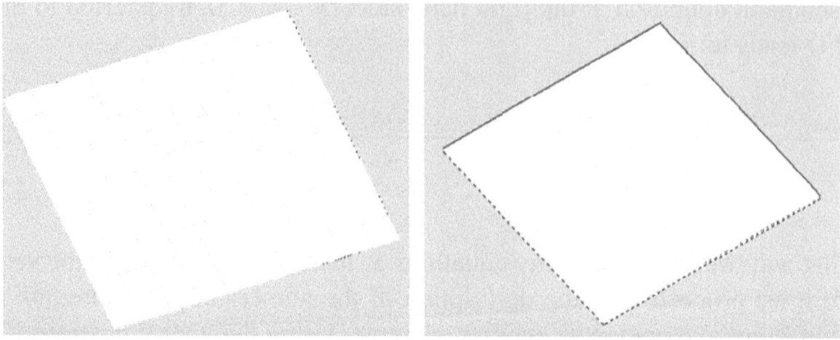

Fig. 21.2 Examples of 2D regular (equidistant) and irregular FD meshes of a square model region

For each position (*x*,*y*) (21.16) and (21.19) state a relation between the (unknown) function value *u*(*x*,*y*) and the (also unknown) function values at the neighboring four nodes. Boundary conditions can be considered as described for the ode-case above.

Altogether, one obtains an equation for each of the unknown nodes. I.e. there are as many equations as there are unknown values – a classical situation for utilizing a mathematical solver for such a problem. Thus the problem of solving a differential equation, ode or pde, is reduced to the problem of solving a system of equations. The latter is a feasible task (in principle), using help offered by mathematical toolboxes, such as MATLAB®.

For the given examples with constant coefficients the described procedure leads to linear systems of equations. For example using canonical numbering of the nodes (*x*-direction first, then *y*-direction) the stencil (21.19) leads to the linear system:

$$\begin{pmatrix} -4 & 1 & 0 & \dots & 1 & 0 & \dots \\ 1 & -4 & 1 & 0 & \dots & \dots & 0 \\ \dots & 1 & -4 & 1 & \dots & \dots & 1 \\ 0 & \dots & \dots & \dots & \dots & \dots & 0 \\ 1 & 0 & \dots & \dots & \dots & 1 & \dots \\ 0 & \dots & \dots & 0 & 1 & -4 & 1 \\ \dots & 0 & 1 & \dots & 0 & 1 & -4 \end{pmatrix} \bullet \mathbf{u} = \mathbf{b} \qquad (21.20)$$

where **u** denotes the vector of unknown values *u*(*x*,*y*) at the nodes and **b** the right hand side vector. The matrix has entries −4 along the main diagonal, and entries 1 in 4 side-diagonals. Two of the side diagonals are just aside the main diagonal; both others are further away depending on the size of the mesh in *x*-direction. Due to the boundary condition several of the 1-entries in the side-diagonals will be missing, and instead of that there will appear non-zero entries in the right hand side vector **b**.

In all cases, independent of the dimension, and of the special discretization method, one obtains matrices, in which the number of non-zero entries is very quite small in relation to the space available in the matrix. Such matrices are called *sparse*, in contrast to dense matrices, in which the majority of elements is non-zero.

MATLAB® has an extra variable type 'sparse array' in order to store and operate with such matrices.

Although the problem of solving a system of equations is feasible, it may be very hard depending on the dimension, the type of the differential equation, the (ir) regularity of the model region and/or the mesh and the structure and dependencies of the parameters.

Let's start with the dimension. In 1D one will usually have some 100s, maybe 1,000 grid points, called *nodes*. The size of the linear system is the lower or equal 1,000. The matrix has only diagonals with non-zero entries. For 2D the situation is surely more severe: there are usually few 100 nodes in each direction, thus we obtain 10,000–100,000 nodes easily. Using the same estimations for a 3D model one easily comes to the order of one million nodes.

Sidebar 21.1: Sparse Matrices

Sparse matrices are stored in a different format in which the location of the non-zero entries is stored and taken into account. Thus it becomes feasible to work with matrices, which have more than a million rows and columns – in most computers that is not possible for dense matrices. There are various operations that work for sparse matrices only, for the transversion of matrices of different type and there are also operations, which can be applied for sparse matrices in the same manner as they are used for usual arrays. Here some examples:

```
sparse(A)
```

converts a matrix into sparse form. Try for the matrix

```
A = [1 0 ; 0 3]
```

for demonstration of the representation (do not use the sparse type for such a simple matrix in your programs). Using `sparse` with a longer list of input parameters it is possible to assign each single element of the matrix. For a complete description of the command see the help. The conversion from a sparse matrix to a full matrix is performed by:

```
full(A)
```

If you want to know, if a matrix is internally stored in sparse form, use:

```
issparse(A)
```

Sparse systems can also be combined from vectors, better: from a matrix containing the diagonals of the sparse matrix in columns. Use the `spdiags` command. Here an example: the command

(continued)

```
A = spdiags([ones(9,1) -2*ones(9,1) ones(9,1)],[-2 0 2],9,9)
```

creates a matrix with three diagonals. The content of the diagonals is determined by the first input parameter, which is a matrix with three columns. The position of the diagonals in the matrix is determined by the second input parameter, which is an integer array. A zero entry in this vector denotes the main diagonal. The other entries are related to that setting, The command above creates a 9×9 matrix with -2 entries in the main diagonal and 1 entries in two off-diagonals. To see the structure of the matrix, use

```
spy(A)
```

and obtain the figure on the right as a result.

All nonzero entries are indicated by a blue star.

Matrix operations, i.e. addition, subtraction, multiplication etc. can be performed as with usual matrices. For example

```
A\ones(9,1)
```

delivers the solution of the system $\mathbf{A} \cdot \mathbf{x} = \mathbf{b}$with right-hand side vector containing one in each entry.

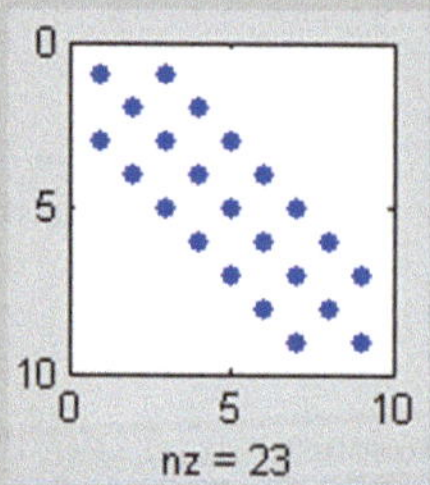

21.3 A Finite Difference Example

In the following example we demonstrate the method for the Poisson equation

$$\nabla^2 u = q \tag{21.21}$$

for the unknown function $u(x,y)$ and right hand side $q(x,y)$.

In the specification part the dimension of the system is assessed. Here the system has the length 5 (`nx*h`) and the width 20 (`ny*h`). Moreover we specify the boundary condition at the upper and lower boundary, and a constant source term.

```
nx = 5; ny = 20;        % dimensions in x- and y-direction
h = 1;                  % grid spacing
btop = 1;               % boundary condition at left side
bbottom = 0;            % boundary condition at right side
q = .1;                 % right hand side (source term)
```

In the main part we compute the matrix and the right hand side, as shown in (21.20) according to the finite difference method. `N` is the total number of unknowns. In the example we compute an approximate solution at all locations of a grid of 5 × 20. For the exact set-up of the matrix (21.20) we use the `spdiags` command (see Sidebar 21.1) before starting the 'real' processing (the solution of the linear system). All commands before that statement prepare the exact entries of the diagonals.

The vector `d` determines the locations of the diagonal. To understand this we have to be aware that the locations of the grid, the nodes, are numbered in canonical order, i.e. starting with the 1 for the (1,1) position we number in x-direction first and in y-direction subsequently. In that order their function values appear in the unknown vector **x**. The left and right (x-direction!) neighbors of the node i then have the indices $i + 1$ and i-1, and the top and bottom (y-direction) neighbor nodes have the indices $i + n_x$ and i-n_x. This determines the locations of the off-diagonals, and is taken into account by the given choice of `d`.

As mentioned above, annlogously to the derivation of formula (21.19) for the Poisson equation we obtain the discretization:

$$u(x + \Delta x, y) + u(x - \Delta x, y) + u(x, y + \Delta y) + u(x, y - \Delta y) - 4u(x, y) = h^2 q \tag{21.22}$$

or

$$\frac{1}{4}(u(x + \Delta x, y) + u(x - \Delta x, y) + u(x, y + \Delta y) + u(x, y - \Delta y) - 4 * u(x, y)) = \frac{h^2 q}{4} \tag{21.23}$$

Thus there are −4 entries in the main diagonal and 1 is the standard entry in the off-diagonal. This we have to consider in the assignment of matrix `B`, in which all diagonals of the matrix `A` are stored. Thus we proceed with the commands:

```
N = nx*ny;
d = [-nx,-1,0,1,nx];
B = [ones(N,2) -4*ones(N,1) ones(N,2)];
b = -q*h*h*ones(N,1);
```

However, there are some exceptions that must be taken into account. For some nodes not all four neighbor nodes are available. For example along the left boundary the left neighbor is missing, and at all nodes on the right side the right neighbor is missing. Thus in the matrix the corresponding off-diagonal entries should be 0 and not 1. In the example this is done in the first two commands of the first loop.

At all nodes at which we require a Neumann condition, the 4 entry in the main diagonal needs to be changed, according to formula (21.18). If we assume a Neumann condition on the left and right boundary, we thus have to change the corresponding diagonal entries to 3. This is done in the last two statements of the first loop.

```
for i = 1:ny
   B(i*nx,2) = 0;
   B(i*nx+1,4) = 0;
   B((i-1)*nx+1,3) = -3;
   B(i*nx,3) = -3;
end
```

At all nodes with Dirichlet condition, according to (21.17) we have to add an additional term on the right hand side, i.e. in the vector `b`. This is done in the next loop, before the matrix `A` finally is set up as intended.

```
for i = 1:nx
    b(i) = b(i)-btop;
    b(N+1-i) = b(N+1-i)-bbottom;
end
A = spdiags(B,d,N,N);
```

After all these preparations we can solve the linear system. The solution is obtained in the vector `U` using the backslash operator (see chapter *). In order to rearrange calculated (before) unknown values into a matrix that corresponds to the format of the grid, we use the `reshape` command.

```
% processing: solution
U = A\b;
U = reshape(U,nx,ny);
```

Finally we check the result. The MATLAB® command `del2` computes the outcome of the right hand side of (21.23). In our example this is a constant value at all nodes. The last command delivers graphical output, as shown in Fig. 21.3.

```
% check & visualize
4*del2(U)
surf(U)
```

The complete code is included in the accompanying software under the name *'Poisson1.m'*

Note that the Dirichlet boundary values, in the example 1 on the left and 0 on the right side, are not taken at the outer nodes. In the given example all nodes are lying in the interior of the model region. The boundaries, for which the conditions are valid, have a distance of h from the outer nodes, as the m-file terms for the boundary blocks correspond with the discrete formulation (21.17). In order to show the solution in the entire model region, a vectors with boundary values would have to be combined with `U`, for example by using:

```
surf([btop*ones(nx,1) U bbottom*ones(nx,1)])
```

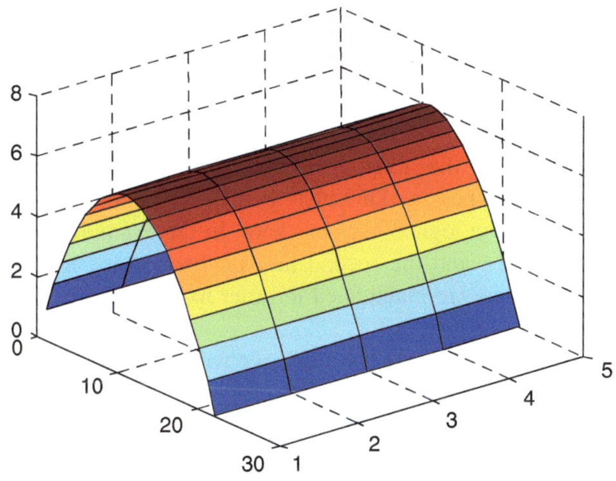

Fig. 21.3 Solution of the example setting for the Poisson equation

21.4 Solution for the 2D Poisson equation

In the previous sub-chapter the numerical solution of the Poisson equation was described as a first example solving a stationary problem on higher dimensional model region. Here we extend the program to deal with heterogeneous boundary conditions and right hand side.

In the previous example Dirichlet-type boundary conditions are given at the constant-x boundaries, while at the constant-y boundaries Neumann-type conditions are required. In order to allow arbitrary combinations of Dirichlet- and Neumann-type conditions, the program has to be extended. We choose to use logical vectors for the indication of the boundary type and double vectors for the values. There are four vectors representing the four sides of the square model region:

```
% boundary type indicators (1=Dirichlet, 0=Neumann no-flow)
ltop = logical(zeros(1,nx));                    % top
lbottom = logical(zeros(1,nx));                 % bottom
lleft = logical([ones(ny/2,1) zeros(ny/2,1)]);  % left
lright = logical([zeros(ny/2,1) ones(ny/2,1)]); % right
```

A 1-entry in the vector indicates a Dirichlet-type condition, while a 0-entry stands for a Neumann-type no flow condition. In the example, given by the four commands above, we have no-flow conditions along the top and bottom boundary, while the left and right boundary a split in two parts, one having Dirichlet- and one Neumann-type conditions.

Corresponding to the logical vectors as boundary-type indicators we introduce double vectors for the eventually needed boundary values:

```
% boundary values (Dirichlet only)
btop = ones(1,nx);                    % top
bbottom = zeros(1,nx);                % bottom
bleft = 2*ones(ny,1);                 % left
bright = zeros(ny,1);                 % right
```

The values given in the four vectors above will be taken only if the corresponding position has a Dirichlet boundary condition, i.e. when the corresponding entry in the logical vector is zero. In the example here only the first half of the vector `bleft` and the second half of the `bright` are relevant for the problem. All other values are maintained in order to allow M-file to be capable to treat the general case.

As a generalization of the right hand side we write

```
q = [ones(nx/2,ny);zeros(nx/2,ny)];     % right hand side (source term)
```

The entries in the matrix `q` represent positions in the model region. If an element is non-zero there is a source (or sink if the value is negative) in the corresponding block of the model region. In the example we have a source term in the upper half of the model region, while there is no source in the lower part – for whatever reason, we don't have to discuss here. The next commands are identical to the former program.

```
N = nx*ny;
d = [-nx,-1,0,1,nx];
B = [ones(N,2) -4*ones(N,1) ones(N,2)];
b = -h*h*reshape(q,N,1);
```

The last command only is modified. One has to take into account that the matrix `q` has to be changed into a vector for the further computations. In the following we modify matrix `B` and right hand side `b` according to the general boundary conditions. In the first loop we treat top and bottom boundary conditions. For the top boundary (indices 1 to n_x) an additional term has to be added to the right hand side in case of a Dirichlet condition or the entry in the main diagonal has to be reset to 3 in case of Neumann condition. These settings were outlined already in the previous subchapter, but for the different boundaries as a whole. Now the distinction has to be made for each location along the boundaries. The bottom boundary blocks then (indices N-n_x + 1 to N) are treated similarly in the second command block.

```
for i = 1:nx
    if ltop(i)
        b(i) = b(i)-btop(i);
    else
        B(i,3) = -3;
    end
```

```
        if lbottom(i)
            b(N-nx+i) = b(N-nx+i)-bbottom(i);
        else
            B(N-nx+i,3) = -3;
        end
    end
```

In the next loop we treat left and right boundaries similarly. However there are certain peculiarities to be considered.

- Some of the entries in the adjacent off-diagonals have to be set to zero. These represent those block neighbors that are outside of the model region, i.e. left neighbors for boundary blocks on the left, and right neighbors for boundary blocks on the right. This is considered below in the first two commands in the outer `if` block. For the top and bottom boundary this was not necessary because there these entries are missing in the matrix as the outer diagonals do not extend over all rows and columns of `B`.
- The diagonal elements must be increased by 1, in contrast to the first loop above, where they could be set to -3 in general. That is because the former assignments in the first loop have to be considered in the second loop.

```
for i = 1:ny
    B(i*nx,2) = 0;
    if i<ny B(i*nx+1,4) = 0; end
    if lleft(i)
        b((i-1)*nx+1) = b((i-1)*nx+1)-bleft(i);
    else
        B((i-1)*nx+1,3) = B((i-1)*nx+1,3)+1;
    end
    if lright(i)
        b(i*nx) = b(i*nx)-bright(i);
    else
        B(i*nx,3) = B(i*nx,3)+1;
    end
end
```

After all these settings to execution and visualisation part is in fact identical to the former M-file:

```
U = A\b;
U = reshape(U,nx,ny);
4*del2(U)
surf(U)
```

The system is solved using the \-operator. The solution has to be reshaped before the output is checked and visualized. The final result is depicted in Fig. 21.4.

The complete code is included in the accompanying software under the name *'Poisson2.m'*

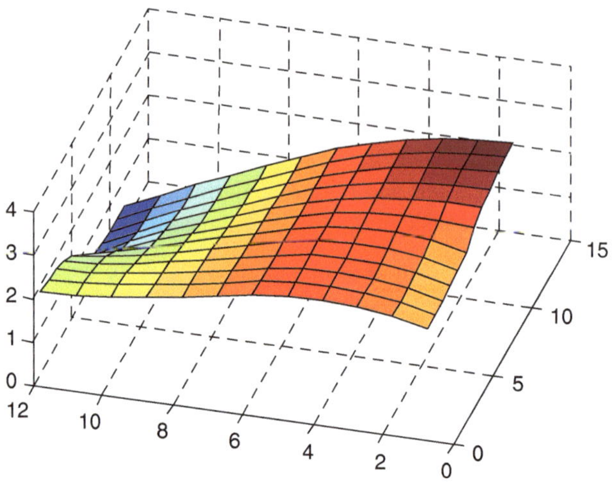

Fig. 21.4 Solution of the example setting for the Poisson equation with general boundary conditions and sources/sinks

21.5 Solution for the 2D Diffusion-Decay Equation

The program M-file from the last sub-chapter can be easily extended to account for decay. This shows the strengths of the numerical methods. The differential equation for the steady state is:

$$D\nabla^2 c - \lambda c = 0 \tag{21.24}$$

(see chapter 5), where the unknown function is again denoted by c to indicate concentration. D is the diffusivity and λ the decay constant.

We can re-write the (21.24) as

$$\nabla^2 c = \frac{\lambda}{D} c \tag{21.25}$$

which now shows a high resemblance with the Poisson (21.21). We see that the difference is only on the right hand side, where we have a function depending on c in this case. In the formalism of the finite difference procedure that is thus not a crucial difference. Following the FD approach the term $(\lambda/D)c$is evaluated at the center position and enters the differential equation for each block. Thus we obtain another term in the main diagonal of the matrix `B`. This is considered by replacing the former initial setting for `B` by:

```
B = [ones(N,2) -(4+lambda/D)*ones(N,1) ones(N,2)];
```

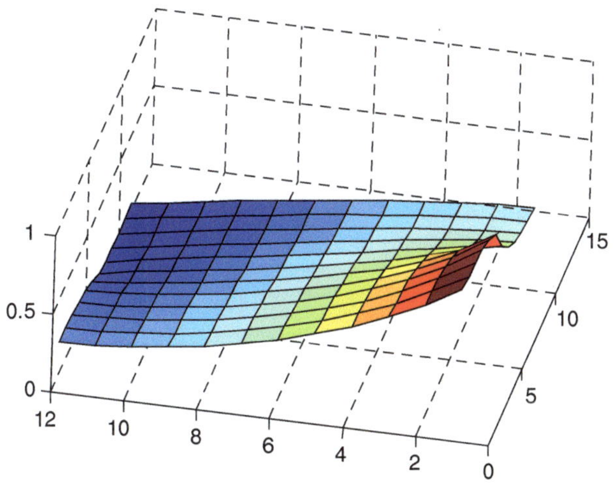

Fig. 21.5 Solution of the example setting for the diffusion-decay equation

That is all we have to do; aside from initializing `D` and `lambda` in the first part of the program. In the example we use the same boundary conditions as in the previous sub-chapter. The result is shown in Fig. 21.5.

We left the general source/sink-term considered so that we can actually solve a diffusion-decay-source/sink problem with the M-file. In the example however, there were no sources, i.e. $q = 0$.

The code can be found in the accompanying software under the name *'DiffDecay2D.m'*

References

Moler CB (2004) Numerical computing with MATLAB. SIAM, Philadelphia

Thomas JW (1995) Numerical partial differential equations: finite difference methods (graduate texts in mathematics). Springer, New York

Supplements

Supplement 1: MATLAB® Data Import

There are several ways to import data into the MATLAB workspace. See topic 'Using Import Functions with Text Data' in the online help for an overview concerning text data. There are special commands for importing spreadsheet data (`csvread`); there is even a special command for importing from Microsoft EXCEL: `xlsread`.

It is not the intention here to go into details. We demonstrate a user-friendly tool, which provides a data preview and several data manipulation tools during importing. The 'Open Import Wizard' interface is called by

```
uiimport
```

from the MATLAB® command window. The functionality is exemplified on one of the most cited data sets, showing the increase of atmospheric CO_2 concentrations within almost 50 years. The data-set, which is measured at the Mauna Loa Observatory in Hawai'i at a height of 3,400 m above sea-level, can be obtained from the internet. For monthly recorded data see: http://www.seattlecentral.org/qelp/sets/078/078.html. All data are given in a single ASCII text file. Typical content is depicted after calling `uiimport` from the command window and opening the file (see Fig. S.1).

After pressing the 'Next' button another similar window appears, which allows some manipulations on the data file. Here it is important to increase the number of text header lines to 15. After that in the right data window the data matrix appears, as depicted in the Fig. S.2.

Click two times 'Next' to move to the final window. Here, choose only to re-name the data variable (use again right mouse button) to 'CO2'. The final 'Finish' (button!) creates a new variable with the chosen name in the workspace (see Fig. S.3). With that operation the command is finished and the wizard disappears.

E. Holzbecher, *Environmental Modeling*,
DOI 10.1007/978-3-642-22042-5, © Springer-Verlag Berlin Heidelberg 2012

Fig. S.1 Data import, first screen

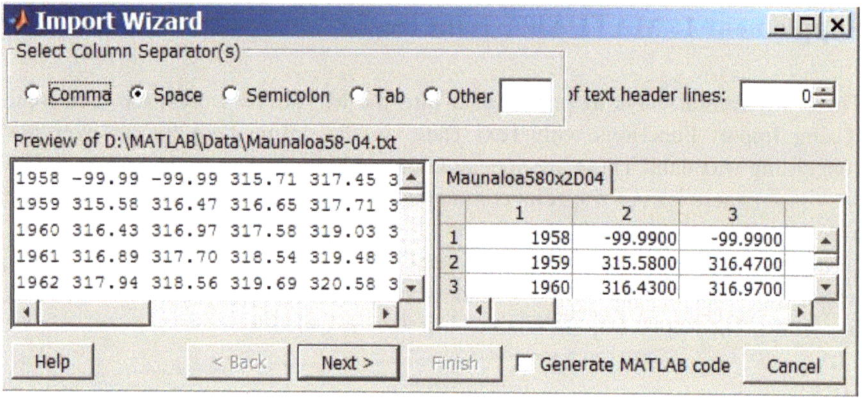

Fig. S.2 Data import, second screen

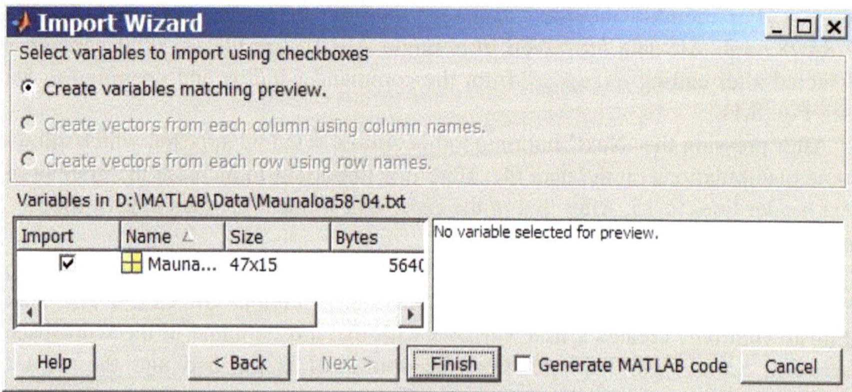

Fig. S.3 Data import, third screen

The year is given in the first column; 14 following columns show month related concentrations, the annual mean and a fitted annual mean. As we liked to take the monthly measurements only, we highlight the corresponding 12 columns by mouse-click in the right data window. Use the right mouse button to obtain a pop-up window, including a copy button. After copying the highlighted columns, use the 'Back' button to return to the first window; and select 'Clipboard' using the corresponding radio button. Now the carbon-dioxide values appear in the data blocks; years are omitted.

In the next step we manually change '-99.99' entries in the data set that is used in the data set for missing values, to 'NaN', which fits with the MATLAB® convention (using copy and paste operations in the array editor. There are only few missing values, which allow the manual operation. For more complex data sets one has to utilize some MATLAB® commands as demonstrated in the following.

In order to plot the data, the matrix is transformed into a row vector. This can be done using the following command sequence:

```
for i=1:47
   for j = 1:12
     co2(12*(i-1)+j)=CO2(i,j);
     t(12*(i-1)+j)=datenum(1957+i,j,15)
   end
end
```

The 2D data set in the variable 'CO2' is converted to a row in the variable 'co2'. In addition another row vector with corresponding times is created. We use the serial date number, which is one of several MATLAB® alternatives to represent date and time (for more information see the online help index 'dates and times'). The `datenum` command converts a date into the serial date number. Called with three numbers, these correspond to year, month and day. There are several more alternative calls of the command, which the user may look up in the help. The aimed plot is finally created by the commands:

```
plot (t,co2)
datetick ('x',11)
xlabel (year); ylabel('Atmospheric CO_2 [ppm]');
```

The following figure results. Corresponding to the time format the datetick command offers several options to display time. There is a list of 28 alternatives which can be applied by the `datetick` command. Here we choose to show the year only. The result is shown in Fig. S.4.

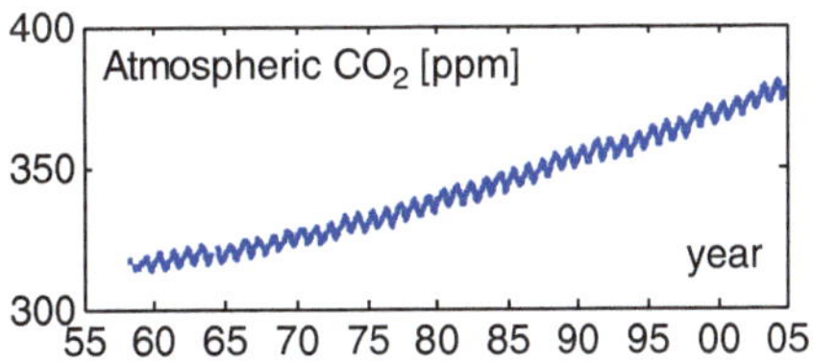

Fig. S.4 Atmospheric CO_2 increase; visualized using MATLAB®

Supplement 2: Data Export

Data are exported by using the `save` command. Let us take the calculated `co2` data from Supplement 1 as an example. The command

```
save ('co2.mat', 'co2')
```

stores the values in the file *'co2.mat'* in the working directory. Make sure that the user has write-permissions on that file. Otherwise change the directory by using the `cd` (change directory) command. Note that MATLAB® has its own data storage format that is the default here. Usually the extension '.mat' is used for files with that data format.

Other data formats can also be stored. Most important is the ASCII format, which is obtained by using:

```
save ('co2.mat', 'co2','-ascii')
```

Also important is the `-append` option for the data to be appended at the end of an existing file.

Supplement 3: Data Presentation in a Histogram

There are various ways to represent environmental data using MATLAB®. The reader may have a look in MATLAB® online command index for the `hist`, `bar` and `bar3` commands.

As an example we show a histogram of concentration measurements of different chemical species at various observation points. Six species were measured at 13 positions. The entire data-set is stored in a matrix `C`.

The histogram is then produced by the bar command:

```
bar(C);
```

The code is included in the accompanying software under the name *'bardemo.m'*

The result of the M-file is depicted in Fig. S.5. Further commands concern the labels of the axes and the legend. Note how Greek characters are introduced in the text, by using the \ operator.

```
xlabel ('observation points');
ylabel ('C [\mug/l]');
title ('measured concentrations');
legend ('Na','Cl','B','HCO_3','F','TOC');
```

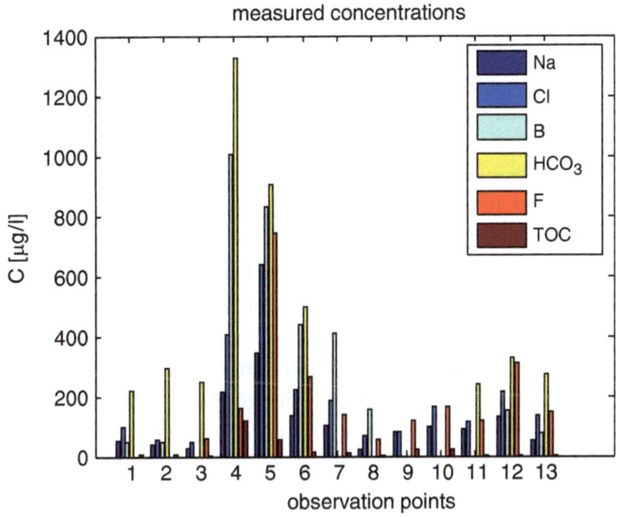

Fig. S.5 Example data representation in a histogram

Epilogue

In 20 chapters, the book shows various applications of MATLAB® in the field of environmental modeling. Numerous MATLAB® commands are introduced and their use demonstrated. Various fields of environmental modeling have been touched.

After 20 chapters, the book remains incomplete. Neither the entire field of environmental modeling is covered, nor is the entire capability of MATLAB® exploited. Of course, either of the mentioned tasks would be too ambitious to be worked out, even within several book volumes.

Is something missing that is important? Probably everyone working in the field of environmental modeling, who does not find her/his special problem set-up, will say, yes. It was already mentioned that the entire field is too vast. Concerning MATLAB®, the important application field of numerical methods for 2D and 3D applications is missing. MATLAB® can be used to implement important numerical approaches, like finite differences, or finite elements. These methods were omitted as a consequence of the decision to focus on core MATLAB®. The easiest way to apply such numerical techniques is to use the partial differential toolbox of MATLAB®. Core MATLAB® could also be used to implement higher-dimensional numerical models, but manual programming skills are required. Only the advanced user would be addressed by this topic, to whom the recently published book of Danaila et al. (2007) can be recommended. Among other numerical topics Quarteroni (2003) outlines methods for the advection diffusion equation using advanced Finite Element modeling techniques and presents MATLAB@ source code for the solution of the 1D steady state.

Concerning the environment, the hydrosphere is surely over-represented in the book, while the atmosphere and the pedosphere appear only sporadically. Among the hydrosphere topics, groundwater has the biggest share. The choice of the topics is surely due to the background of the author, who in the past mainly worked in the favored fields. However, the mathematical concepts that were introduced are mostly independent from the environmental compartment and thus applicable in several environmental areas. The given applications should be viewed as examples for the mathematical techniques.

It was the purpose of the book to give a first introduction. I hope that goal is reached. Aside from that, some novel approaches have been introduced and examined which are beyond state-of-the-art. Some of these approaches turn out to be simple and useful and will hopefully find their way into the practice of environmental modeling. If that really happens, is due to the reader and her/his conception of the book. In that sense, I wish the book to find understanding readers who make these concepts work.

References

Danaila I, Joly P, Kaber SM, Postel M (2007) Twelve computational projects solved with MATLAB, Springer, New York, 294p

Quarteroni A (2003) Modellistica Numerica per Problemi Differenziali, Springer, Milan, 332p (in Italian)

MATLAB®Command Index

The following list gathers the MATLAB® commands, which are mentioned in the book. For each command find the corresponding chapter and sub-chapter numbers within the book. For frequently appearing commands the most important occurrences are listed, only.

DOI 10.1007/978-3-642-22042-5,

Companion Software List

The accompanying CD contains the M-files which are described in the book, sometimes extended versions. The files can also be downloaded from the MATLAB® central file exchange (www.mathworks.com/matlabcentral/fileexchange).

advection.m
analtrans.m
analtrans_s1.m
analtrans_s2.m
analtrans_s3.m
analtrans_s4.m
analtransnodim.m
AnElements.m
animation.mpg
bardemo.m
boudreau_westrich.m
comparts.m
compspec.m
cplxPot.m
DiffDecay2D
diffusion.m
dipole.m
DischargePotential.m
Fracture.m
GaussianPlume.m
GaussianPuff.m
GdDTPA.m
georef.m
gw_flow.m
ierfc.m
kinetics.m
logistic.m
lorenza.m
MichaelisMenten.m
newtondemo.m
nuclides.m
numdemo.m
OpenChannel.m
par_est.m
par_esta.m
par_estb.m
par_estc.m
par_est2.m
par_est2a.m
pdepetrans.m
phasediag.m
Poisson1.m
Poisson2.m
predprey.m
redoxsteady.m
retention.m
richards.m
simpltrans.m

DOI 10.1007/978-3-642-22042-5,

slowsorp.m
Speciation.m
StreeterPhelps.m
sttransanal.m
thiem_test.m
ThreeD_flow.m
transport.m
viscosity_dyn.m
welldrawdown.m
wellvortex.m

Index

E. Holzbecher, *Environmental Modeling*,
DOI 10.1007/978-3-642-22042-5, © Springer-Verlag Berlin Heidelberg 2012

MIX
Papier aus verantwortungsvollen Quellen
Paper from responsible sources
FSC® C105338

If you have any concerns about our products,
you can contact us on
ProductSafety@springernature.com

In case Publisher is established outside the EU,
the EU authorized representative is:
Springer Nature Customer Service Center GmbH
Europaplatz 3, 69115 Heidelberg, Germany

Printed by Libri Plureos GmbH
in Hamburg, Germany